# In Vitro Models of Tissue and Organ Regeneration

# In Vitro Models of Tissue and Organ Regeneration

Editors

**Patrick C. Baer**
**Ralf Schubert**

MDPI

Basel • Beijing • Wuhan • Barcelona • Belgrade • Novi Sad • Cluj • Manchester

*Editors*

| | |
|---|---|
| Patrick C. Baer | Ralf Schubert |
| Universitätsklinikum | Universitätsklinikum |
| J. W. Goethe University | J. W. Goethe University |
| Frankfurt/Main, Germany | Frankfurt/Main, Germany |


*Editorial Office*
MDPI
St. Alban-Anlage 66
4052 Basel, Switzerland

This is a reprint of articles from the Special Issue published online in the open access journal *International Journal of Molecular Sciences* (ISSN 1422-0067) (available at: https://www.mdpi.com/journal/ijms/special_issues/models_TOR).

For citation purposes, cite each article independently as indicated on the article page online and as indicated below:

Lastname, A.A.; Lastname, B.B. Article Title. *Journal Name* **Year**, *Volume Number*, Page Range.

**ISBN 978-3-0365-9218-3 (Hbk)**
**ISBN 978-3-0365-9219-0 (PDF)**
**doi.org/10.3390/books978-3-0365-9219-0**

Cover image courtesy of Patrick C. Baer

# Contents

# About the Editors

**Patrick C. Baer**

Patrick C. Baer is a cell biologist and Associate Professor of Experimental Medicine at the Hospital of the Goethe University in Frankfurt/M. He completed his studies in biochemistry at the Technical University of Darmstadt and received his doctorate at the Goethe University of Frankfurt/M. P.C.B. has currently published over 100 research articles, including three book chapters and three patents. P.C.B. has been working with cell culture models of proximal and distal tubular epithelial cells of the kidney for more than 25 years. The research areas of P.C.B. also focus on the isolation, culture and differentiation of mesenchymal stromal/stem cells (MSCs) and the transplantation of MSCs or their derivatives (conditioned medium, extracellular vesicles) to improve organ regeneration.

**Ralf Schubert**

Ralf Schubert Ph.D., is an Associate Professor with tenure at the Department of Pediatrics at the University Hospital Frankfurt, Goethe University and is the Head of the Pneumological/ Immunological Laboratory. He graduated from the Technische Universität Darmstadt with a Ph.D. degree in Immunology and recieved his postdoctral training at the Medical School at the University of California, San Diego. He has been working in the field of immunology for more than 20 years and the research activity of his group is focused on the investigation of cellular and molecular mechanisms of inflammatory lung diseases such as bronchial asthma. There work is especially concerned with rare diseases such as Ataxia telangiectasia, and bronchiolitis obliterans. He has authored more than 180 scientific publications in international journals on this topic and has received numerous national and international grants. Current projects focus on the regulatory role of microRNAs in the resolution of inflammation in in vivo and in vitro models.

International Journal of
*Molecular Sciences*

MDPI

*Editorial*

# In Vitro Models of Tissue and Organ Regeneration

**Patrick C. Baer * and Ralf Schubert ***

Division of Allergology, Pneumology and Cystic Fibrosis, Department for Children and Adolescents, University Hospital, Goethe-University, 60596 Frankfurt/M., Germany
* Correspondence: p.baer@em.uni-frankfurt.de or pcbaer@arcor.de (P.C.B.); ralf.schubert@kgu.de (R.S.); Tel.: +49-69-6301-83611 (R.S.); Fax: +49-69-6301-83349 (R.S.)

The recovery of cells after tissue and organ injury is a complex process. To understand the underlying molecular biological mechanisms, more detailed insights into the cellular processes of repair and regeneration are urgently needed. Based on this knowledge, this Special Issue focuses on current in vitro systems exploring repair and regeneration mechanisms. Experimental research approaches to investigate the mechanisms involved and laboratory methods to establish and optimise models for tissue and organ repair and regeneration, as well as theoretical modelling and computational models, but also review papers are included here. Eleven articles are published in the Special Issue, which deals with various tissue and organ regeneration questions or the modelling or summary of the research models used in this process.

Shyam et al. comprehensively summarise various methods involved in developing 3D cell culture systems, emphasising the differences between 2D and 3D systems and methods involved in recapitulating the organ-specific 3D microenvironment [1]. They also discuss the latest developments in 3D tissue model fabrication techniques, microfluidics-based organ-on-a-chip, and imaging as a characterisation technique for 3D tissue models. Lieto et al. summarise current research to accurately evaluate ocular toxicity and drug effectiveness [2]. The recent achievements in tissue engineering of in vitro 2D, 2.5D, 3D, organoid and organ-on-chip ocular models and in vivo and ex vivo ocular models were discussed in terms of their advantages and limitations. Another review by Young et al. looks in detail at different strategies for fatty liver treatment in non-alcoholic fatty liver disease [3]. They discuss various defatting strategies, including in vitro use of pharmacologic agents, machine perfusion of extracted livers, and genomic approaches targeting specific proteins. Another work by Hentabli et al. deals with modelling a neural network for bioactivity prediction [4]. This paper describes a novel technique based on a deep learning convolutional neural network for predicting chemical compounds' bioactivity. The authors explain the importance of this work by stating that determining and modelling the possible behaviour and effects of molecules requires the study of the basic structural features and physicochemical properties that determine their behaviour in chemical, physical, biological and environmental processes.

Two original works use in vitro models of mesenchymal stromal/stem cells (MSCs) to investigate regenerative purposes. Barbon and coworkers use an in vitro conditioning regimen of MSCs towards the endothelial lineage to stimulate coagulation factor VIII production [5]. The background of this work was the development of a cell therapy for the treatment of Haemophilia A and, therefore, for future pre-clinical investigation using preconditioned MSCs. Leppik and coworkers demonstrate a new perspective on bone tissue engineering [6]. The work shows that MSCs survive cryopreservation on scaffolds and, after thawing, could be released as ready-to-use products for permanent implantation during surgery.

In addition, two other in vitro studies use epithelial cell systems to investigate their differentiation or their involvement in inflammatory processes. Primary alveolar epithelial cells' main limitation is the difficulty of maintaining the type II phenotype in culture.

**Citation:** Baer, P.C.; Schubert, R. In Vitro Models of Tissue and Organ Regeneration. *Int. J. Mol. Sci.* **2023**, *24*, 14592. https://doi.org/10.3390/ijms241914592

Received: 7 September 2023
Revised: 13 September 2023
Accepted: 22 September 2023
Published: 26 September 2023

Marhuenda and coworkers show that culturing primary alveolar epithelial cells on lung extracellular matrix-derived hydrogels facilitated the prolonged culturing of these cells and enhanced the preservation of the type II phenotype [7]. Baer and coworkers characterised the mRNA expression of renal proximal tubular epithelial cells and the cargo in extracellular vesicles in an inflammatory microenvironment [8]. This study demonstrates the altered miRNA expression of epithelial cells and their released vesicles during induced inflammation, with only three miRNAs overlapping between cells and vesicles. The background to this study is that understanding the precise molecular and cellular mechanisms that lead to inflammation is the most important way to identify targets for the prevention or treatment of inflammation.

Steyn-Ross and coworkers describe the ex vivo quantification of tissue oxygen consumption by measuring oxygen partial pressure as a function of probe depth using thin slices of cortical brain tissue [9]. The authors confirm that a previously published diffusion-consumption model provides an excellent description of the oxygen-tension distribution in a thin slice of active tissue.

Finally, this Special Issue contains two publications addressing issues using in vivo models. Azam and coworkers investigated the prevention of neuroinflammation in vitro and in vivo using an herbal extract and purified dioscin [10]. The in vitro study demonstrates protection against lipopolysaccharide-activated inflammatory responses in microglial cells. The following in vivo study shows that dioscin upregulates brain-derived neurotrophic factor and cAMP-response element binding protein phosphorylation in the cerebral cortex and hippocampus regions of the mouse brain. The authors conclude that dioscin protects against neurotoxicity. Esteves-Monteiro and coworkers evaluated changes in ileum and colon histomorphometry and Angiotensin II reactivity in a rat model of diabetes mellitus. They showed the structural remodelling of the gut wall with a decreased contractile response to Angiotensin II [11]. They summarise that these findings may help to explain diabetic dysmotility.

**Conflicts of Interest:** The authors declare no conflict of interest.

## References

1. Shyam, R.; Reddy, L.V.K.; Palaniappan, A. Fabrication and Characterization Techniques of In Vitro 3D Tissue Models. *Int. J. Mol. Sci.* **2023**, *24*, 1912. [CrossRef] [PubMed]
2. Lieto, K.; Skopek, R.; Lewicka, A.; Stelmasiak, M.; Klimaszewska, E.; Zelent, A.; Szymański, Ł.; Lewicki, S. Looking into the Eyes-In Vitro Models for Ocular Research. *Int. J. Mol. Sci.* **2022**, *23*, 9158. [CrossRef] [PubMed]
3. Young, E.N.; Dogan, M.; Watkins, C.; Bajwa, A.; Eason, J.D.; Kuscu, C.; Kuscu, C. A Review of Defatting Strategies for Non-Alcoholic Fatty Liver Disease. *Int. J. Mol. Sci.* **2022**, *23*, 11805. [CrossRef] [PubMed]
4. Hentabli, H.; Bengherbia, B.; Saeed, F.; Salim, N.; Nafea, I.; Toubal, A.; Nasser, M. Convolutional Neural Network Model Based on 2D Fingerprint for Bioactivity Prediction. *Int. J. Mol. Sci.* **2022**, *23*, 13230. [CrossRef] [PubMed]
5. Barbon, S.; Stocco, E.; Rajendran, S.; Zardo, L.; Macchi, V.; Grandi, C.; Tagariello, G.; Porzionato, A.; Radossi, P.; de Caro, R.; et al. In Vitro Conditioning of Adipose-Derived Mesenchymal Stem Cells by the Endothelial Microenvironment: Modeling Cell Responsiveness towards Non-Genetic Correction of Haemophilia A. *Int. J. Mol. Sci.* **2022**, *23*, 7282. [CrossRef] [PubMed]
6. Leppik, L.; Gempp, A.; Kuçi, Z.; Kuçi, S.; Bader, P.; Bönig, H.; Marzi, I.; Henrich, D. A New Perspective for Bone Tissue Engineering: Human Mesenchymal Stromal Cells Well-Survive Cryopreservation on β-TCP Scaffold and Show Increased Ability for Osteogenic Differentiation. *Int. J. Mol. Sci.* **2022**, *23*, 1425. [CrossRef] [PubMed]
7. Marhuenda, E.; Villarino, A.; Narciso, M.L.; Camprubí-Rimblas, M.; Farré, R.; Gavara, N.; Artigas, A.; Almendros, I.; Otero, J. Lung Extracellular Matrix Hydrogels Enhance Preservation of Type II Phenotype in Primary Alveolar Epithelial Cells. *Int. J. Mol. Sci.* **2022**, *23*, 4888. [CrossRef] [PubMed]
8. Baer, P.C.; Neuhoff, A.-K.; Schubert, R. microRNA Expression of Renal Proximal Tubular Epithelial Cells and Their Extracellular Vesicles in an Inflammatory Microenvironment In Vitro. *Int. J. Mol. Sci.* **2023**, *24*, 11069. [CrossRef] [PubMed]
9. Steyn-Ross, D.A.; Steyn-Ross, M.L.; Sleigh, J.W.; Voss, L.J. Determination of Krogh Coefficient for Oxygen Consumption Measurement from Thin Slices of Rodent Cortical Tissue Using a Fick's Law Model of Diffusion. *Int. J. Mol. Sci.* **2023**, *24*, 6450. [CrossRef] [PubMed]

10. Azam, S.; Kim, Y.-S.; Jakaria, M.; Yu, Y.-J.; Ahn, J.-Y.; Kim, I.-S.; Choi, D.-K. *Dioscorea nipponica* Makino Rhizome Extract and Its Active Compound Dioscin Protect against Neuroinflammation and Scopolamine-Induced Memory Deficits. *Int. J. Mol. Sci.* **2022**, *23*, 9923. [CrossRef]
11. Esteves-Monteiro, M.; Menezes-Pinto, D.; Ferreira-Duarte, M.; Dias-Pereira, P.; Morato, M.; Duarte-Araújo, M. Histomorphometry Changes and Decreased Reactivity to Angiotensin II in the Ileum and Colon of Streptozotocin-Induced Diabetic Rats. *Int. J. Mol. Sci.* **2022**, *23*, 13233. [CrossRef]

International Journal of
*Molecular Sciences*

MDPI

*Review*

# Fabrication and Characterization Techniques of In Vitro 3D Tissue Models

**Rohin Shyam [1,2], L. Vinod Kumar Reddy [3] and Arunkumar Palaniappan [2,*]**

1 School of Bio Sciences and Technology (SBST), Vellore Institute of Technology (VIT), Vellore 632014, Tamil Nadu, India
2 Centre for Biomaterials, Cellular, and Molecular Theranostics (CBCMT), Vellore Institute of Technology (VIT), Vellore 632014, Tamil Nadu, India
3 Department of Biomedical Engineering, University of North Texas, Denton, TX 76207, USA
* Correspondence: arunkumar.p@vit.ac.in

**Abstract:** The culturing of cells in the laboratory under controlled conditions has always been crucial for the advancement of scientific research. Cell-based assays have played an important role in providing simple, fast, accurate, and cost-effective methods in drug discovery, disease modeling, and tissue engineering while mitigating reliance on cost-intensive and ethically challenging animal studies. The techniques involved in culturing cells are critical as results are based on cellular response to drugs, cellular cues, external stimuli, and human physiology. In order to establish in vitro cultures, cells are either isolated from normal or diseased tissue and allowed to grow in two or three dimensions. Two-dimensional (2D) cell culture methods involve the proliferation of cells on flat rigid surfaces resulting in a monolayer culture, while in three-dimensional (3D) cell cultures, the additional dimension provides a more accurate representation of the tissue milieu. In this review, we discuss the various methods involved in the development of 3D cell culture systems emphasizing the differences between 2D and 3D systems and methods involved in the recapitulation of the organ-specific 3D microenvironment. In addition, we discuss the latest developments in 3D tissue model fabrication techniques, microfluidics-based organ-on-a-chip, and imaging as a characterization technique for 3D tissue models.

**Keywords:** in vitro models; 2D and 3D cell cultures; 3D tissue models; 3D bioprinting; confocal microscopy

**Citation:** Shyam, R.; Reddy, L.V.K.; Palaniappan, A. Fabrication and Characterization Techniques of In Vitro 3D Tissue Models. *Int. J. Mol. Sci.* **2023**, *24*, 1912. https://doi.org/10.3390/ijms24031912

Academic Editors: Patrick C. Baer and Ralf Schubert

Received: 27 November 2022
Revised: 30 December 2022
Accepted: 1 January 2023
Published: 18 January 2023



## 1. Introduction

In vitro two-dimensional (2D) cell culture methods are a widely used tool for understanding biological functions such as cellular interaction, mechanisms of disease initiation and progression, production of proteins, cellular biology, and, more recently, the development of engineered tissue mimics. In a 2D environment, cells are grown as a monolayer over a flat plastic surface, where they adhere and spread. However, the simplicity of this model makes the depiction and simulation of complex tissue structures challenging. Two-dimensional monolayer cultures have been used for decades to study the cellular responses to biochemical and biophysical cues. These systems do not always mimic human physiological conditions despite providing significant advancements in the understanding of cellular behavior [1], thereby resulting in non-predictive results.

In recent years, the paradigm has shifted towards three-dimensional (3D) cell cultures. Increasing research-based evidence suggests that 3D tissue models are a better option for mimicking complex tissue or organ architecture (cell–cell and cell–matrix interactions) and physiology [2]. These models are gaining importance from basic research to advanced application-based research such as drug testing/screening and other translational purposes. In human tissue, cells are encapsulated within extracellular matrix (ECM) proteins in a 3D environment [3]. The ECM function under defined biophysical and biochemical signals, which regulate cellular functions such as proliferation, adhesion, migration, differentiation, and morphogenesis and maintain homeostasis [4]. Hence, different 3D models are evolving with the combination of cells and proteins to recapitulate native organs and the cellular microenvironment. This aids in understanding the various organs and tissue functions under a controlled laboratory setting and offer the possibility to generate organ-specific and personalized drug testing platforms [5].

Recent advances in microfabrication techniques and tissue engineering technology have influenced the development of complex culture systems and biomimetic microfluidic platforms to capture the structural and functional complexity of the native physiological environment. Tissue engineering is a subfield of regenerative medicine that aims to repair, replace, or regenerate tissues or organs. This is achieved through the translation of fundamental principles of physics, chemistry, and biology combined with the principles of materials engineering and cell transplantation. The goal of this approach is to mimic native tissues that can function as medical devices with therapeutic benefits to regenerate damaged tissues, function as a platform to study drug cytotoxicity at a cellular and molecular level, and model disease under laboratory conditions [6,7]. With these advancements, 3D models with ECM-mimicking proteins could recapitulate the microarchitecture and functional cellular environment of the native organ. In recent years, organ-on-the-chip technology has been gaining prominence due to its ability to simulate organ-level physiology by recreating the multicellular connections and interfaces, vascular perfusion, mechanical cues, and chemical gradient under highly controlled environments.

The bioengineering and designing of complex biomimetic tissue for model systems involve considering several design characteristics and parameters. A 3D tissue model system can be generated through the fabrication of spheroids and organoids; however, while being able to provide a 3D microenvironment, a critical challenge with these systems is the lack of vasculature, which is essential in providing oxygen and nutrients while removing metabolic waste from cells. Alternatively, a scaffold that mimics the ECM is generated via techniques such as 3D bioprinting, electrospinning, and solvent casting/particulate leaching (SCPL) to create porous structures that house the cells, growth factors, vasculature, and transcription factors. The choice of biomaterial to generate the ECM is critical. There are a variety of natural and synthetic biomaterials available, with each having its own benefits and limitations. There has been an increased interest in the combination of biomaterials to generate hybrid biomaterials, which enhance the structural and biological properties of biomaterials. Another consideration is the choice of cells, which is dependent on the tissue being modeled. Stem cells are state-of-the-art in tissue engineering due to their differentiation potential into any cell lineage. Figure 1 summarizes the essential considerations in the realm of 3D tissue models. In the following sections, we discuss the types of 3D tissue models, types of biomaterials and their key characteristics, techniques used in the mimicking of tissue architectures and generation of porous scaffold structures, the types of cells used in 3D models, and their advantages and disadvantages, concluding with the imaging modalities of tissue architecture.

It is often challenging to directly study and observe the complex mechanisms of human development and disease due to a lack of experimental accessibility to biological processes. As a result, the use of model systems that recapitulate these functions ex vivo has been of primary interest to researchers. Two-dimensional culture systems have been established as standard protocols to observe cellular behavior; however, these systems do not completely recapitulate the cellular microenvironment. For instance, the epithelia of the small intestine is an active and rapidly renewing tissue that can undergo tissue widening and form compact folds, invaginations, evaginations, and wavy morphologies [8]. Similarly, in the cardiovascular system, myocardial fibrillar proteins form a 3D complex structure that changes orientation during systole and diastole, resulting in the cardiac tissue undergoing cyclic stress, torsion, and compression [9,10]. These microenvironments represent a major challenge to replicate in vitro. The replication of tissue-specific conditions within a 3D model can offer the ability to study these complex mechanisms and enable a deeper understanding of the role in human development and disease progression and as a platform for drug testing. The key differences between 2D and 3D cell cultures are tabulated in Table 1. There are several strategies to fabricate and characterize 3D tissue models. In this review, we have explored the various types of 3D model strategies involved in the fabrication of complex models. Importantly, we have also explained unique imaging techniques involved in the characterization of 3D tissue models.

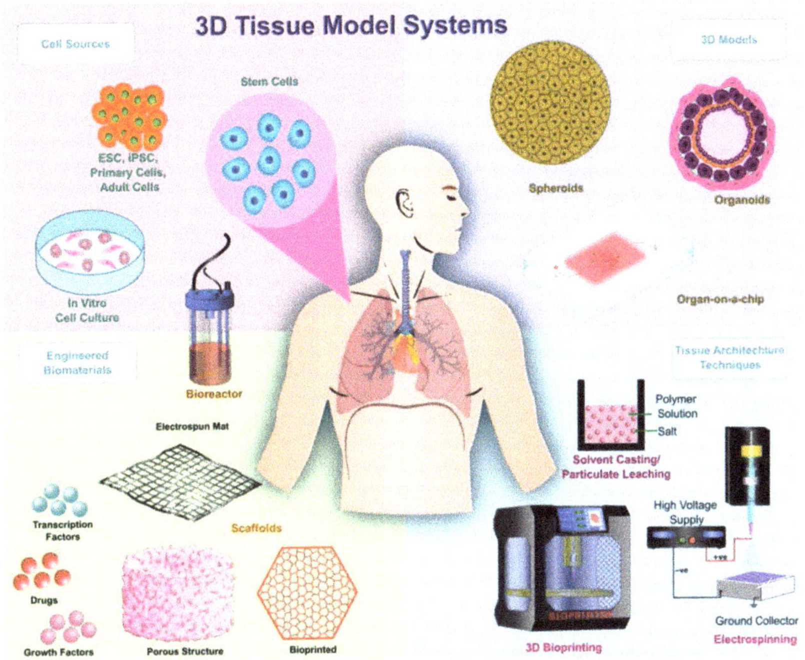


**Figure 1.** Schematic representation of considerations in the fabrication of 3D tissue models.

**Table 1.** Key differences between 2D and 3D cell cultures for modeling in vivo conditions.

| S.No | Characteristics | 2D Cell Cultures | 3D Cell Cultures | Ref |
|---|---|---|---|---|
| 1 | Cell morphology | Cell shape is elongated and grows on a flat 2D surface. | Natural cell shape is preserved with spheroids or organoids structures and with other 3D models | [11] |
| 2 | Cell proliferation | Cell growth in 2 dimensions is rapid and does not mimic in vivo | Cell growth is realistic under 3D culture conditions | [12] |
| 3 | Cell and ECM interactions | Growing on a flat surface is not mimicking the native tissue environment. There is no cell and ECM interactions. | Cells and ECM interact with each other and make a 3D environment such as the existing interactions in native tissues. These models reduce the cost of in vivo testing. | [1,13] |
| 4 | Cell–Cell Interactions | Multi-cell interactions cannot mimic the native organ environment. | Multi-cell interactions in different 3D tissue models can mimic the native environment. | [14] |
| 5 | Cell differentiation | Less resemblance to the native tissue. | Mimics native tissue-like differentiation and markers expression are close to the native tissues. | [15] |

**Table 1.** *Cont.*

| S.No | Characteristics | 2D Cell Cultures | 3D Cell Cultures | Ref |
| --- | --- | --- | --- | --- |
| 6 | Vasculature | 2D co-culture vasculature studies do not mimic the native vascular system. | Ability to incorporate complex vasculature in the 3D model | [16] |
| 7 | Protein and gene expression | Lack of 3D culture conditions, the expression levels may not show much resemblance to in vivo | Show resemblance with in vivo environment | [3] |
| 8 | Drug response efficacy | Low or sometimes not predictable due to cells growing on plastic substrate | Predictable as structure is—in vivo environment. | [3,17] |
| 9 | Apoptosis and viability (tumor models) | Sensitive to study the target drugs | High resistance to the anti-cancer drugs, which replicates the in vivo environment. | [12,18] |
| 10 | Mechanical stimulation | Mechanical stimulation may not mimic the native tissue | We can apply mechanical stimulus according to the native environment and it is an accurate representation of cells in vivo. | [14] |
| 11 | Physiological relevance | Highly non-relevant to the physiological environment. | Feasible to make physiologically relevant nutritional and oxygen conditions. | [1] |
| 12 | Exposure to culture condition | All cells receive nutrients and growth supplements equally. | The core site of the models will not obtain enough nutrients. Different approaches are explored to improve upon the nutrient and oxygen diffusion to the core site. | [13] |
| 13 | Experimentation and analysis | Easy to handle and highly reproducible. | Handling is difficult when compared to 2D cultures, less reproducible, and difficult to handle. | [19] |
| 14 | Characterizations | Easy to characterize the cells for experiment with any instrument. Well-established characterization techniques are available. | Difficult to characterize the 3D models. Specific instrumentation is required. Time consuming and not so well-established. | [20] |
| 15 | Cost | Inexpensive and well established | Expensive and requires further standardization. | [18] |

## 2. Types of 3D Tissue Models

### *2.1. Anchorage Independent (Non-Scaffold Based) 3D Tissue Models*

#### 2.1.1. Spheroids

Spheroids are perfectly spherical cellular aggregates in suspension generated from primary cell types and cell lines. The term was coined in 1970 by Sutherland et al. when the group dissociated Chinese Hamster V79 lung cells which formed spherical aggregates [21]. There are various techniques involved in the fabrication of spheroid, including the hanging drop technique, microwell hanging drop technique, liquid overlay technique, microwell array from micropatterned agarose wells, rotating wall vessel, and magnetic levitation (Figure 2A(i–vi)) [22,23]. Microfluidic technology and 3D bioprinting have also been utilized in the generation of spheroids [24–26]. The most common applications of spheroids are in creating tumor models, stem cell research, tissue engineering, and transplantation therapy. The key advantages of using this method as a 3D tissue model are that it facilitates cell–cell and cell–matrix interactions providing a physiochemical environment similar to in vivo while maintaining intrinsic phenotypic properties and improving the viability and proliferation of cells [23].

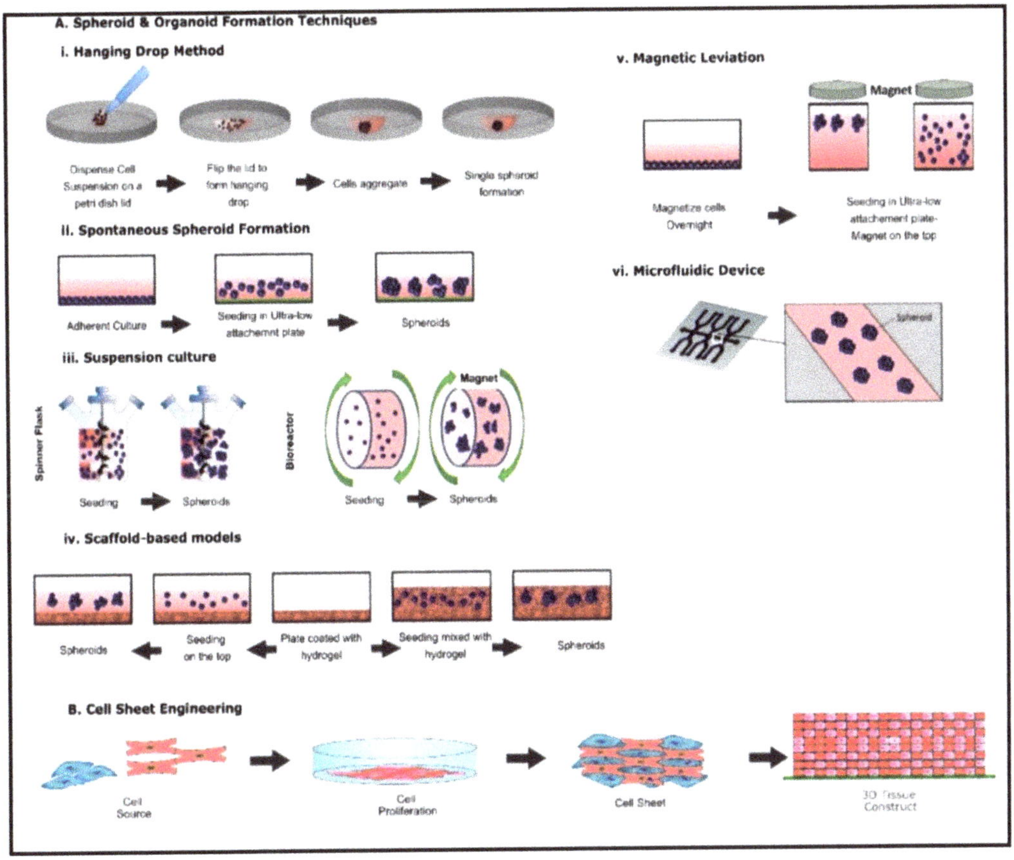


**Figure 2.** Common fabrication techniques used for the creation of spheroids, organoids, and cell sheet (**A**) (i) Hanging drop method (ii) Spontaneous spheroid formation (iii) suspension culture (iv) ECM method (v) Magnetic levitation method, (vi) Microfluidic device method. Altered and reproduced with permission from [27] under Open Access CC BY 4.0. MDPI (**B**) Schematic representation of cell sheet engineering.

However, there are several drawbacks to this method. Due to the lack of vasculature within the aggregates, the supply of nutrients to the core of spheroids is limited, and this limitation becomes pronounced with larger spheroid aggregates as it forms a diffusion gradient [23]. Additionally, despite the various techniques utilized in spheroid formation, each has its own unique challenges. For example, the hanging drop method is a simple method to implement and provides uniform spheroid shapes with greater control over spheroid shapes. However, it is tedious to handle and time-consuming, and inefficient due to low throughput. In other spheroid formation techniques, long-term survivability and tedious media exchange are the key challenges [28]. Despite exhibiting a 3D structure, inherently, spheroids lack the complex architecture of tissues in vivo and, therefore, cannot completely recapitulate the physiological environment.

### 2.1.2. Organoids

Organoids are 3D self-aggregating assemblies containing multiple cell types arranged spatially, such as cells in a tissue, recapitulating cellular and molecular stages in early organ development [29,30]. They have been used as tissue models to explore mechanisms of organ development. Organoids are increasingly being used in medical research, specifically in preclinical studies and in 3D tissue models, to study cellular interactions and drug-toxicology, pharmacology, and microbiology [29]. The 3D architectural and functional similarities to the tissue of origin make organoids an excellent model for studying complex cell–cell interactions and tissue development. The fabrication of organoid models is similar to the processes involved in the generation of spheroids (Figure 2A(i–vi)) [27,31]. However, the key difference is that in organoid formation, pluripotent stem cells and embryonic stem cells are given specific signaling cues that act as instructions to form 3D organoids of a variety of tissues [31]. Organoids have been employed in the generation of optical cups, liver, brain, lung, and heart [32–36]. They have also been used to model disease conditions to study disease development and progression. For example, in a recent study by Richards et al., cardiac organoids with oxygen-diffusion gradients were fabricated to model the human heart after myocardial infarction while recapitulating the hallmarks of myocardial infarction [36]. Yang et al. developed a mice 3D testicular organoid using testicular cells from BALB/c mice to investigate Zika-virus-induced mammalian testicular damage [37]. The key challenge of using organoids is the lack of vasculature. Optimization of the conditions for incorporating more than one type of cells to mimic in vivo structure is required [38]. Additionally, the effect of ECM composition and cell–matrix interaction requires further investigation to develop robust model systems. While there have been significant advances to overcome this challenge, research into multi-organ communication requires further investigation.

#### 2.1.3. Cell Sheet Engineering

Cell sheet engineering is a form of tissue engineering methodology that does not require a scaffold. In this method, cells are grown in vitro by placing a single-type cell on a stimuli-sensitive polymer (Figure 2B). In a culture environment suitable for cell growth, cells are grown till a three-dimensional cell sheet is generated. By inducing a stimulus such as heat, the polymer becomes hydrophilic, enabling the detachment of the cell sheet from the polymer base [39]. Cell sheet engineering has applications among various organs such as the heart, cornea, bladder, liver, and bone. The key advantage of using cell sheet engineering is the ability to co-culture cells and generate a vasculature network. For example, Sakaguchi et al. observed that endothelial cells within cell sheets spontaneously form blood vessel networks as in vivo capillaries [40]. Wu et al. investigated the therapeutic benefits of cell sheets derived from umbilical cord mesenchymal stem cells on rat models with induced ischemic heart failure [41]. The authors subjected H9C2 cardiomyocytes under hypoxia conditions and starvation to observe cell apoptosis as a 2D model, and an ischemic model was made by subjecting rats with Left Anterior descending artery (LAD) ligation to induce ischemic conditions [41]. The study observed that the cell sheets improved cell retention in the myocardium affected by ischemic heart failure, improved cardiac function, attenuated cardiac fibrosis, and induced neovascularization [41]. While recent research indicates that cell sheet engineering may pose a viable therapeutic solution, a major drawback of this method is the generation of hypoxic conditions within thicker cell sheets. Additionally, the lack of well-developed vascular networks within the cell sheet at the time of the generation of sheets poses further translational limitations [42].

### 2.2. *Anchorage Dependent (Scaffold Based) 3D Tissue Models*

3D tissue models offer the versatility of generating mini-organs that mimic in vivo physiology of a specific tissue. However, these models do not completely recapitulate the characteristics of the tissue. Spheroids and organoids have major drawbacks, such as poor mechanical strength and closed 3D geometry. This results in decreased oxygen and nutrients delivery to the center and hampers the use of conventional assays and instrumentation for screening studies such as nutrient and oxygen transport, absorption kinetics of drugs, and cell–cell interactions [43,44]. The paradigm of tissue engineering involves the conglomeration of living cells within bioartificial support to generate a 3D living structure with mechanical, structural, and functional properties equivalent to human tissue [45].

While the generation of artificial constructs is primarily for regenerative purposes, artificial tissues are being developed to replace reliance on animal models, which are dissimilar to human physiology and do not provide accurate predictions for human tissue responses. The conventional methods, from the perspective of tissue engineering for regenerative purposes, rely on the generation of support structures that act as a temporary scaffold to aid tissue regeneration while gradually degrading and being replaced by autologous tissues [46]. However, from the perspective of modeling, tissue replication should be designed to recapitulate the specific conditions being mimicked. This process is extremely complex due to several factors involved in the mimicking of tissue. Specifically, each tissue exhibits varying features such as porosity, ECM composition, cell phenotypes, and signaling pathways [47]. Ergo, the fundamental elements to consider in the designing of artificial tissue are the material for scaffolds, the cell source, the chemical stimuli, and the method for generating the correct tissue architecture. Additionally, it is pertinent that the choice of material is significantly dependent on the tissue being mimicked as the material will form the ECM, and therefore, the scaffold must meet the specific mechanical, chemical, physical, and biological requirements to achieve cell diffusion, proliferation, viability, and functionality [46]. The key modalities used in the generation of scaffolds for 3D tissue models are Solvent Casting Particulate Leaching (SCPL), Electrospinning, and 3D Bioprinting. Figure 3 provides a schematic representation of the methods, and Table 2 highlights the advantages and disadvantages each method has to offer.

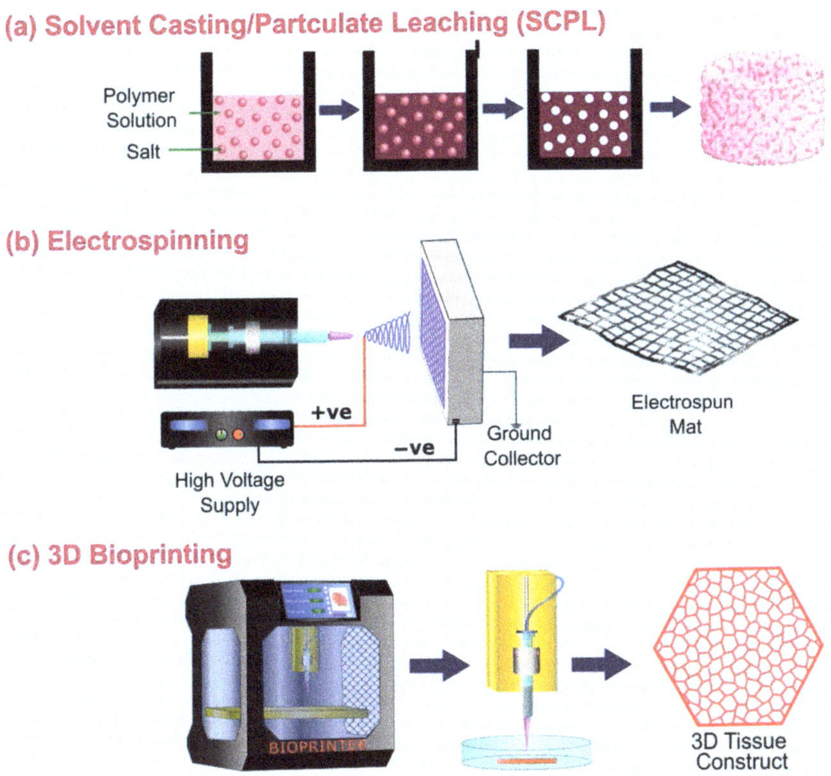


**Figure 3.** Schematic representation of the methods used in the generation of 3D tissue architecture.

**Table 2.** The advantages and disadvantages of fabrication methods used in creating 3D architectures.

| Method | SCPL | Electrospinning | 3D Bioprinting |
|---|---|---|---|
| Advantages | Control over pore size and density, ease of fabrication | Highly efficient and well-understood process, in-expensive, controllable fiber dimensions, high resolution | Moderate resolution, diverse choice of biomaterials, dimensional control, cheap, variation in geometry, spatial control of cells deposition. |
| Disadvantages | Residual solvents and salts, generally isotropic properties, weaker mechanical integrity | 3D architecture is challenging, toxic solvents, lower mechanical properties, inhomogeneous cell distribution, time consuming. | Finding the right bio-inks with the right sterilizable and cross-linkable properties. Limited devices can print whole organ, application dependent |

#### 2.2.1. Solvent Casting Particulate Leaching (SCPL)

SCPL is a popular technique used in the fabrication of highly porous polymer scaffolds for hard tissues such as bone and teeth. In this method, a salt that is insoluble in the polymer is admixed in a polymer solution followed by an evaporation process to remove the solvent, resulting in a salt-polymer composite. The composite matrix is then submerged in water to leach out the salt resulting in a highly porous structure (Figure 3a) [48]. Through this method, 50–90% porosity is achieved [49]. A key advantage of this method is the relative ease and low cost associated with the fabrication of highly porous and tunable pore size that enables the migration of cells within the scaffolds [50]. Similar processes that are employed in the generation of highly porous structures are freeze-drying [51,52], thermal-induced phase separation (TIPS) [53], and gas foaming [54]. The advantages and disadvantages of this method are covered in Table 2.

#### 2.2.2. Electrospinning

The term is derived from electrostatic spinning and is a method that utilizes a high-voltage electric field to draw charged threads of ultrafine nanometric scale fibers from polymer melts or solutions [55,56]. The technique is complicated and involves a process where a charged droplet of polymer in a liquid phase under high voltage results in an electrostatic repulsion counteracting surface tension and elongation of the droplet to a critical point of liquid stream eruption termed a Taylor cone [56]. As shown in Figure 3b, a standard electrospinning system consists of a syringe pump, a metallic needle, a high-voltage DC supply, and a grounded collector. In the process of electrospinning, solvents evaporate, and the resulting fibers are solidified to form nonwoven fibrous membranes. Typically, cells suspended in cell culture media are seeded on electrospun mats in tissue culture well plates to cultivate cells within the scaffold [57]. Recently, there have been advances in incorporating cells within the polymer solution as a cell-laden bioink to generate cell-laden fibrous structures [39]. This technique was first introduced by Townsend-Nicholson et al., who used a coaxial system to encapsulate cells in a bio-suspension within an outer core of PDMS [40]. The key material and process parameters that need to be considered in the generation of either electrospinning or cell-electrospinning are viscosity, applied electric field, feed rate, and the distance between the nozzle and collector plate, along with environmental factors such as room temperature, relative humidity [39]. Table 2 summarizes the advantages and disadvantages of using such a system.

#### 2.2.3. Bioprinting

3D bioprinting is the layer-by-layer deposition of cell-laden biomaterials in 3D space based on a predetermined geometry. Complex geometries and shapes are designed through computer-aided design (CAD) software or geometries extracted from medical images. The main modalities of 3D bioprinting are based on the delivery system of the cell-laden biomaterials termed bio-inks and include extrusion-based (extrusion can be achieved via pneumatic, piston, or screw), inkjet (thermal or piezoelectric), and laser-assisted [58] (Figure 4). In a typical extrusion-based 3D bioprinting system, bioink is extruded via a needle, and based on the pattern generated in a CAD file, a 3D structure in a bottom-up approach is generated (Figure 4A). Three-dimensional bioprinting is a rapidly evolving technology employed to print a variety of tissue structures of various organs, and the frontier of 3D bioprinting is the printing of a complete artificial whole organ, which was most recently achieved by Mirdamadi et al. [59] using a novel technique termed Freeform Reversible Embedding of Suspended Hydrogels (FRESH). In the study, the authors modified an extrusion-based bioprinter and embedded alginate in a support bath comprised of gelatin microparticles suspended in a calcium chloride solution [59]. The core principle is that the gelatin microparticles act as a support bath with multiple crosslinking strategies to gel the different types of hydrogels while providing support for embedded hydrogels that would normally collapse in conventional additive manufacturing processes as they are being printed (Figure 4F) [60]. Senior et al. modified the FRESH bioprinting approach to generate stable hydrogels with low viscosity, termed Suspended Layer Additive Manufacturing (SLAM) [61]. In their study, bioinks with low viscosity in the liquid phase prior to gelation were extruded in an agarose gel that exhibited shear thinning property as the material was extruded and regained its structure upon removal of the shear force entrapping the suspended hydrogel [61]. A crosslinker was then allowed to diffuse through the agarose fluid gel, which resulted in the hydrogel forming stable structures and could be easily removed from the fluid gel [61] (Figure 4G). While microgel support baths have been used to demonstrate full organ printing [59], a shift from this paradigm is the bioprinting of a sacrificial bioink within a slurry-support bath comprised of cellular spheroids in a technique termed sacrificial writing into functional tissue (SWIFT) (Figure 5). Skylar-Scott et al. reported the use of this technique to generate a living matrix primarily composed of tightly compacted tissue-specific organ building blocks from iPSC-derived embryoid bodies, multicellular spheroids, or organoids [62]. Within the living matrix, a sacrificial ink is patterned and embedded via 3D printing, which, when removed, yields perusable branching channels and conduits, thereby resembling vascularized networks (Figure 5) [62,63]. While Support bath systems with extrusion-based bioprinting could be an effective platform in the fabrication of microtissues, however, controlling the position within the 3D space is a challenge [63]. Novel methods to circumvent these challenges are emerging within the scientific community and have been covered elsewhere [63].

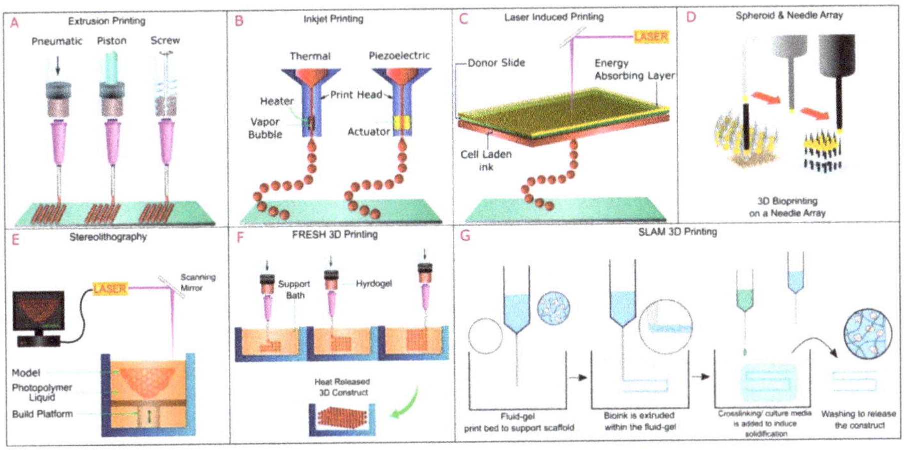


**Figure 4.** Modalities of 3D Bioprinting (**A**) Extrusion based printing (**B**) Inkjet Printing (**C**) Laser Induced Printing (**D**) Kanzen Spheroid and needle array (**E**) Stereolithography (**F**) FRESH 3D printing method (**G**) SLAM 3D Printing. (**D**) reproduced under the terms and conditions of the Creative Commons CC BY 4.0 License [64] Copyright 2017, The Authors. Published by Springer-Nature Publishing. G reproduced under the terms and conditions of the Creative Commons CC BY 4.0 License [61] Copyright 2019, The Authors. Published by Advanced functional materials.

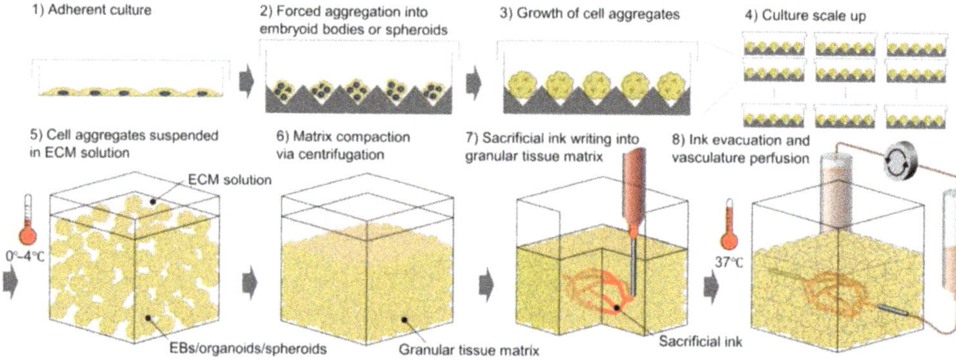


**Figure 5.** Schematic representation of the SWIFT process. Copyright © 2023 The Authors [62], some rights reserved, exclusive licensee American Association for the Advancement of Science. Distributed under a Creative Commons Attribution Non-Commercial License 4.0.

3D bioprinting offers versatility in controlling essential parameters such as bioink composition, printing speed, needle gauge, extrusion pressure, and scaffold geometry. However, despite the plethora of biomaterials available for this technique (Table 3), each biomaterial has unique properties that must be optimized to generate suitable constructs. A major benefit of this method is the ability to generate a complex vasculature network via bioinks laden with endothelial cells, as shown in recent research by Noor et al. [65]. Despite the large library of biomaterials that can be used, not all materials have gelling properties required to hold the shape fidelity of the final printed structure and need to be modified to enhance mechanical strength along with chemical, physical, and biological properties. In certain circumstances, bioinks are stabilized through post-processing crosslinking mechanisms via photon activation through UV light in the presence of a photoinitiator or via ionic crosslinking in the presence of divalent cations. While 3D bioprinting offers vast opportunities, it is severely limited by the availability of printers capable of printing whole organs [66]. Additionally, further research into improving the print resolution of the printed construct and encapsulation of cell densities from a clinical translation outlook remains a challenge [67].

#### 2.2.4. Organ-on-a-Chip

The process of developing novel drugs and medical interventions requires the use of in vitro modeling, followed by animal studies, to test the safety and efficacy of newly developed drugs before testing on humans. However, animal models do not provide accurate predictions for human responses. Clinical trials are time-consuming and not cost-effective in the long run. Most novel drugs fail in clinical trials, and therefore, there is a need to develop a system or model that mimics human physiology, remains cost-efficient, and has the capability to provide accurate data. In contrast to biological approaches to generate 3D tissue models, organ-on-a-chip (OOC) systems are used to recapitulate tissue and organ structure by leveraging microfluidic physics along with microfabrication engineering techniques and biomaterials to create micro-physiological systems that model tissue structure and disease conditions. Research into the development of microfluidic channels to study signal pathways, drug responses, and tissue functions is ongoing [68]. For example, Zhao et al. employed OOC to create a platform to generate chamber-specific cardiac tissue and disease modeling to measure contractile force in ventricles and atriums and their response in the presence of drugs [69]. Similarly, Parsa et al. developed a platform to study mechanisms of cardiac hypertrophy with low cell volume [70].

There is a wide range of organ systems that have been modeled on an OOC platform, including the heart [71], kidney [72], brain [73], lung [74], intestine [75], liver [76], and eyes [77]. Additionally, OOC has been employed to study tissue-specific diseases.

Costa et al. reported the use of microfluidics to mimic arterial thrombosis in vitro [78]. The study was designed to replicate a three-dimensional architecture of coronary arteries under healthy and stenotic conditions by modeling healthy and stenotic arteries to create a microfluidic chip with inlets and outlets to allow perfusion through the system [78]. This enabled the authors to study the effect of shear rates within arteries and enable a better understanding of arterial thrombosis [78]. Microfluidic technology has also been leveraged as a tool to generate spheroids and organoids [26], study drug pharmacokinetics, and the generation of micro bioreactors where 3D bio-printed tissue constructs can receive oxygen and nutrients under laminar flow conditions. While most microfluidic systems use a design-based approach and leverage fluid behavior on a microscale, the lack of ECM or an in-vivo-like microenvironment is a drawback in OOC. OOC technology is based on the use of soft lithography to generate molds of microchannels with the use of polydimethylsiloxane (PDMS) as a substrate material. The high resolution offered by stereolithography and the ability to miniaturize the microenvironment enables researchers to study complex diseases and their behavior in a heterogeneous environment. In designing and production of OOC, the selection of cells and biomaterials must be given extensive consideration. In order to improve the relevance of OOC, it is crucial to include vascular networks that can provide efficient nutrient and oxygen diffusion across the tissue or microfluidic channels. There has been a focus on the incorporation of scaffolds or hydrogels into microfluidic systems to overcome this drawback. The presence of an ECM-like matrix to house cells provides both biophysical and chemical cues that aid in the development of a more in-vivo-like microenvironment. Figure 6 provides a schematic representation of the microfluidic system integrated with hydrogels to generate 3D in vitro models to study disease. For example, Shang et al. used 3D bioprinting to generate biomimetic hollow blood capillaries [79]. The authors created microchannels using 3D printing and injected a composite of GelMA and Alginate incorporated with human umbilical cord endothelial cells (HUVECS), the hydrogel being crosslinked with either barium or calcium chloride Figure 7i and studied the proliferation of cells in the hollow chamber [79]. Hong et al. used 3D bioprinting to fabricate cancer spheroids for evaluating the drug resistance of cancer cells [80]. In their study to evaluate the efficacy of drug resistance of cancer cells, the authors printed 3D mini-wells using poly (lactic acid) in a grid structure. The authors then embedded drug-resistant MCF-7 breast cancer cells in a gelatin–alginate hydrogel bioink and 3D bioprinted into the mini-wells to encourage single spheroid formation [80]. The use of hydrogels encapsulated within microfluidic devices to provide a more comprehensive in-vivo-like environment could change the research field towards lesser reliance on animal models.

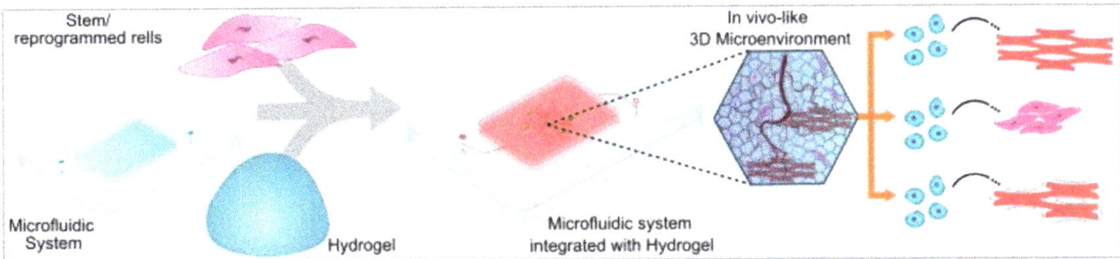


**Figure 6.** Microfluidic system integrated with hydrogels and cells to provide in-vivo-like 3D microenvironment with biochemical and biophysical cues that result in enhanced differentiation of stem cells or reprogrammed cells, generate functionally mature tissue specific cells and enable a structurally organized microenvironment.

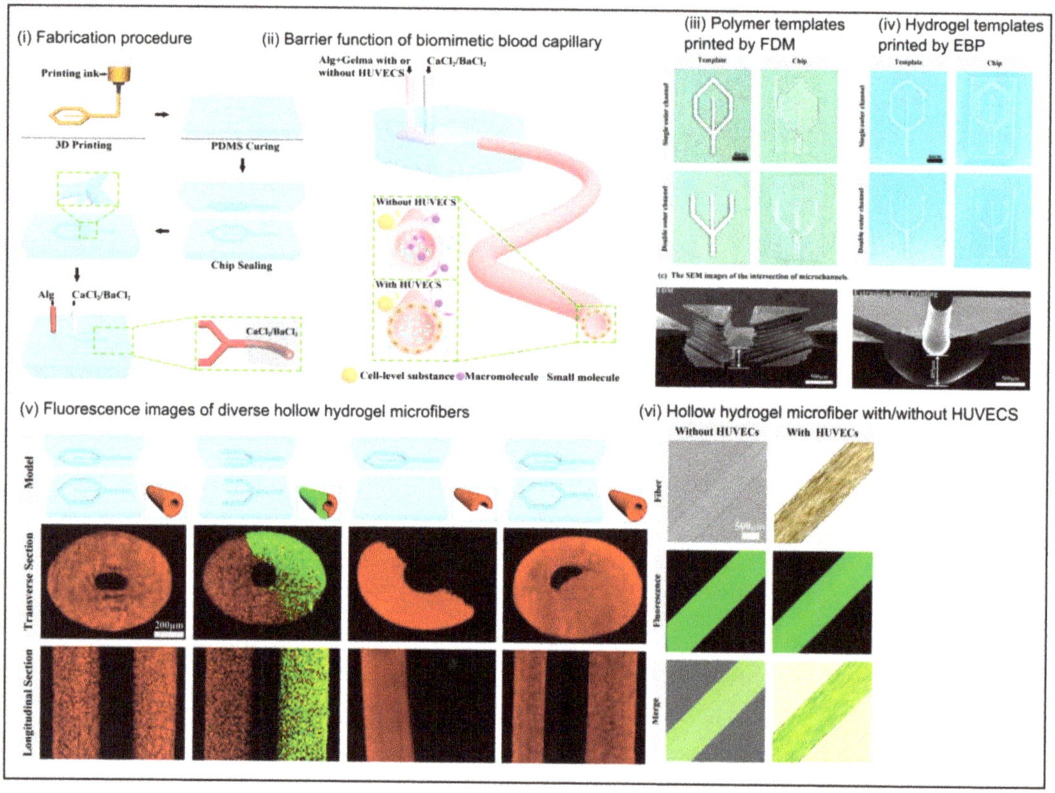


**Figure 7.** Fabrication of microfluidic chip to generate an in vitro model to simulate hollow biomimetic capillary using 3D bioprinting and hydrogel. (**i**) fabrication of templates and microfluidic device (**ii**) simulating hollow blood capillary. Comparison of templated printed using 3D printing, with (**iii**) fused deposition model (FDM) and (**iv**) using hydrogels and extrusion-based bioprinting. (**v**) fabrication and characterization using fluorescence microscopy of diverse hollow structures (**vi**) barrier function of hollow hydrogel microfiber with cells. Image reprinted with permission from [79] Copyright © 2023 by the authors ACS Biomaterials Science and Engineering.

Despite the significant advantages this technology has to offer, there are several challenges that need to be addressed. For example, PDMS is the most common material used as a substrate to build a microfluidic device. However, it is known that PDMS absorbs small molecules such as drugs and may have an impact on drug bioactivity in OOC devices designed to study cell behavior and drug efficiency [81]. While there are other materials that can be used in the generation of microfluidic devices, PDMS is one of the most predominantly used materials that is used in the generation of microfluidic devices, and these materials have been thoroughly reviewed elsewhere [82]. The lack of multi-organ interaction and communication is a drawback of this technology. However, researchers have reported the generation of multi-organ/human-on-a-chip. Abaci et al. reported a conceptual study on the design parameters and considerations in developing such a model [83]. The benefits offered by OOC technology outweigh the drawbacks, which have resulted in the continued development of this technology. With the incorporation of novel biomaterials and nanotechnology, OOC platforms are expected to evolve with technological advancements in the future.

## 3. Biomaterials for 3D Tissue Modelling

Advances in research have led to the development of improved 3D tissue models for in vitro studies. Cells in nature reside in a molecular matrix composed of protein, glycosaminoglycan, and glycoconjugate, termed the extracellular matrix (ECM). The ECM provides physical scaffolding, biochemical cues, and mechanical stability to cells and is necessary for morphogenesis and homeostasis [84]. The engineering of ECM that mimics native tissue matrix begins with the identification of a biomaterial that is critical in the formation of a scaffold. The choice of biomaterial is dependent on the tissue being modeled. Biomaterials are based on three categories (a) Polymers, (b) metallic, and (c) ceramics. Factors that influence the choice of materials are the type of tissue being mimicked, structural integrity, adequate mechanical environment, bioactivity, biocompatibility, and biodegradability [84]. The biomaterial should provide structural support for cellular attachment, growth, proliferation, and migration while consisting of adequate mechanical properties and an environment native tissue matrix provide to cells. Materials should be bioactive and biocompatible to provide bioactive cues and growth factors while reducing the risk of immunological response in the presence of an artificial scaffold. Additionally, the scaffold or matrix should act as a support structure facilitating correct localization and retention at the site of tissue damage [85]. While biodegradability is key for the formation of the vascular network and allows for patients' own ECM to replace the scaffold and degrade over time without any cytotoxic effects [86], this factor is organ-specific. For example, in regenerative medicine for hard tissues such as bone or teeth, materials are engineered from metallic or ceramic biomaterials to reduce the rate of biodegradability. Table 3 provides a summary of the various biomaterials and their pros and cons.

**Table 3.** List of Biomaterials, both natural and synthetic employed in tissue engineering and their advantages and disadvantages.

| Biomaterial | Type | Pros | Cons | Ref |
|---|---|---|---|---|
| **Collagen** | Natural | High biocompatibility, biodegradable, high cell adhesion, and cell remodeling. Has high printability, is biocompatible, low immunogenicity | Poor mechanical properties, unpredictable degradation in vivo, high thrombogenic potential | [87] |
| **Gelatin** | Natural | Cheap, biocompatible, easy to modify, good proliferation, biodegradable | Brittleness, low mechanical properties, fast degradation | [88] |
| **Chitosan** | Natural | Biocompatible, biodegradable, high cell proliferation | Lower mechanical properties, immunogenic | [89] |
| **Fibrin** | Natural | High cell adhesion and viability, quick gelation and good cell migration, and vascularization | low printability, biocompatibility, low mechanical strength | [90] |
| **Hyaluronic Acid** | Natural | Biocompatible, biodegradable, high cell proliferation and viability, high printability | low mechanical strength | [91] |
| **Alginate** | Natural | Biocompatible, biodegradable, sustained release, adoptable mechanical strength with cell growth, rapid gelation | low cell adhesion | [92] |
| **Pectin** | Natural | Cheap, biocompatible, can be modified, plant derived, good cell proliferation, biodegradable | Poor mechanical properties, Slower gelation time | [93] |
| **Decellularized ECM** | Natural | Keeps vasculature network intact | Variation caused by different decellularization methods, | [94] |

**Table 3.** *Cont.*

| Biomaterial | Type | Pros | Cons | Ref |
|---|---|---|---|---|
| **Starch** | Natural | Cheap, biocompatible, versatile rheology, | Poor mechanical properties, slower gelation time, needs high temperature (70–90 °C) to gelatinize, at higher temperatures, phase separation between composite materials may occur | [95] |
| **Fucoidan** | Natural | Good bioactive properties, biocompatible, biodegradable, used to enhance properties of other natural biomaterials | Does not gel on its own, crosslinking strategies need to be optimized, high synthesis cost | [96] |
| **Silk Fibroin** | Natural | Biocompatible, good mechanical properties | High cost of production, | [97] |
| **Hydroxyapatite** | Natural/Synthetic Synthesis | Bioactive, biocompatible, hydrophilic, | brittleness, low tensile strength and fracture toughness | [98,99] |
| **Polycaprolactone (PCL)** | Synthetic | Moderate mechanical strength. Biocompatible | Slow degradation, lower cell adhesion/aggregation, hydrophobic, inflammation due to acid degradation products | [100] |
| **Poly Lactic-co-Glycolic Acid (PLGA)** | Synthetic | Biocompatible, biodegradable, immunogenic | Brittle and relatively hard, lower cell adhesion/aggregation, inflammation due to acid degradation products | [101] |
| **Poly(itaconate-co-citrate-cooctanediol) (PICO)** | Synthetic | Biocompatible, biodegradable, cheap, good mechanical properties, fast crosslinking, non-cytotoxic to cells | UV cross linking | [102] |
| **Poly (ethylene glycol) (PEG)** | Synthetic | Biocompatible, biodegradable, can be modified with various functional groups | Moderate mechanical strength, low printability, difficulty in scalability, Lower cell adhesion | [92] |
| **Polyphosphazenes** | Synthetic | Biocompatible, good mechanical properties, slow degradation (hard tissues) | Slow degradation (soft tissues) | [97,103] |
| **Polyurethanes** | Synthetic | Good mechanical properties, good rheological properties | Poor degradability, copolymerization is required | [104] |
| **Polyanhydrides** | Synthetic | Good flexibility, controllable degradation rates | Weak mechanical properties | [104] |
| **Poly(propelene-fumarate)** | Synthetic | Good processability, good ductility, biocompatibility, easily forms covalent polymer networks | Challenging to handle the material due to high viscosity, increased cytotoxicity and acute inflammation, variation in molecular weight between crosslinking agents | [105,106] |
| **Metals** | Synthetic | Biocompatible with good mechanical properties, low degradability (Tissue dependent) | Subject to oxidation, low degradability (Tissue dependent), may be cytotoxic due to release of free metal ions | [107] |
| **Ceramics** | Synthetic | Osteoinductive and osteoconductive in bioactive ceramics, low toxicity, biocompatible, angiogenetic potential, | High brittleness, weak, low bioactivity | [107] |

*Characterization and Optimization of Biomaterials*

There is a plethora of biomaterials available in the generation of ECM, such as structures, and the choice of biomaterial is highly dependent on the tissue of interest. On the formation of stable tissue-like constructs through any of the biofabrication techniques, the constructs should be subjected to various characterization techniques to ensure that they meet the parameters as close as possible to native tissue. Table 4 highlights the fundamental properties and the quantitative methods utilized in the characterization of these properties of biomaterials.

**Table 4.** List of quantitative methods utilized in the characterization of the fundamental properties of biomaterials.

| Characterization | | Properties | Method |
| --- | --- | --- | --- |
| **Physical** | Swelling Ratio | Fraction increase in weight of hydrogel due to water absorption | Weighing difference |
| | Degradation rate | Fractional decrease in material to facilitate tissue growth | Collagenase and Weight measurement |
| | Porosity and Morphology | Determination of porous structure to facilitate cellular impregnation and proliferation | Scanning Electron Microscope (SEM), Brunauer, Emmett and Teller (BET) technique |
| **Chemical** | FTIR Spectroscopy | Investigate formation of Chemical Bonds | Standard FTIR Protocol |
| | H-NMR | Investigate the Molecular Structure | Standard NMR Protocol |
| | Degree of Functionalization | Quantify functional groups | Habeeb Method |
| **Mechanical** | Mechanical Loading | Determine elasticity of biomaterials | Youngs Modulus, Tension, Compression, Shear, Torsion, Yield Strength, Ultimate Yield Strength |
| | Rheology | Determine viscoelastic characteristics such as Shear Thinning, Viscosity, Storage and Loss Modulus | Rheometer, Viscometer |
| | Printability | Determine optimum parameters to enable efficient printing e.g., Needle Gauge, Print speed, Extrusion Pressure, Geometry, laser power, UV Crosslinking | Based on the tissue architecture equipment |
| **Biological** | Cell Volume | Optimize quantity of cells required for functional models | Cell Culture methods |
| | Cell Viability and Proliferation | Determine and monitor the response and health of cells, survivability, and spread tissue model | CCK-8, MTT, XTT, Live/Dead |
| | Cytotoxicity, Adhesion | Determine toxicity of biomaterials on cells and how well cells adhere to surface of tissue models. | Fluorescence microscopy, Confocal Microscopy |
| | Immunostaining | Identification and assessment of the topographical distribution of cells, proteins, and detect antigen levels. Eg F-Actin/DAPI | Immunohistochemistry, Flow cytometry, Western Blotting, ELISA, Immuno-electron Microscopy |

## 4. Cell Sources

The incorporation of cells is essential in the generation of functional 3D tissue models. In general, cells can either be seeded on an existing carrier matrix or can be encapsulated within a biomaterial [108]. The factor governing cell incorporation is dependent on the tissue architecture fabrication method. While there are various methods to create scaffolds for 3D tissue models, the choice of the cell is highly dependent on the tissue being modeled. Primary cells closely mimic in vivo physiological state of the tissue or organ of interest; however, not all organs or tissues have primary cells in sufficient quantities or have limited proliferative potential. Figure 8 provides a schematic representation of the various cell sources [108–112].

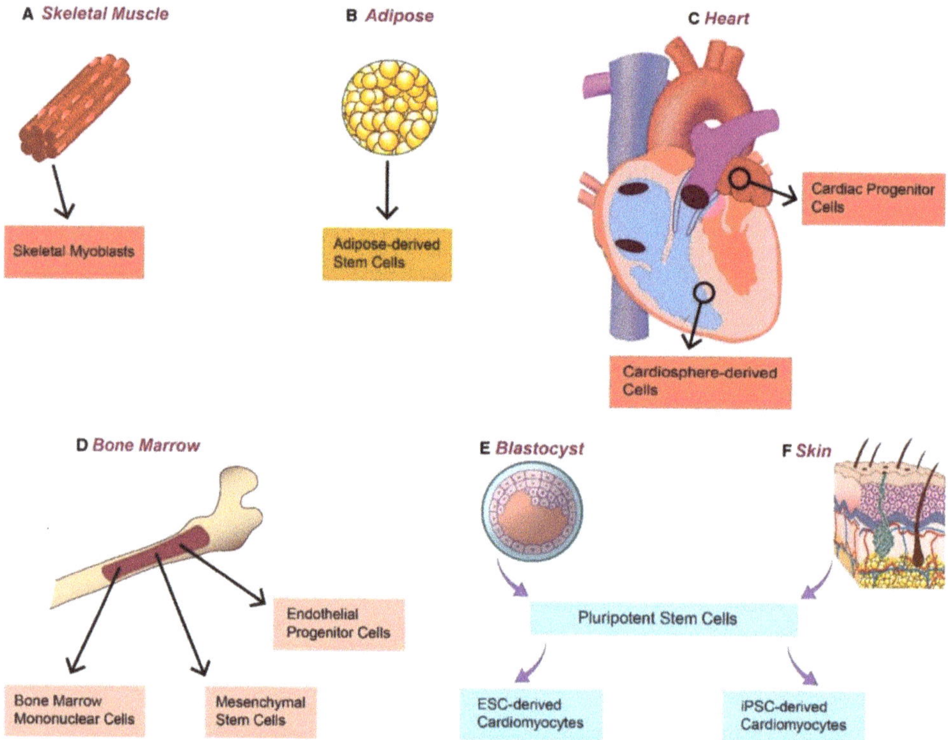


**Figure 8.** Schematic representation of the Cell Sources. (**A**) skeletal myoblasts, (**B**) adipose-derived stem cells, (**C**) cardiac 'progenitor' cells and cardio sphere-derived cells (**D**) bone marrow-derived stem cells (**E**) Embryonic stem cells derived from the blastocyst (**F**) induced pluripotent stem cells derived from skin. Modified and adapted with permission from [113] under creative commons license CC BY 4.0. Copyright 2015 The Authors. Journal of Cellular and Molecular Medicine published by John Wiley & Sons Ltd. and Foundation for Cellular and Molecular Medicine.

Recent findings on the differentiation of stem cells towards any tissue-specific lineages have led to significant advancements in tissue modeling and have been reviewed exhaustively elsewhere [108,111,112]. The conglomeration of novel biomaterials fabrication strategies, advances in stem cell biology, and 3D bioprinting has evolved as a next-generation technology for in vitro tissue model development. Table 5 provides a summary of the advantages and disadvantages of the various cells used in 3D bioprinting.

**Table 5.** Advantages and Disadvantages of the various types of stem cells found within the human body.

| Cell category | Pros | Cons | Reference |
|---|---|---|---|
| **Mesenchymal Stem Cells** | Can differentiate into many types of cells | Limited in quantity, differentiation capacity diminishes with age | [113,114] |
| **Adult stem cell** | Can differentiate into cells of the lineage it belongs to, main function is to repair the organ they are found in | Limited in quantity, expensive Limited capacity to divide | [85] |
| **Adipose Derived Stem Cells** | Multipotent, easily isolated, easily available | Low survival | [115] |
| **Embryonic stem cells** | Can differentiate into any cells in the right conditions, | Ethical concerns, allogenic and hence require immunosuppressants | [114] |
| **Induced Pluripotent Stem Cells** | Can be reprogrammed to embryonic stem cell-like state, can differentiate into any cell, | teratomas formation | [114] |

## 5. Imaging Modalities of 3D Tissue Models

There are a variety of methods used in the generation of 3D tissue models which have been discussed. While characterization methods are employed to ensure that the physical, mechanical, chemical, and biological parameters are met, these methods are often destructive and do not provide an insight into what is happening in the tissue once it is constructed. Therefore, additional methods are required to characterize and ensure that the final tissue model works as intended. In 2D cell culture and model systems, imaging, molecular, and immunohistochemistry techniques are commonplace. However, in a 3D system, advanced techniques are essential. Imaging techniques allow observation of the live-cell morphology and other organelles within the cells from 3D tissue models. Scanning electron microscopy analysis helps to find the cell morphology, migration, attachment, and cell–cell and cell–matrix interaction. Recent techniques have shown the real-time analysis of biological parameters in 3D cell/tissue models. Ruslan et al. used polymer-conjugated nanoparticles to identify $O_2$ in cells present in the 3D tissue models [116]. Muller et al. used nanoparticle-based fluroionophore to study live analysis of $K^+$ flux in 3D tissue models and animals [117]. Cell density can be analyzed with the presence of nucleated cells with H&E-stained histologic section photographs by using the ImageJ tool [118]. Table 6 provides a summary of the common advanced imaging techniques used for the analysis and examination of 3D tissue models, and Figure 9 provides the use of Optical Coherence Tomography (OCT) to characterize hydrogels.

**Table 6.** Common advanced imaging techniques used to analyze 3D tissue models and live cells within scaffolds.

| Imaging Modalities | Characteristics | Application | Ref |
|---|---|---|---|
| **Fluorescence** | Cells are marked with fluorescence markers and a sample is irradiated with wavelengths between visible and ultraviolet to reveal fluorescent species. | Cell viability, proliferation | [119] |
| **Confocal Imaging** | Advanced version of fluorescence resulting in high-resolution images by collecting light from a single plane of focus and eliminating out-of-focus light | Cellular Structure, viability, live imaging, 3D reconstruction | [120] |
| **Scanning Electron Microscope (SEM)** | A technique used to produce high-resolution images of surface topography by scanning with electrons on the surface | Surface morphology | [121] |

**Table 6.** *Cont.*

| Imaging Modalities | Characteristics | Application | Ref |
|---|---|---|---|
| **Transmission Electron Microscope (TEM)** | A technique in which electrons pass through ultrathin samples to generate high-resolution images | Characterization of pore structure, nano structures | [122] |
| **Fluorescence Recovery After Photobleaching (FRAP)** | A high-intensity laser causes bleaching in a region of interest (ROI) and gradual recovery of fluorescence from the surrounding environment to the bleached area is observed. | Used to study oxygen and nutrient diffusion across cells and tissue structures. | [119] |
| **Fluorescence Loss in Photobleaching (FLIP)** | Involves repeated bleaching of an ROI and measuring fluorescence intensity outside the bleached area where a drop in fluorescence intensity due to bleached non-fluorescent molecules provides quantitative data on molecular mobility. | Molecular mobility, exchange of molecules between cell compartments | [123] |
| **Fluorescence localization after photobleaching (FLAP)** | Involves labeling molecules with two fluorescent labels: one to be bleached locally and the second is a reference label that remains intact. By measuring the difference between bleached and unbleached signals gives an absolute FLAP signal which can be used to track the labeled molecule. | Ability to identify molecules and populations that have varying speeds and have dissimilar dynamics. | [123] |
| **Fluorescence Resonance Energy Transfer (FRET)** | A physical process in which a molecular fluorophore is excited and a nonradiative energy transfer occurs to another fluorophore through intermolecular long-range dipole-dipole coupling. This process is highly dependent on the distance between two fluorophores. | Live -cell analysis of cell biology, cellular interaction in 3D scaffolds, Protein–protein interaction, receptor activation, intramolecular distances | [124,125] |
| **Fluorescence Lifetime Imaging Microscopy (FLIM)** | A method in which the fluorescence decay time is measured. In combination with FRET, this method can be used to map spatial distribution to indirectly measure bimolecular interactions, concentration, and conformational changes. | Measuring intramolecular distances, evaluate therapeutic efficacy in drug screening | [119] |
| **Phosphorescence Lifetime Imaging Microscopy (PLIM)** | Similar to FLIM; however, this process images phosphorescence quenching. | Measuring partial oxygen concentration and identify hypoxic environment. | [119] |
| **Optical Coherence Tomography (OCT)** | Measures optical backscatter from different microstructural features within materials and tissues to generate high-resolution images of the cross-sections of tissue. | Quantify changes in porosity of scaffold, pore size, pore interconnectivity, cell dynamics and tissue development. | [126,127] |
| **Micro-Computerized Tomography (MCT)** | Involves exploiting variations in X-ray absorption, refraction, and scattering to form contrast alterations resulting in spatial distribution of material densities and providing 3D images of the internal structure. | Largely used in bone tissue models, as this technique offers the ability to form contrasts between soft and hard tissues. | [119] |

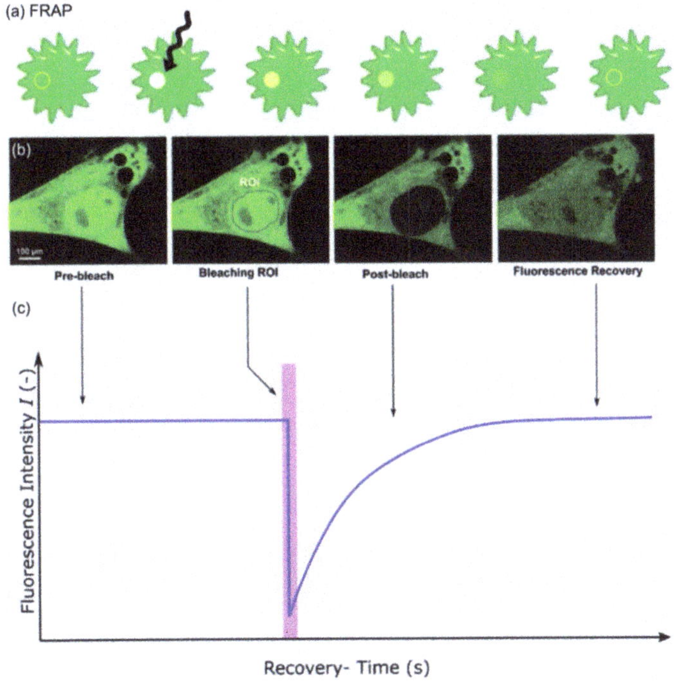


**Figure 9.** (**a**) Schematic representation of FRAP method, (**b**) Example of a FRAP experiment (**c**) Anatomy of a typical FRAP curve. Modified and reprinted with permission from [123] under creative commons CC BY 4.0. Copyright © 2023 by the authors; licensee MDPI, Basel, Switzerland.

### *5.1. Fluorescence Recovery after Photobleaching (FRAP) Using Confocal Microscopy*

Confocal fluorescence microscopy is an advanced method of fluorescence microscopy where high-resolution images can be obtained by the introduction of a spatial pinhole before the light source and the detector [128]. The aperture of the pinhole can be controlled to limit diffraction and thereby eliminate out-of-focus light from the sample. FRAP is a method used to study the movement of molecules that have been doped with a fluorescent dye (Figure 9) [123]. In FRAP, mobile fluorescent molecules are bleached by a high-intensity laser source. The bleached molecules are exchanged with fluorescent molecules from the surrounding area resulting in a recovery of fluorescent intensity. This information is plotted on a recovery curve and can be used to study the behavior of the molecules (Figure 9) [129]. The key advantage of using FRAP with confocal microscopy is that a small region in high resolution can be observed. For example, it is possible to study oxygen diffusion in a scaffold. Lee et al. used this method to examine the microscale diffusion of oxygen in scaffolds generated via electrospinning [130]. By introducing simulated cell concentrations, the study reports the ability to predict the efficiency of the scaffold. However, this technique requires further standardization protocols to be established as a viable method to characterize 3D tissue models.

### *5.2. Optical Coherence Tomography (OCT)*

OCT is a type of imaging modality that performs high-resolution, cross-sectional imaging of microstructures in biological materials by measuring optical backscatter from different microstructural features within materials and tissues [126]. OCT can be used to observe the spatial and temporal changes of these features in real-time and in three dimensions, allowing the screening, identification, and optimization of parameters that govern the usability of tissue [131]. A key feature of OCT is capturing details in high resolution between 15–20 μm depths, thereby allowing the ability to observe scaffold architecture in intricate details [131]. The characteristics of scaffold architecture include parameters such as porosity, pore size, and degree of pore interconnectivity, which influence cellular activity, including cell adhesion, distribution, and proliferation [119,131]. A non-destructive method, OCT imaging, can be used to quantify changes in porosity as the scaffold degrades and cellular growth profile. For example, Zheng et al. used OCT to demonstrate the importance of OCT in the reconstruction of scaffold architecture and cell adhesion by capturing high-resolution images of two scaffolds with different seeding densities of human embryonic kidney cells [131]. Their study concluded that OCT is a viable method that can be used to optimize the parameters of scaffolds. More recently, Wang et al. used OCT to capture high-resolution images of the inner microstructures of cell-laden 3D-printed scaffolds. The study incorporated C3A cells in the gelatin–alginate hydrogel with varying pore sizes and utilized OCT to quantify morphological features, including pore size, pore shape factor, volume porosity, and the interconnectivity of the pores, as shown in Figure 10 [132]. Ultimately, this imaging modality has the capability to improve the understanding of the intricate structures, thereby leading to improved scaffold architecture designs, efficiently mimicking in vivo architecture and improving the efficacy of 3D tissue models.

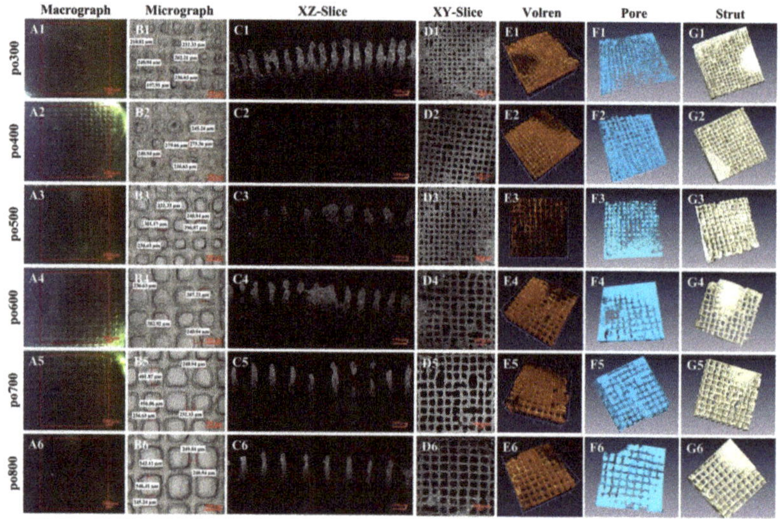


**Figure 10.** Cell-laden 3D bioprinted structures with varying pore size characterized using OCT (**A1–A6**) Macrographs (**B1–B6**) Micrographs, (**C1–C6**) C OCT Cross-sectional images to a depth of 3 mm. (**D1–D6**) en-face OCT images. (**E1–E6**) rendering in 3D and (**F1–F6**,**G1–G6**) 3D reconstruction of hydrogel exhibiting variation in pore size. Image reprinted with permission from [132] under the creative commons license CC BY 4.0. Copyright © 2023 by the authors; Scientific Reports [132].

## 6. Other Imaging Modalities

While FRAP and OCT are imaging modalities that can be utilized to perform characterization on 3D tissue models, confocal microscopy imaging provides other methods to characterize and analyze 3D tissue models. Such modalities include Fluorescence Loss in Photobleaching (FLIP), Fluorescence localization after photobleaching (FLAP), Fluorescence Resonance Energy Transfer (FRET), Fluorescence Lifetime Imaging Microscopy (FLIM), Phosphorescence Lifetime Imaging Microscopy (PLIM), and Micro-Computerized Tomography (MCT), and a summary of their characteristics and application can be found in Table 6. Figure 11 provides a workflow of the modalities. Ishikawa-Ankerhold et al. have provided an exhaustive review of the same [123].

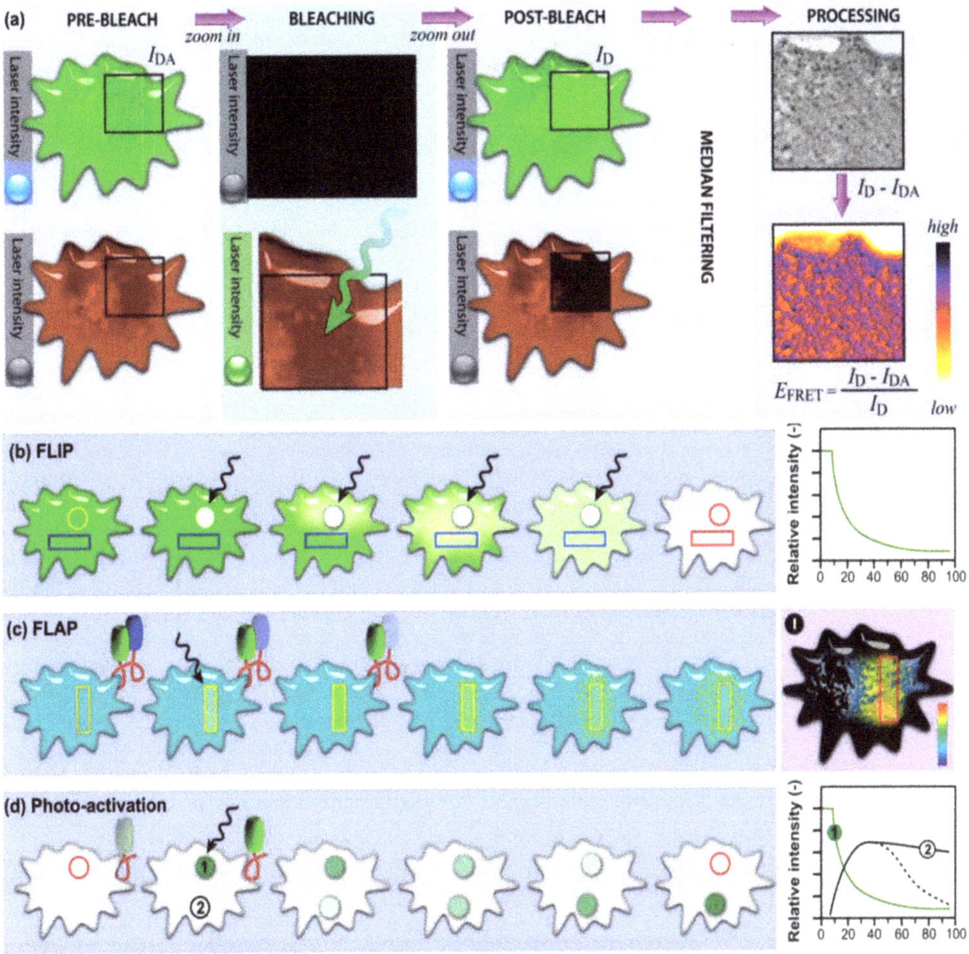


**Figure 11.** (**a**) workflow of FRET (**b**) Workflow of FLIP, (**c**) workflow of FLAP, (**d**) workflow of Photoactivation. Images reprinted with permission from [123] under creative commons CC BY 4.0. Copyright © 2023 by the authors; licensee MDPI, Basel, Switzerland.

## 7. Conclusions and Future Perspective

This review highlights the vast potential of 3D in vitro models for the generation of tissue mimics, disease modeling, and assessment of innovative drugs toward personalized medicine over 2D models. While 2D modeling is a traditional and established method, it lacks the capability to replicate human physiology and diseased conditions. In the context of tissue engineering, the various methods used in the generation of artificial constructs, along with their advantages and disadvantages, are discussed. The potential role of these methods in regenerative medicine is also highlighted. Biomaterials play an important role in the generation of such constructs and models. The choice of biomaterials that have the capability to closely replicate human physiology and promote cellular functions within artificial constructs is critical when considering modeling. With the advent of various stem-cell types, specifically iPSCs (induced pluripotent stem cells), research in disease modeling and personalized medicine has taken an innovative direction. The key advantage of employing the strategy of using 3D in vitro model systems is a reduced dependence on animal models, which are dissimilar to human physiology.

Most reviews discuss the state-of-the-art in tissue engineering research and regenerative medicine; however, methods used in the assessment of artificially generated constructs are a key area that is often neglected. An important aspect of 3D models and tissue engineering is to ensure that the artificial construct has the capability to replicate physiological conditions as closely as possible. Methods such as FTIR, mechanical testing, and biological activity assays to determine cell proliferation and survivability, to name a few, enable researchers to establish artificial tissues as efficient models and maintains standardization from a regulatory perspective. This review provides an exhaustive analysis of the various characterization methods used to evaluate artificially constructed 3D models along with various imaging modalities. Imaging has the capability to provide researchers with a tool to observe the functioning of cells at a microscopic level. It provides a platform where researchers can develop a deeper understanding of the attributes involved in the development and progression of the disease through direct observation. Methods such as optical coherence tomography are used in observing the structure of scaffolds in 3D, while FRAP and FRET can be employed to observe cellular functions.

A key challenge with 3D in vitro modeling is that while it has the capability to closely mimic human physiological conditions, it is an incomplete model, hence the reliance on animal models. Towards the future (Figure 12), it is imperative to focus on research towards the development of models that completely considers and mimics various factors and functions within the human body. The choice of biomaterials to have the right cellular microenvironment, appropriate mechanical properties as that of the relevant tissues of interest, the right orientation of cell/s, vascular networks, immune cells, the spatio-temporal release of necessary factors needed for the differentiation or growth of cells, and other factors unique to the tissues of interests, such as conduction properties in case of cardiac and neural tissues. This will allow researchers to work with improved 3D models, develop an improved understanding of diseases, and provide targeted solutions which are easy to manufacture, economically viable, and safe to administer. Furthermore, with the recent implementation of FDA Modernization Act 2.0, we envision that more emphasis will be given to complex and more sophisticated human physiology-relevant 3D in vitro tissue models for drug testing applications in the near future.

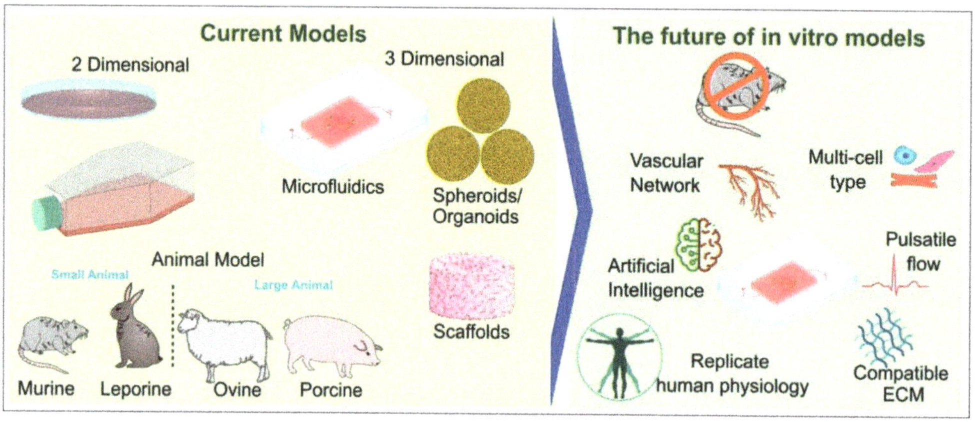


**Figure 12.** The future of in vitro models: a perspective.

**Funding:** Arunkumar Palaniappan would like to kindly acknowledge the financial support from the Science and Engineering Research Board (SERB), Department of Science and Technology, Government of India through its start-up research grant scheme (SRG/2020/001115).

**Institutional Review Board Statement:** Not applicable.

**Informed Consent Statement:** Not applicable.

**Data Availability Statement:** All the data and materials that support the results or analyses presented in the paper will be made available upon request.

**Conflicts of Interest:** The authors declare no competing interest.

## References

1. Duval, K.; Grover, H.; Han, L.; Mou, Y.; Pegoraro, A.; Fredberg, J.; Chen, Z. Modeling physiological events in 2D vs. 3D cell culture. *Physiology* **2017**, *32*, 266–277. [CrossRef]
2. Edmondson, R.; Broglie, J.; Adcock, A.; Yang, L. Three-dimensional cell culture systems and their applications in drug discovery and cell-based biosensors. *Assay Drug Dev. Technol.* **2014**, *32*, 266–277. [CrossRef]
3. Langhans, S. Three-dimensional in vitro cell culture models in drug discovery and drug repositioning. *Front. Pharmacol.* **2018**, *9*, 6. [CrossRef] [PubMed]
4. Kusindarta, D.; Wihadmadyatami, H. The Role of Extracellular Matrix in Tissue Regeneration. In *Tissue Regeneration*; InTech: Singapore, 2018. [CrossRef]
5. Gibot, L. 3D tissue models to bridge the gap between cell culture and tissue in assessing electroporation. In *Handbook of Electroporation*; Springer: Cham, Germany, 2017. [CrossRef]
6. Ruan, J.; Tulloch, N.; Razumova, M.V.; Saiget, M.; Muskheli, V.; Pabon, L.; Reinecke, H.; Regnier, M.; Murry, C. Mechanical Stress Conditioning and Electrical Stimulation Promote Contractility and Force Maturation of Induced Pluripotent Stem Cell-Derived Human Cardiac Tissue. *Circulation* **2016**, *134*, 1557–1567. [CrossRef] [PubMed]
7. Di Silvio, L. Bone tissue engineering and biomineralization. In *Tissue Engineering Using Ceramics and Polymers*; Elsevier Ltd.: Amsterdam, The Netherlands, 2007; pp. 319–331. [CrossRef]
8. Torras, N.; García-Díaz, M.; Fernández-Majada, V.; Martínez, E. Mimicking epithelial tissues in three-dimensional cell culture models. *Front. Bioeng. Biotechnol.* **2018**, *18*, 197. [CrossRef] [PubMed]
9. Oberman, R.; Bhardwaj, A. *Physiology, Cardiac*; StatPearls Publishing: Vienna, Austria, 2019.
10. Omar, A.; Vallabhajosyula, S.; Sengupta, P. Left Ventricular Twist and Torsion. *Circ. Cardiovasc. Imaging* **2015**, *8*, e003029. [CrossRef]
11. Baker, B.; Chen, C. Deconstructing the third dimension-how 3D culture microenvironments alter cellular cues. *J. Cell Sci.* **2012**, *125*, 3015–3024. [CrossRef]
12. Ravi, M.; Paramesh, V.; Kaviya, S.; Anuradha, E.; Solomon, F.P. 3D cell culture systems: Advantages and applications. *J. Cell. Physiol.* **2015**, *230*, 16–26. [CrossRef]
13. Dhaliwal, A. Three Dimensional Cell Culture: A Review. *Mater. Methods* **2012**, *2*. [CrossRef]
14. Jensen, C.; Teng, Y. Is It Time to Start Transitioning From 2D to 3D Cell Culture? *Front. Mol. Biosci.* **2020**, *7*, 33. [CrossRef]
15. Griffanti, G.; Rezabeigi, E.; Li, J.; Murshed, M.; Nazhat, S. Rapid Biofabrication of Printable Dense Collagen Bioinks of Tunable Properties. *Adv. Funct. Mater.* **2020**, *30*, 1903874. [CrossRef]
16. Mastrullo, V.; Cathery, W.; Velliou, E.; Madeddu, P.; Campagnolo, P. Angiogenesis in Tissue Engineering: As Nature Intended? *Front. Bioeng. Biotechnol.* **2020**, *8*, 188. [CrossRef]
17. Fang, Y.; Eglen, R. Three-Dimensional Cell Cultures in Drug Discovery and Development. *Adv. Sci. Drug Discov.* **2017**, *22*, 456–472. [CrossRef]
18. Costa, E.; Moreira, A.; de Melo-Diogo, D.; Gaspar, V.; Carvalho, M.; Correia, I. 3D tumor spheroids: An overview on the tools and techniques used for their analysis. *Biotechnol. Adv.* **2016**, *34*, 1427–1441. [CrossRef]
19. Kapałczyńska, M.; Kolenda, T.; Przybyła, W.; Zajączkowska, M.; Teresiak, A.; Filas, V.; Ibbs, M.; Bliźniak, R.; Łuczewski, Ł.; Lamperska, K. 2D and 3D cell cultures—A comparison of different types of cancer cell cultures. *Arch. Med. Sci.* **2018**, *14*, 910–919. [CrossRef] [PubMed]
20. De Hoogt, R.; Estrada, M.; Vidic, S.; Davies, E.; Osswald, A.; Barbier, M.; Santo, V.; Gjerde, K.; Van Zoggel, H.; Blom, S.; et al. Data descriptor: Protocols and characterization data for 2d, 3d, and slice-based tumor models from the predect project. *Sci. Data* **2017**, *4*, 170170. [CrossRef] [PubMed]
21. Sutherland, R.; McCredie, J.; Inch, W. Growth of multicell spheroids in tissue culture as a model of nodular carcinomas. *J. Natl. Cancer Inst.* **1971**, *46*, 113–120. [CrossRef] [PubMed]
22. Fennema, E.; Rivron, N.; Rouwkema, J.; van Blitterswijk, C.; De Boer, J. Spheroid culture as a tool for creating 3D complex tissues. *Trends Biotechnol.* **2013**, *31*, 108–115. [CrossRef] [PubMed]
23. Ryu, N.; Lee, S.; Park, H. Spheroid Culture System Methods and Applications for Mesenchymal Stem Cells. *Cells* **2019**, *8*, 1620. [CrossRef] [PubMed]

24. Lim, W.; Park, S. A microfluidic spheroid culture device with a concentration gradient generator for high-throughput screening of drug efficacy. *Molecules* **2018**, *23*, 3355. [CrossRef]
25. Utama, R.; Atapattu, L.; O'Mahony, A.; Fife, C.; Baek, J.; Allard, T.; O'Mahony, K.; Ribeiro, J.; Gaus, K.; Kavallaris, M.; et al. A 3D Bioprinter Specifically Designed for the High-Throughput Production of Matrix-Embedded Multicellular Spheroids. *IScience* **2020**, *23*, 101621. [CrossRef] [PubMed]
26. Moshksayan, K.; Kashaninejad, N.; Warkiani, M.; Lock, J.; Moghadas, H.; Firoozabadi, B.; Saidi, M.; Nguyen, N. Spheroids-on-a-chip: Recent advances and design considerations in microfluidic platforms for spheroid formation and culture. *Sens. Actuators B Chem.* **2018**, *263*, 151–176. [CrossRef]
27. Hoarau-Véchot, J.; Rafii, A.; Touboul, C.; Pasquier, J. Halfway between 2D and animal models: Are 3D cultures the ideal tool to study cancer-microenvironment interactions? *Int. J. Mol. Sci.* **2018**, *19*, 181. [CrossRef] [PubMed]
28. Benien, P.; Swami, A. 3D tumor models: History, advances and future perspectives. *Futur. Oncol.* **2014**, *10*, 1311–1327. [CrossRef] [PubMed]
29. Davies, J. Organoids and mini-organs: Introduction, history, and potential. In *Organoids and Mini-Organs*; Elsevier Inc.: Amsterdam, The Netherlands, 2018; pp. 3–23. [CrossRef]
30. Augustyniak; Bertero, A.; Coccini, T.; Baderna, D.; Buzanska, L.; Caloni, F. Organoids are promising tools for species-specific in vitro toxicological studies. *J. Appl. Toxicol.* **2019**, *39*, 1610–1622. [CrossRef]
31. Velasco, V.; Shariati, S.; Esfandyarpour, R. Microtechnology-based methods for organoid models. *Microsyst Nanoeng.* **2020**, *6*, 1–13. [CrossRef]
32. Eiraku, M.; Takata, N.; Ishibashi, H.; Kawada, M.; Sakakura, E.; Okuda, S.; Sekiguchi, K.; Adachi, T.; Sasai, Y. Self-organizing optic-cup morphogenesis in three-dimensional culture. *Nature* **2011**, *472*, 51–56. [CrossRef]
33. Takebe, T.; Sekine, K.; Enomura, M.; Koike, H.; Kimura, M.; Ogaeri, T.; Zhang, R.; Ueno, Y.; Zheng, Y.; Koike, N.; et al. Vascularized and functional human liver from an iPSC-derived organ bud transplant. *Nature* **2013**, *499*, 481–484. [CrossRef]
34. Lancaster, M.; Knoblich, J. Generation of cerebral organoids from human pluripotent stem cells. *Nat. Protoc.* **2014**, *9*, 2329–2340. [CrossRef]
35. Miller, A.; Dye, B.; Ferrer-Torres, D.; Hill, D.; Overeem, A.; Shea, L.; Spence, J. Generation of lung organoids from human pluripotent stem cells in vitro. *Nat. Protoc.* **2019**, *14*, 518–540. [CrossRef]
36. Richards, D.; Li, Y.; Kerr, C.; Yao, J.; Beeson, G.; Coyle, R.; Chen, X.; Jia, J.; Damon, B.; Wilson, R.; et al. Human cardiac organoids for the modelling of myocardial infarction and drug cardiotoxicity. *Nat. Biomed. Eng.* **2020**, *4*, 446–462. [CrossRef] [PubMed]
37. Yang, W.; Zhang, C.; Wu, Y.-H.; Liu, L.-B.; Zhen, Z.-D.; Fan, D.-Y.; Song, Z.-R.; Chang, J.-T.; Wang, P.-G.; An, J. Mice 3D testicular organoid system as a novel tool to study Zika virus pathogenesis. *Virol. Sin.* 2022, *in press*. [CrossRef]
38. Joseph, J.S.; Malindisa, S.T.; Ntwasa, M. Two-Dimensional (2D) and Three-Dimensional (3D) Cell Culturing in Drug Discovery. In *Cell Culture*; IntechOpen: Singapore, 2019. [CrossRef]
39. Hong, J.; Yeo, M.; Yang, G.; Kim, G. Cell-electrospinning and its application for tissue engineering. *Int. J. Mol. Sci.* **2019**, *20*, 6208. [CrossRef] [PubMed]
40. Townsend-Nicholson, A.; Jayasinghe, S. Cell electrospinning: A unique biotechnique for encapsulating living organisms for generating active biological microthreads/scaffolds. *Biomacromolecules* **2006**, *7*, 3364–3369. [CrossRef] [PubMed]
41. Wu, Y.; Zhang, H.; Wang, S.; Li, L.; Wang, R.; Jiang, S. Human umbilical cord-derived stem cell sheets improve left ventricular function in rat models of ischemic heart failure. *Eur. J. Pharmacol.* **2022**, *925*, 174994. [CrossRef] [PubMed]
42. Owaki, T.; Shimizu, T.; Yamato, M.; Okano, T. Cell sheet engineering for regenerative medicine: Current challenges and strategies. *Biotechnol. J.* **2014**, *9*, 904–914. [CrossRef] [PubMed]
43. Lin, R.; Chang, H. Recent advances in three-dimensional multicellular spheroid culture for biomedical research. *Biotechnol. J.* **2008**, *3*, 1172–1184. [CrossRef] [PubMed]
44. Wilson, S.; Tocchi, A.; Holly, M.; Parks, W.; Smith, J. A small intestinal organoid model of non-invasive enteric pathogen-epithelial cell interactions. *Mucosal Immunol.* **2015**, *8*, 352–361. [CrossRef]
45. Kim, M.; Evans, D. Tissue Engineering: The Future of Stem Cells. *Top. Tissue Eng.* **2005**, *2*, 1–21.
46. Eltom, A.; Zhong, G.; Muhammad, A. Scaffold Techniques and Designs in Tissue Engineering Functions and Purposes: A Review. *Adv. Mater. Sci. Eng.* **2019**, *2019*, 1–13. [CrossRef]
47. Caddeo, S.; Boffito, M.; Sartori, S. Tissue engineering approaches in the design of healthy and pathological in vitro tissue models. *Front. Bioeng. Biotechnol.* **2017**, *5*, 40. [CrossRef] [PubMed]
48. Sola, A.; Bertacchini, J.; D'Avella, D.; Anselmi, L.; Maraldi, T.; Marmiroli, S.; Messori, M. Development of solvent-casting particulate leaching (SCPL) polymer scaffolds as improved three-dimensional supports to mimic the bone marrow niche. *Mater. Sci. Eng. C* **2019**, *96*, 153–165. [CrossRef] [PubMed]
49. Li, Z.; Xie, M.B.; Li, Y.; Ma, Y.; Li, J.; Dai, F. Recent progress in tissue engineering and regenerative medicine. *J. Biomater. Tissue Eng.* **2016**, *6*, 755–766. [CrossRef]
50. Sanz-Herrera, J.; García-Aznar, J.; Doblaré, M. On scaffold designing for bone regeneration: A computational multiscale approach. *Acta Biomater.* **2009**, *5*, 219–229. [CrossRef] [PubMed]
51. Brougham, C.; Levingstone, T.; Shen, N.; Cooney, G.; Jockenhoevel, S.; Flanagan, T.; O'Brien, F. Freeze-Drying as a Novel Biofabrication Method for Achieving a Controlled Microarchitecture within Large, Complex Natural Biomaterial Scaffolds. *Adv. Healthc. Mater.* **2017**, *6*, 1700598. [CrossRef] [PubMed]

52. Anandan, D.; Stella, S.M.; Nambiraj, N.A.; Vijayalakshmi, U.; Jaiswal, A. Development of mechanically compliant 3D composite scaffolds for bone tissue engineering applications. *J. Biomed. Mater. Res. Part A* **2018**, *106*, 3267–3274. [CrossRef]
53. Martínez-Pérez, C.A.; Olivas-Armendariz, I.; Castro-Carmona, J.S.; García-Casillas, P.E. Scaffolds for Tissue Engineering Via Thermally Induced Phase Separation. In *Advances in Regenerative Medicine*; IntechOpen: Singapore, 2011. [CrossRef]
54. Dehghani, F.; Annabi, N. Engineering porous scaffolds using gas-based techniques. *Curr. Opin. Biotechnol.* **2011**, *22*, 661–666. [CrossRef]
55. Zhong, W. Nanofibres for Medical Textiles. In *Advances in Smart Medical Textiles: Treatments and Health Monitoring*; Woodhead Publishing: Southen, UK, 2016. [CrossRef]
56. Zheng, Y. Fabrication on bioinspired surfaces. In *Bioinspired Design of Materials Surfaces*; Elsevier: Amsterdam, The Netherlands, 2019. [CrossRef]
57. Singh, R.; Eitler, D.; Morelle, R.; Friedrich, R.; Dietel, B.; Alexiou, C.; Boccaccini, A.; Liverani, L.; Cicha, I. Optimization of cell seeding on electrospun PCL-silk fibroin scaffolds. *Eur. Polym. J.* **2020**, *134*, 109838. [CrossRef]
58. Li, J.; Chen, M.; Fan, X.; Zhou, H. Recent advances in bioprinting techniques: Approaches, applications and future prospects. *J. Transl. Med.* **2016**, *14*, 1–15. [CrossRef]
59. Mirdamadi, E.; Tashman, J.; Shiwarski, D.; Palchesko, R.; Feinberg, A. FRESH 3D bioprinting a full-size model of the human heart. *ACS Biomater. Sci. Eng.* **2020**, *6*, 6453–6459. [CrossRef]
60. Lee, A.; Hudson, A.; Shiwarski, D.; Tashman, J.; Hinton, T.; Yerneni, S.; Bliley, J.; Campbell, P.; Feinberg, A. 3D bioprinting of collagen to rebuild components of the human heart. *Science* **2019**, *365*, 482–487. [CrossRef] [PubMed]
61. Senior, J.; Cooke, M.; Grover, L.; Smith, A.; Senior, J.; Smith, A.; Cooke, M.; Grover, L. Fabrication of Complex Hydrogel Structures Using Suspended Layer Additive Manufacturing (SLAM). *Adv. Funct. Mater.* **2019**, *29*, 1904845. [CrossRef]
62. Skylar-Scott, M.; Uzel, S.; Nam, L.; Ahrens, J.; Truby, R.; Damaraju, S.; Lewis, J. Biomanufacturing of organ-specific tissues with high cellular density and embedded vascular channels. *Sci. Adv.* **2019**, *5*, eaaw2459. [CrossRef]
63. Burdis, R.; Kelly, D. Biofabrication and bioprinting using cellular aggregates, microtissues and organoids for the engineering of musculoskeletal tissues. *Acta Biomater.* **2021**, *126*, 1–14. [CrossRef] [PubMed]
64. Ong, C.; Fukunishi, T.; Zhang, H.; Huang, C.; Nashed, A.; Blazeski, A.; Disilvestre, D.; Vricella, L.; Conte, J.; Tung, L.; et al. Biomaterial-Free Three-Dimensional Bioprinting of Cardiac Tissue using Human Induced Pluripotent Stem Cell Derived Cardiomyocytes. *Sci. Rep.* **2017**, *2017*, 1–11. [CrossRef]
65. Noor, N.; Shapira, A.; Edri, R.; Gal, I.; Wertheim, L.; Dvir, T. 3D Printing of Personalized Thick and Perfusable Cardiac Patches and Hearts. *Adv. Sci.* **2019**, *6*, 1900344. [CrossRef]
66. Loai, S.; Kingston, B.R.; Wang, Z.; Philpott, D.N.; Tao, M.; Cheng, H.-L.M. Clinical Perspectives on 3D Bioprinting Paradigms for Regenerative Medicine. *Regen. Med. Front.* **2019**, *1*, e190004. [CrossRef]
67. Murphy, S.V.; Atala, A. 3D bioprinting of tissues and organs. *Nat. Biotechnol.* **2014**, *32*, 773–785. [CrossRef]
68. Aziz, A.; Geng, C.; Fu, M.; Yu, X.; Qin, K.; Liu, B. The role of microfluidics for organ on chip simulations. *Bioengineering* **2017**, *4*, 39. [CrossRef]
69. Zhao, Y.; Rafatian, N.; Feric, N.; Cox, B.; Aschar-Sobbi, R.; Wang, E.; Aggarwal, P.; Zhang, B.; Conant, G.; Ronaldson-Bouchard, K.; et al. A Platform for Generation of Chamber-Specific Cardiac Tissues and Disease Modeling. *Cell* **2019**, *176*, 913.e18–927.e18. [CrossRef]
70. Parsa, H.; Wang, B.; Vunjak-Novakovic, G. A microfluidic platform for the high-throughput study of pathological cardiac hypertrophy. *Lab Chip* **2017**, *17*, 3264–3271. [CrossRef]
71. Zhao, Y.; Rafatian, N.; Wang, E.; Feric, N.; Lai, B.; Knee-Walden, E.; Backx, P.; Radisic, M. Engineering microenvironment for human cardiac tissue assembly in heart-on-a-chip platform. *Matrix Biol.* **2019**, *85-86*, 189–204. [CrossRef]
72. Jang, K.; Suh, K. A multi-layer microfluidic device for efficient culture and analysis of renal tubular cells. *Lab Chip* **2010**, *10*, 36–42. [CrossRef]
73. Kilic, O.; Pamies, D.; Lavell, E.; Schiapparelli, P.; Feng, Y.; Hartung, T.; Bal-Price, A.; Hogberg, H.; Quinones-Hinojosa, A.; Guerrero-Cazares, H.A. Levchenko, Brain-on-a-chip model enables analysis of human neuronal differentiation and chemotaxis. *Lab Chip* **2016**, *16*, 4152–4162. [CrossRef]
74. Grigoryan, B.; Paulsen, S.; Corbett, D.; Sazer, D.; Fortin, C.; Zaita, A.; Greenfield, P.; Calafat, N.; Gounley, J.; Ta, A.; et al. Multivascular networks and functional intravascular topologies within biocompatible hydrogels. *Am. Assoc. Adv. Sci.* **2019**, *364*, 458–464. [CrossRef]
75. Sontheimer-Phelps, A.; Chou, D.; Tovaglieri, A.; Ferrante, T.; Duckworth, T.; Fadel, C.; Frismantas, V.; Sutherland, A.; Jalili-Firoozinezhad, S.; Kasendra, M.; et al. Human Colon-on-a-Chip Enables Continuous In Vitro Analysis of Colon Mucus Layer Accumulation and Physiology. *Cell. Mol. Gastroenterol. Hepatol.* **2020**, *9*, 507–526. [CrossRef]
76. Bhise, N.; Manoharan, V.; Massa, S.; Tamayol, A.; Ghaderi, M.; Miscuglio, M.; Lang, Q.; Zhang, Y.; Shin, S.; Calzone, G.; et al. A liver-on-a-chip platform with bioprinted hepatic spheroids. *Biofabrication* **2016**, *8*, 014101. [CrossRef]
77. Seo, J.; Byun, W.; Alisafaei, F.; Georgescu, A.; Yi, Y.; Massaro-Giordano, M.; Shenoy, V.; Lee, V.; Bunya, V.; Huh, D. Multiscale reverse engineering of the human ocular surface. *Nat. Med.* **2019**, *25*, 1310–1318. [CrossRef]

78. Costa, P.; Albers, H.; Linssen, J.; Middelkamp, H.; Van Der Hout, L.; Passier, R.; Van Den Berg, A.; Malda, J.; Van Der Meer, A. Mimicking arterial thrombosis in a 3D-printed microfluidic: In vitro vascular model based on computed tomography angiography data. *Lab Chip* **2017**, *17*, 2785–2792. [CrossRef]
79. Lan, D.; Shang, Y.; Su, H.; Liang, M.; Liu, Y.; Li, H.; Feng, Q.; Cao, X.; Dong, H. Facile Fabrication of Hollow Hydrogel Microfiber via 3D Printing-Assisted Microfluidics and Its Application as a Biomimetic Blood Capillary. *ACS Biomater. Sci. Eng.* **2021**, *7*, 4971–4981. [CrossRef]
80. Hong, S.; Song, J. 3D bioprinted drug-resistant breast cancer spheroids for quantitative in situ evaluation of drug resistance. *Acta Biomater.* **2022**, *138*, 228–239. [CrossRef] [PubMed]
81. van Meer, B.; de Vries, H.; Firth, K.; van Weerd, J.; Tertoolen, L.; Karperien, H.; Jonkheijm, P.; Denning, C.; IJzerman, A.; Mummery, C. Small molecule absorption by PDMS in the context of drug response bioassays. *Biochem. Biophys. Res. Commun.* **2017**, *482*, 323–328. [CrossRef] [PubMed]
82. Campbell, S.; Wu, Q.; Yazbeck, J.; Liu, C.; Okhovatian, S.; Radisic, M. Beyond polydimethylsiloxane: Alternative materials for fabrication of organ on a chip devices and microphysiological systems. *ACS Biomater. Sci. Eng.* **2020**, *7*, 2880–2899. [CrossRef] [PubMed]
83. Abaci, H.; Shuler, M. Human-on-a-chip design strategies and principles for physiologically based pharmacokinetics/pharmacodynamics modeling. *Integr. Biol.* **2015**, *7*, 383–391. [CrossRef] [PubMed]
84. Kim, Y.; Ko, H.; Kwon, I.; Shin, K. Extracellular matrix revisited: Roles in tissue engineering. *Int. Neurourol. J.* **2016**, *20*, S23–S29. [CrossRef]
85. NIH Stem Cell, NIH Stem Cell Information Home Page—Stem Cell Basics, In Stem Cell Information. Available online: https://stemcells.nih.gov/ (accessed on 31 December 2022).
86. Doss, M.; Sachinidis, A. Current Challenges of iPSC-Based Disease Modeling and Therapeutic Implications. *Cells* **2019**, *8*, 403. [CrossRef]
87. Ricklefs, M.; Korossis, S.; Haverich, A.; Schilling, T. Polymeric Scaffolds for Bioartificial Cardiovascular Prostheses. In *Scaffolds in Tissue Engineering—Materials, Technologies and Clinical Applications*; IntechOpen: Singapore, 2017. [CrossRef]
88. Wang, X.; Ao, Q.; Tian, X.; Fan, J.; Tong, H.; Hou, W.; Bai, S. Gelatin-based hydrogels for organ 3D bioprinting. *Polymers* **2017**, *9*, 401. [CrossRef]
89. Xu, B.; Li, Y.; Deng, B.; Liu, X.; Wang, L.; Zhu, Q. Chitosan hydrogel improves mesenchymal stem cell transplant survival and cardiac function following myocardial infarction in rats. *Exp. Ther. Med.* **2017**, *13*, 588–594. [CrossRef]
90. Wang, Z.; Lee, S.; Cheng, H.; Yoo, J.; Atala, A. 3D bioprinted functional and contractile cardiac tissue constructs. *Acta Biomater.* **2018**, *70*, 48–56. [CrossRef]
91. Gaetani, R.; Doevendans, P.; Metz, C.; Alblas, J.; Messina, E.; Giacomello, A.; Sluijter, J. Cardiac tissue engineering using tissue printing technology and human cardiac progenitor cells. *Biomaterials* **2012**, *33*, 1782–1790. [CrossRef]
92. Maiullari, F.; Costantini, M.; Milan, M.; Pace, V.; Chirivì, M.; Maiullari, S.; Rainer, A.; Baci, D.; Marei, H.; Seliktar, D.; et al. A multicellular 3D bioprinting approach for vascularized heart tissue engineering based on HUVECs and iPSC-derived cardiomyocytes. *Sci. Rep.* **2018**, *8*, 13532. [CrossRef]
93. Mehrali, M.; Thakur, A.; Kadumudi, F.; Pierchala, M.; Cordova, J.; Shahbazi, M.; Mehrali, M.; Pennisi, C.; Orive, G.; Gaharwar, A.; et al. Pectin Methacrylate (PEMA) and Gelatin-Based Hydrogels for Cell Delivery: Converting Waste Materials into Biomaterials. *ACS Appl. Mater. Interfaces* **2019**, *11*, 12283–12297. [CrossRef]
94. Kc, P.; Hong, Y.; Zhang, G. Cardiac tissue-derived extracellular matrix scaffolds for myocardial repair: Advantages and challenges. *Regen. Biomater.* **2019**, *6*, 185–199. [CrossRef] [PubMed]
95. Dong, D.; Li, J.; Cui, M.; Wang, J.; Zhou, Y.; Luo, L.; Wei, Y.; Ye, L.; Sun, H.; Yao, F. In Situ "clickable" Zwitterionic Starch-Based Hydrogel for 3D Cell Encapsulation. *ACS Appl. Mater. Interfaces* **2016**, *8*, 4442–4455. [CrossRef]
96. Reys, L.; Silva, S.; Da Costa, D.S.; Oliveira, N.; Mano, J.; Reis, R.; Silva, T. Fucoidan Hydrogels Photo-Cross-Linked with Visible Radiation As Matrices for Cell Culture. *ACS Biomater. Sci. Eng.* **2016**, *2*, 1151–1161. [CrossRef] [PubMed]
97. Chen, Q.; Zhu, C.; Thouas, G.A. Progress and challenges in biomaterials used for bone tissue engineering: Bioactive glasses and elastomeric composites. *Prog. Biomater.* **2012**, *1*, 2. [CrossRef]
98. Family, R.; Solati-Hashjin, M.; Nik, S.; Nemati, A. Surface modification for titanium implants by hydroxyapatite nanocomposite. *Casp. J. Intern. Med.* **2012**, *3*, 460.
99. Lee, H.; Byun, S.; Cho, S.; Yang, B. Past, present, and future of regeneration therapy in oral and periodontal tissue: A review. *Appl. Sci.* **2019**, *9*, 1046. [CrossRef]
100. Ho, C.; Mishra, A.; Lin, P.; Ng, S.; Yeong, W.; Kim, Y.; Yoon, Y. 3D Printed Polycaprolactone Carbon Nanotube Composite Scaffolds for Cardiac Tissue Engineering. *Macromol. Biosci.* **2017**, *17*, 1600250. [CrossRef] [PubMed]
101. Mironov, A.V.; Grigoryev, A.; Krotova, L.; Skaletsky, N.; Popov, V.; Sevastianov, V. 3D printing of PLGA scaffolds for tissue engineering. *J. Biomed. Mater. Res. A* **2017**, *105*, 104–109. [CrossRef]
102. Savoji, H.; Huyer, L.D.; Mohammadi, M.; Lai, B.L.; Rafatian, N.; Bannerman, D.; Shoaib, M.; Bobicki, E.; Ramachandran, A.; Radisic, M. 3D Printing of Vascular Tubes Using Bioelastomer Prepolymers by Freeform Reversible Embedding. *ACS Biomater. Sci. Eng.* **2020**, *6*, 1333–1343. [CrossRef] [PubMed]

103. Ulery, B.; Nair, L.; Laurencin, C. Biomedical applications of biodegradable polymers. *J. Polym. Sci. Part B Polym. Phys.* **2011**, *49*, 832–864. [CrossRef] [PubMed]
104. Kunduru, K.; Basu, A.; Domb, A. Biodegradable Polymers: Medical Applications. In *Encyclopedia of Polymer Science and Technology*; John Wiley & Sons, Inc.: Hoboken, NJ, USA, 2016. [CrossRef]
105. Kasper, F.; Tanahashi, K.; Fisher, J.; Mikos, A. Synthesis of poly(propylene fumarate). *Nat. Protoc.* **2009**, *4*, 518–525. [CrossRef] [PubMed]
106. Kinard, L.; Kasper, F.; Mikos, A. Synthesis of oligo(Poly(ethylene glycol) fumarate). *Nat. Protoc.* **2012**, *7*, 1219–1227. [CrossRef]
107. Bahraminasab, M.; Sahari, B.; Edwards, K.; Farahmand, F.; Arumugam, M. Aseptic loosening of femoral components—Materials engineering and design considerations. *Mater. Des.* **2012**, *44*, 155–163. [CrossRef]
108. Ma, X.; Liu, J.; Zhu, W.; Tang, M.; Lawrence, N.; Yu, C.; Gou, M.; Chen, S. 3D bioprinting of functional tissue models for personalized drug screening and in vitro disease modeling. *Adv. Drug Deliv. Rev.* **2018**, *132*, 235–251. [CrossRef]
109. Kengla, C.; Kidiyoor, A.; Murphy, S.V. Bioprinting Complex 3D Tissue and Organs. In *Kidney Transplantation, Bioengineering, and Regeneration: Kidney Transplantation in the Regenerative Medicine Era*; Elsevier Inc.: Amsterdam, The Netherlands, 2017; pp. 957–971. [CrossRef]
110. Skardal, A. Bioprinting essentials of cell and protein viability. In *Essentials of 3D Biofabrication and Translation*; Elsevier Inc.: Amsterdam, The Netherlands, 2015; pp. 1–17. [CrossRef]
111. Salaris, F.A. Rosa, Construction of 3D in vitro models by bioprinting human pluripotent stem cells: Challenges and opportunities. *Brain Res.* **2019**, *1723*, 146393. [CrossRef]
112. Augustine, R.; Kalva, S.; Ahmad, R.; Zahid, A.; Hasan, S.; Nayeem, A.; McClements, L.; Hasan, A. 3D Bioprinted cancer models: Revolutionizing personalized cancer therapy. *Transl. Oncol.* **2021**, *14*, 101015. [CrossRef]
113. Chen, C.; Sereti, K.; Wu, B.; Ardehali, R. Translational aspects of cardiac cell therapy. *J. Cell. Mol. Med.* **2015**, *19*, 1757–1772. [CrossRef]
114. Liao, S.-Y.; Tse, H.-F. Multipotent (adult) and pluripotent stem cells for heart regeneration: What are the pros and cons? *Stem Cell Res. Ther.* **2013**, *4*, 151. [CrossRef]
115. Gálvez-Montón, C.; Prat-Vidal, C.; Roura, S.; Soler-Botija, C.; Bayes-Genis, A. Cardiac Tissue Engineering and the Bioartificial Heart. *Rev. Española De Cardiol. (Engl. Ed.)* **2013**, *66*, 391–399. [CrossRef]
116. Dmitriev, R.; Borisov, S.; Düssmann, H.; Sun, S.; Müller, B.; Prehn, J.; Baklaushev, V.; Klimant, I.; Papkovsky, D. Versatile conjugated polymer nanoparticles for high-resolution $O_2$ imaging in cells and 3D tissue models. *ACS Nano* **2015**, *9*, 5275–5288. [CrossRef]
117. Müller, B.; Zhdanov, A.V.; Borisov, S.; Foley, T.; Okkelman, I.; Tsytsarev, V.; Tang, Q.; Erzurumlu, R.; Chen, Y.; Zhang, H.; et al. Nanoparticle-Based Fluoroionophore for Analysis of Potassium Ion Dynamics in 3D Tissue Models and In Vivo. *Adv. Funct. Mater.* **2018**, *28*, 1704598. [CrossRef]
118. Goliwas, K.; Richter, J.; Pruitt, H.; Araysi, L.; Anderson, N.; Samant, R.; Lobo-Ruppert, S.; Berry, J.; Frost, A. Methods to Evaluate Cell Growth, Viability, and Response to Treatment in a Tissue Engineered Breast Cancer Model. *Sci. Rep.* **2017**, *7*, 14167. [CrossRef]
119. Bardsley, K.; Deegan, A.; El Haj, A.; Yang, Y. Current state-of-the-art 3D tissue models and their compatibility with live cell imaging. *Adv. Exp. Med. Biol.* **2017**, *1035*, 3–18. [CrossRef]
120. Elliott, A. Confocal Microscopy: Principles and Modern Practices. *Curr. Protoc. Cytom.* **2020**, *92*, e68. [CrossRef]
121. Fischer, E.; Hansen, B.; Nair, V.; Hoyt, F.; Dorward, D. Scanning electron microscopy. *Curr. Protoc. Microbiol.* **2012**, *25*, 2B.2.1–2B.2.47. [CrossRef]
122. Misof, B.; Roschger, P.; Fratzl, P. Imaging mineralized tissues in vertebrates. In *Comprehensive Biomaterials*; Elsevier: Amsterdam, The Netherlands, 2011; pp. 407–426. [CrossRef]
123. Ishikawa-Ankerhold, H.; Ankerhold, R.; Drummen, G. Advanced fluorescence microscopy techniques-FRAP, FLIP, FLAP, FRET and FLIM. *Molecules* **2012**, *17*, 4047–4132. [CrossRef]
124. Donius, A.; Bougoin, S.V.; Taboas, J. FRET imaging in three-dimensional hydrogels. *J. Vis. Exp.* **2016**, *2016*, 54135. [CrossRef]
125. Sekar, R.; Periasamy, A. Fluorescence resonance energy transfer (FRET) microscopy imaging of live cell protein localizations. *J. Cell Biol.* **2003**, *160*, 629–633. [CrossRef]
126. Fujimoto, J.; Pitris, C.; Boppart, S.; Brezinski, M. Optical coherence tomography: An emerging technology for biomedical imaging and optical biopsy. *Neoplasia* **2000**, *2*, 9–25. [CrossRef] [PubMed]
127. Wang, L.; Xu, M.; Zhang, L.; Zhou, Q.; Luo, L. Automated quantitative assessment of three-dimensional bioprinted hydrogel scaffolds using optical coherence tomography. *Biomed. Opt. Express* **2016**, *7*, 894–910. [CrossRef]
128. Wang, S.; Larina, I.V. High-resolution imaging techniques in tissue engineering. In *Monitoring and Evaluation of Biomaterials and Their Performance In Vivo*; Elsevier Inc.: Amsterdam, The Netherlands, 2017; pp. 151–180. [CrossRef]
129. Meddens, M.; de Keijzer, S.; Cambi, A. High Spatiotemporal Bioimaging Techniques to Study the Plasma Membrane Nanoscale Organization. In *Fluorescence Microscopy: Super-Resolution and Other Novel Techniques*; Elsevier Inc.: Amsterdam, The Netherlands, 2014; pp. 49–63. [CrossRef]
130. Lee, S.; Lee, B.; Lee, J.; Kim, S.; Kim, J.; Jeong, Y.; Jin, S. Microscale diffusion measurements and simulation of a scaffold with a permeable strut. *Int. J. Mol. Sci.* **2013**, *14*, 20157–20170. [CrossRef]

131. Zheng, K.; Rupnick, M.; Liu, B.; Brezinski, M. Three Dimensional OCT in the Engineering of Tissue Constructs: A Potentially Powerful Tool for Assessing Optimal Scaffold Structure. *Open Tissue Eng. Regen. Med. J.* **2009**, *2*, 8–13. [CrossRef]
132. Wang, L.; Xu, M.; Luo, L.; Zhou, Y.; Si, P. Iterative feedback bio-printing-derived cell-laden hydrogel scaffolds with optimal geometrical fidelity and cellular controllability. *Sci. Rep.* **2018**, *8*, 1–13. [CrossRef]

International Journal of
*Molecular Sciences*

MDPI

*Review*

# Looking into the Eyes—In Vitro Models for Ocular Research

**Krystyna Lieto [1,†], Rafał Skopek [2,†], Aneta Lewicka [3], Marta Stelmasiak [4], Emilia Klimaszewska [5], Arthur Zelent [2], Łukasz Szymański [2,*] and Sławomir Lewicki [1,2,4,*]**

1 Department of Regenerative Medicine, Military Institute of Hygiene and Epidemiology, Kozielska 4, 01-163 Warsaw, Poland
2 Department of Molecular Biology, Institute of Genetics and Animal Biotechnology, Polish Academy of Sciences, Postępu 36A, 05-552 Jastrzębiec, Poland
3 Military Centre of Preventive Medicine Modlin, 05-100 Nowy Dwór Mazowiecki, Poland
4 Department of Medicine, Faculty of Medical Sciences and Health Sciences, Kazimierz Pulaski University of Technology and Humanities in Radom, 26-600 Radom, Poland
5 Department of Cosmetology, Faculty of Medical Sciences and Health Sciences, Kazimierz Pulaski University of Technology and Humanities in Radom, 26-600 Radom, Poland
* Correspondence: l.szymanski@igbzpan.pl (Ł.S.); s.lewicki@igbzpan.pl (S.L.)
† These authors contributed equally to this work.

**Abstract:** Animal research undoubtedly provides scientists with virtually unlimited data but inflicts pain and suffering on animals. Currently, legislators and scientists alike are promoting alternative in vitro approaches allowing for an accurate evaluation of processes occurring in the body without animal sacrifice. Historically, one of the most infamous animal tests is the Draize test, mainly performed on rabbits. Even though this test was considered the gold standard for around 50 years, the Draize test fails to mimic human response mainly due to human and rabbit eye physiological differences. Therefore, many alternative assays were developed to evaluate ocular toxicity and drug effectiveness accurately. Here we review recent achievements in tissue engineering of in vitro 2D, 2.5D, 3D, organoid and organ-on-chip ocular models, as well as in vivo and ex vivo models in terms of their advantages and limitations.

**Keywords:** in vitro eye models; 3D eye models; tissue engineering; ocular toxicity; eye irritation; corneal equivalents

**Citation:** Lieto, K.; Skopek, R.; Lewicka, A.; Stelmasiak, M.; Klimaszewska, E.; Zelent, A.; Szymański, Ł.; Lewicki, S. Looking into the Eyes—In Vitro Models for Ocular Research. *Int. J. Mol. Sci.* **2022**, 23, 9158. https://doi.org/10.3390/ijms23169158

Academic Editors: Patrick C. Baer and Ralf Schubert

Received: 13 June 2022
Accepted: 11 August 2022
Published: 15 August 2022



## 1. Introduction

The number of factors that can damage human tissues increases every year. For example, smog, substances contained in cosmetics, unnatural food additives, and UV radiation have a harmful effect on our skin and eyes. Moreover, every year industry delivers thousands of new chemical substances which are necessary for new medicines, chemicals, or food additives. Therefore, biocompatibility assessments of each new compound, especially one that involves animal testing, is impossible. Moreover, the number of studies performed on animals has to be limited in Europe by the law (i.e., Directive 2010/63/EU).

Optic neuropathies, such as glaucoma, anterior ischaemic optic neuropathy (AION), traumatic optic neuropathies, optic neuritis, etc., need new treatment options, which in turn require the development of disease models [1]. However, the pathophysiological mechanisms of these diseases are not fully understood; therefore, developing an animal model is a tough challenge. Moreover, due to the physiological differences, animal models differ significantly from human diseases [2]. For example, rodents' eyes do not have maculae or foveae, and 85–90% of their optic nerve axons decussate to the other side of the brain [1]. On the other hand, monkeys' anatomy of the retina and optic nerve is almost identical to that of human eyes. Still, monkey breeding is complicated, very expensive, and time-consuming; therefore, the number of tests performed on individual animals is limited. As a result, monkeys are often used in the stage just before clinical trials on humans [1].

Historically, one of the most popular experiments performed on animal eyes is the Draize test. Developed in 1944 by American toxicologists John H. Draize and Jacob M. Spines, it was widely used to study cosmetics and other chemicals. However, the test arouses many controversies due to the lack of reliable and objective results. In fact, the test was never correctly validated. Briefly, the test is based on applying the test substance directly to the eye, but the exposition time is not well defined. After observing the eye reaction for some time, the substance is washed from the eye, and the animal is observed for another two weeks. The result is subjectively assessed by the operator [3]. Moreover, the test is considered incorrect mainly due to anatomical and biochemical differences between the human and the animal (mostly the rabbit) eye. Therefore, currently, the Draize test is not performed. Instead, chemicals are usually tested using the EpiOcular eye irritation test, in vitro cytotoxicity assay, and irritation tests on the rabbit's skin. Each of these tests has advantages and disadvantages, but none of them allows for multifactorial compound-eye interaction evaluation. The EpiOcular eye irritation test is an in vitro alternative which allows for the assessment of acute eye irritation in response to the topical administration of chemicals onto the EpiOcular cornea epithelial model. The test makes cytotoxic effect measurement possible and provides a tool for eye-hazardous chemical identification.

For these reasons, it is crucial to develop new in vitro tissue models to study all substances for ocular treatment and to understand the development and molecular causes of eye diseases [4]. Here, we summarized all cellular and tissue-specific animal models used in in vitro eye studies.

## 2. Eye Structure

The eye is a highly complex biological machine (Table 1). The human eye has the shape of a sphere about 24 mm in diameter. It is filled with a vitreous body that allows the shape of the ball to be maintained. It is located in the eye socket, which reduces the risk of mechanical damage. The eye is divided into two parts: external and internal. The eye's outside layers are tough, elastic structures, with a white sclera and transparent cornea which provide eye shape [5]. The second inner layer is the vascular membrane, including the iris, ciliary body, and choroid. The third layer is the retina. It consists of light receptors, cells, nerve fibers, and blood vessels originating from the central retinal artery [6]. The primary function of the eye is to convert light pulses into electrical signals, which are transmitted to the brain and converted into images. Light is refracted by the cornea and the lens, which results in a sharp, inverted, reduced image formed on the retina. The amount of light reaching the receptors is regulated by the iris, which changes pupil diameter [6].

The retina in vertebrates is characterized by light-sensitive structures and covers 60% of the back of the eyeball. It is located above the choroid. It consists of 10 cell layers and contains photoreceptors called rod cells and cone cells [7]. These cells contain a visual pigment which is located in the cell membrane. Rod cells allow the recognition of shapes and motion at low light intensity [8]. Cone cells are responsible for seeing color and detail in a more intense light than rod cells. Individual cone cells differ in sensitivity at wavelength, which allows us to distinguish colors. In humans, cone cells are concentrated in the macula—a small pocket of the retina center. The optic nerve's axons that leave the retina at the ONH form a blind spot. There are no photoreceptors in that spot [9].

The eyeball is surrounded by the connective tissue—the sclera. It is a hard layer protecting the inner structures of the eye. In addition, it stiffens the eyeball like a bag of collagen and elastic fibers [10]. The sclera is thinner, more permeable to light, and creates a transparent cornea on the front. It refracts the light rays so that they fall on the lens. It is susceptible to pain and able to partially regenerate. It is avascular and consists of six layers (epithelial, Bowmann, stroma, Dua's, Descemet, and endothelial).

The conjunctiva lines the inner surface of the eyelid. It contains a lot of mucous cells, which ensures the constant humidity of the eyeball—it produces mucus and tears. It covers the eyeball up to the edge of the cornea. It has a very high regeneration ability. The conjunctiva is sensitive to any irritation, such as smoke, dust, or chemical substances.

These factors can lead to conjunctivitis. During this type of inflammation, blood vessels are firmly filled with blood, causing redness and swelling of the eye [11]. Glands are dispersed in the conjunctiva, and the lids secrete mucus, water, and lipids, forming a tear film whose primary function is to moisturize and cleanse the eye from undesired foreign bodies if needed. There are many glands dispersed in the conjunctiva and the lids that secrete mucus, water, and lipids, thus forming the tear film. However, the main lacrimal gland (responsible for emotional tears) is outside the eye structure.

**Table 1.** Layers of the eyeball and their functions.

| Membrane | Part of Membrane | Structure and Function |
|---|---|---|
| Fibrous tunic | Sclera | ✓ composed of collagen fibers<br>✓ surrounds 80% of the eye<br>✓ opaque<br>✓ protects against mechanical injuries<br>✓ place of muscle attachment<br>✓ maintains the oval shape of the eyeball |
| | Cornea | ✓ avascular, consists of six layers<br>✓ surrounds 20% of the eyeball (from the front)<br>✓ transparent, permeable to light<br>✓ protects against mechanical injuries<br>✓ refracts the rays of light<br>✓ directs the light on the lens<br>✓ sensitive to pain<br>✓ the ability to regenerate from the limbus |
| Vascular anteriorly | Iris | ✓ round, can change diameters<br>✓ around the pupil<br>✓ regulates the amount of light entering the inside of the eye<br>✓ contains a pigment in the epithelial layer on top of the muscular layers |
| | Ciliary body | ✓ the muscle surrounding the lens<br>✓ changes the convexity of the lens (accommodation)<br>✓ connects the choroid with the iris<br>✓ responsible of the secretion of the aqueous humor in the anterior chamber |
| | Choroid | ✓ covers the retina<br>✓ consists of a dense network of capillaries of large diameter<br>✓ nourishes the cells of the eye<br>✓ it absorbs excess light |
| Nervous | Retina | ✓ pigmented, photosensitive layer of the eye<br>✓ it determines the reception of visual impressions<br>✓ contains photosensitive cells (cone cells, rod cells)<br>✓ contains the macula of the retina and the blind spot (wherein the optic nerve head (ONH) is located) |

Even though the eye is a very specialized organ, there is significant progress in the development of tissue engineering, and newer and more suitable in vitro models are emerging. The search for such models is caused, among others, by increasing awareness of the welfare of animals used in experiments, including toxicological effects [12].

## 3. In Vitro Ocular Models

Over the years, many alternative assays were developed to accurately evaluate ocular toxicity and drug effectiveness. Here we present recent achievements in tissue engineering of various ocular models in terms of their advantages and limitations (see also Figures 1 and 2).

### 3.1. 2D Eye Models

Currently, 2D models are the most popular ones in ocular research. Two-dimensional cell line culture is an inexpensive, well-established model providing results that are easy to compare with the vast literature. However, the unquestionable drawback of these culture systems is the lack of predictivity in research connected with the fact that cells growing on a flat surface are not an equal representation of the cell environment in the organism.

#### 3.1.1. Pigment Epithelium Cell Lines

One of the most common 2D models is immortalized retinal pigment epithelium (RPE) cell lines. Primary cultures of retinal cells are challenging to handle. Obtaining a homogenous cell line that is not contaminated with other eye cells is challenging. Furthermore, isolated cells often quickly change their properties. For example, cells can lose keratin-containing intermediate filaments [13]. Cell transformation using the SV40 virus managed to obtain a line that retains the characteristics of retinal cells [14]. These cells are characterized by appropriate polarization and monocellular epithelial cell formation.

In 1995, RPE cells were first isolated by Davis et al. from a patient. However, the RPE cell line was only used in toxicity tests because the cells lost the characteristics of normal metabolism, adequate cytoskeleton polarization, and enzyme activity [14]. In the literature, primary models of RPE cell culture obtained from mice (i.e., Mouse Retinal Pigment Epithelial Cells-Hpv16 E6/E7, Immortalized) [15], rats (RPE primary cells isolated from PVG rats susceptible to experimental uveitis development; RPE isolated from Long Evans rats) [16,17], chickens (primary RPE cells isolated from domestic chickens embryos at stages 29–31 of development) [18], bovines (primary RPE cells) [19], and frogs (Xenopus laevis isolated primary RPE cells) [20] have also been described.

The human cell line ARPE-19 has structural and functional properties characteristic of RPE cells in vivo (in rats, RPE-J) [11]. This line is essential because the number of tissue donors is limited [21]. Studies on ARPE19 showed several features confirming the usefulness of this line for retinal pigment epithelial examination, such as expression of characteristic RPE cell markers, CRALBP, and RPE65, secretion of IL-6 and IL-8, as well as morphological polarization in monolayers, and ability form tight-junctions [22,23].

RPE-340 are primary cells isolated from humans which have epithelial morphology, but after several passages, their ability to replicate is limited [24]. Human RPE cells are good models for pharmacodynamic and physiological evaluation of a drug's effect on the choroid-RPE-photoreceptor, but after 40–60 population doublings, they go into a senescence state [24]. To develop a cell line with an extended lifespan, RPE-340 was transfected with a plasmid expressing the human telomerase reverse transcriptase subunit (hTERT), creating a new cell line—hTERT-RPE-1 (human retinal pigment epithelial RPE-1) [25]. This way, the lifespan of hTERT-RPE-1 is extended without any alterations in the population, doubling time and RPE-340 characteristic features [26]. hTERT-RPE-1 is reported to be an excellent model for epigenetic regulation studies [27–29]. Unfortunately, this line still has its limitations. The handling lasts 20 passages longer than RPE-340, but after this period, the cells change their morphology and function under the phenomenon called deadaptation [30,31]. Due to the low availability of primary human RPE cultures, validating and comparing this cell line with immortalized cell lines is challenging. The perfect line of human RPEs has yet to be developed.

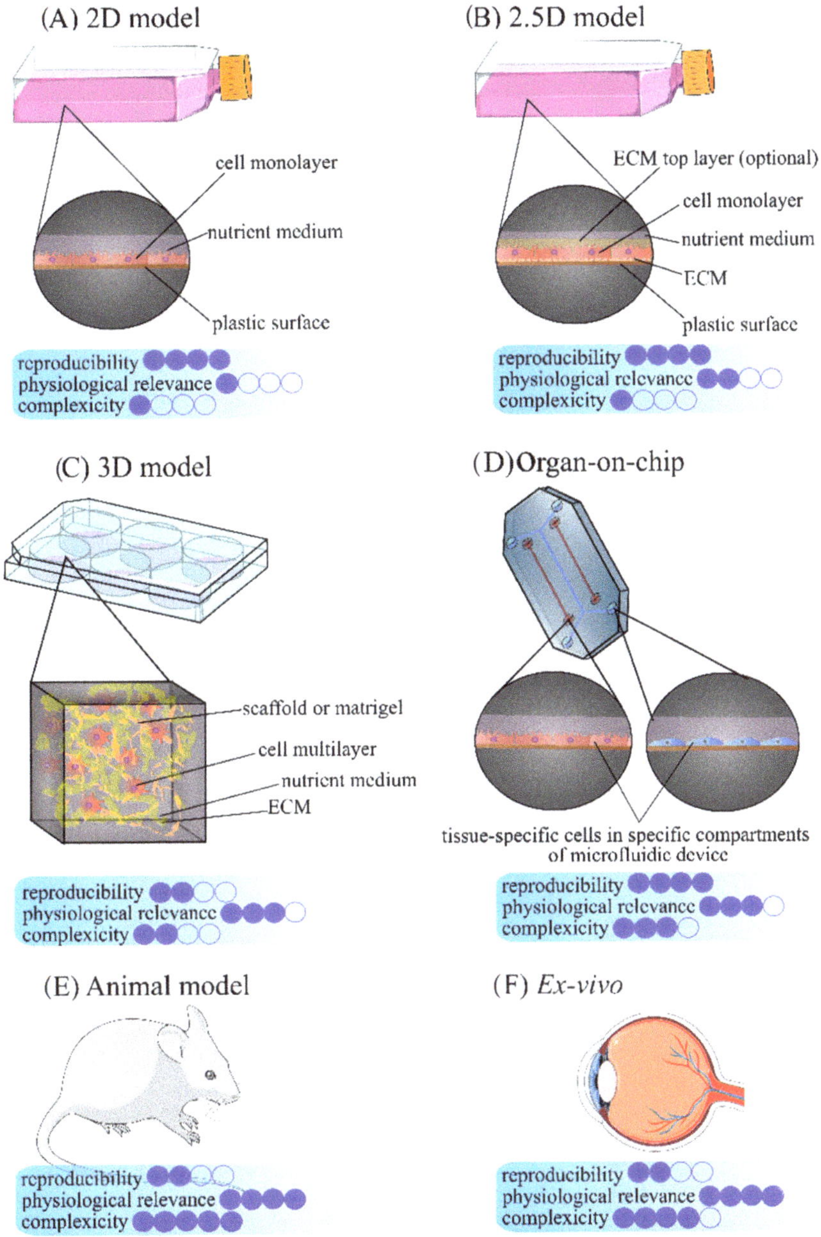


**Figure 1.** Schematic diagram of of eye models types. The figure was created using SMART (Servier Medical ART) modified graphics, licensed under a Creative Commons Attribution 3.0. Generic License.

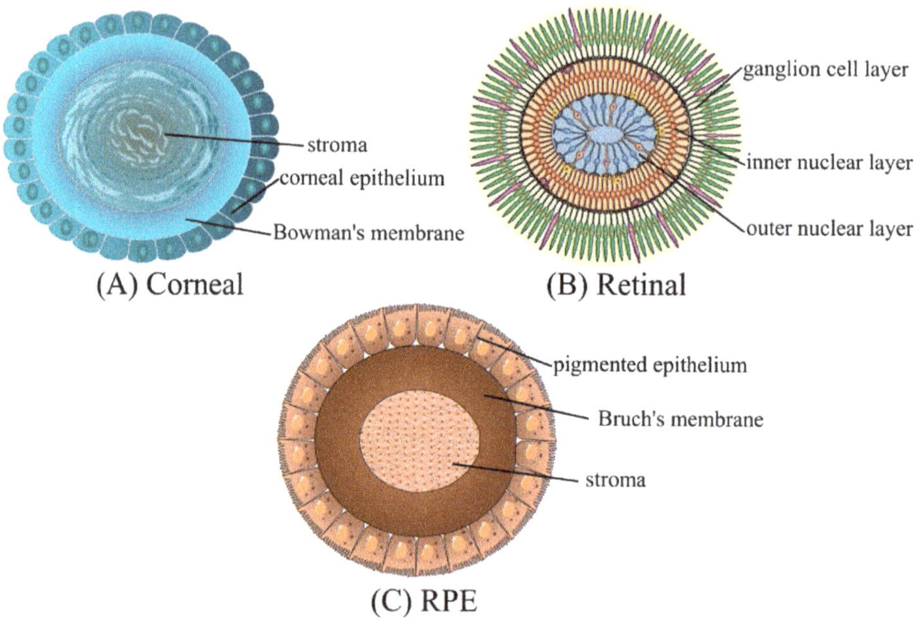


**Figure 2.** Schematic diagram of eye organoids. The figure was created using SMART (Servier Medical ART) modified graphics, licensed under a Creative Commons Attribution 3.0. Generic License.

The R28 immortalized retinal precursor cell line originating from postnatal day 6 rat retinal culture has been frequently used in in vitro and in vivo studies [32]. R28 provides an important system for understanding retinal cell behavior aspects such as differentiation, cytotoxicity, light stimulation, and neuroprotection. Although R28 originated from single clones, they remained highly heterogenous, suggesting the precursor character of these cells [32]. This cell model has been used in various toxicity experiments in vitro [33–37]. In addition, R28 exerts a high potential for studying the neuroprotective properties of chemical compounds [38–40]. Latanoprost was one of the drugs validated on R28 under the angle of cytoprotective properties [33].

RGC-5 was previously described as a rat-derived, transformed retinal ganglion cell line and is widely used in glaucoma research [41]. After more than 220 published papers worldwide involving the use of the RGC-5 cell line, it was reported that these cells are in fact 661W, a mouse SV-40 T antigen transformed photoreceptor cell [41,42]. The 661W cell line was present in the laboratory of origin of RGC-5; therefore, the most probable scenario was the cross-contamination of the newly developed cell line with 661W. This incident has shown how crucial the proper culture protocols and DNA profiling of newly-developed cell lines are. 661W is a model of cone photoreceptor cells. This cell line was widely used as a model for research on macular degeneration, but studying retinal ciliopathies such as retinitis pigmentosa is believed to be possible [43]. 661W shows properties of both retinal ganglion and photoreceptor cells, providing a functional photoreceptor model [43,44]. Moreover, 661W are believed to be an alternative model to the hTERT-RPE-1 cell line previously used for small molecule screening to identify new treatments for retinal ciliopathies [45]. 661W shows potential in studying ciliopathy disease genes not expressed or expressed at a low level in hTERT-RPE-1 cells [43].

### 3.1.2. Cornea Cells

The primary cornea cultures on which the individual in vitro models were developed come mainly from rabbits. Rabbit corneal epithelial cells (RbCEpC) help assess drug safety, pharmaceutical effects, corneal development, pathology, glaucoma, viral infections, keratitis, ocular hypertension, and even special contact lenses that provide sustained, extended-release of ophthalmic drugs [31]. Human corneal epithelial cells (HCEpC) have been used as models for studying corneal damage and reconstruction, re-epithelialization of the eye following surgery, and the effects of degradative enzymes. Corneal research mainly focuses on developing a model of drug permeation through this structure [46]. The models of corneal culture used in cellular research concern simple monolayers, the multilamellar epithelium, and very complex three-dimensional (3D) tissues resembling the functional cornea. HCEpC was used to create a single cellular layer later used for transplantation [47].

Several commercially available in vitro cornea models are destined to be cultured in 2D models (monolayers). One such model is HCE-T, in which cells are grown on the collagen membrane and are located at the air-liquid interface with the serum-free medium. The cells have the features of the primary cell line and form a stratified epithelium whose morphology can be modulated with calcium. Moreover, the cells expressed specific corneal epithelial cell markers such as epidermal growth factor (EGF), EGF receptor, basic fibroblast growth factor (basic FGF), transforming growth factor-beta 1 (TGF-beta 1), and interleukin-1 alpha (IL-1 alpha) [48].

#### 3.1.3. Corneal Endothelial Cells

The role of corneal endothelial cells (CECs) is to control corneal transparency. Unfortunately, the cells exhibit limited proliferative capability; therefore, their dysfunction may be one of the causes of blindness. One of the gold standards in treatment of corneal endothelial dysfunction is the donor isolated corneal transplant [49]. The first culture of human corneal endothelial cells from donors was established by Pistov et al. in 1988 [50], and from that time, plenty of protocols and newly designed biomaterials for the propagation of these cells were developed [51]. However, due to the low proliferation rate of primary CECs culture, protocols for immortalized cells were established [52,53]. Currently, several immortalized human CECs are available in the market and are used mainly to understand corneal endothelial cell dysfunctions. Two clonal cell lines derived from the immortalization of human corneal endothelial cells (obtained from the donor) were described by Valtink et al. [54]: B4G12 and H9C1 cells. B4G12 cells are polygonal, strongly adherent cells, which form a strict monolayer and H9C1 cells are less adherent and formed floating spheres. Both cell lines exhibited the characteristic expression of corneal endothelial cell markers; however, on different levels. Therefore, the authors concluded that the B4G12 cell line is a good model of differentiated CECs, and H9C1 is a good model for developing or transitional CECs. Alternatives for donor corneal endothelial cells or cornea endothelial cell lines may be pluripotent stem cell-derived corneal endothelial cells. The cells generated from a cryopreserved human embryonic stem cell (hESC) are stable, express corneal endothelial cell markers, and have an improved proliferation rate compared to primary CECs [55,56].

#### 3.1.4. Conjunctival Cells

The most popular eye conjunctival test model is the rabbit conjunctiva. For the first time in 1996, Saha et al. isolated the primary culture of the rabbit's conjunctiva. The model created by him represents a tight epithelial barrier [57]. This model is constantly being improved. The most significant difficulty was adapting the cells of this model to contact with air, just like in the natural eye. The use of additional filters (for example, the Transwell filter) allowed cell growth at the air-liquid interface. The layers of the conjunctival epithelial cells showed transepithelial resistance and a difference in potential [58]. The conjunctival epithelial cells are polygonal with many microvilli [59]. The primary culture of conjunctival cells was also obtained from bovines [59] and rats. The immortalized rat conjunctival (CJ4.1A) cell line was created by transfection of SV 40 [60]. CJ4.1A expresses the SV40 T antigen, conjugal cytokeratin 4, and cytokeratin specific for goblet cells 7, but not the cytokeratin 12. The line's lifespan is very long—line cells can be cultured for over 60 passages, and the population doubling times were 22 ± 7 h [60].

The development of methods for obtaining the primary culture of conjunctival cells has contributed to the development of transplantation techniques for heterotopic or allogeneic grafts. After severe damage to the conjunctiva, it is possible to restore its function by taking a piece of epithelium from a healthy eye, multiplying it in a cell culture, and implanting it in the affected conjunctiva [61]. Besides the primary cell culture, several established conjunctiva cell culture lines also exist. An example would be two human immortalized conjunctival cell lines: HCjE [62] and IOBA-NHC [63]. These cells have a typical epithelial morphology of the human epithelium, and after exposure of the cells to inflammatory mediators (IFN$\gamma$ and/or TNF$\alpha$), they increase the expression of the intercellular adhesion molecule (ICAM)-1 and MHC class II cell surface receptor (HLA-DR) [63].

The above-mentioned 2D models have their advantages but also their limitations. Firstly, these models are exceptionally delicate, and their manipulation must be meticulous. For example, the layer is easily damaged and dried. In addition, the models do not take into account cell to cell communication and the influence of immunological factors, which probably have a tremendous impact on the regeneration of this structure. Finally, based on these models' results, it is impossible to recapitulate all the processes occurring in the cornea in the human eye [64]. Therefore, the researchers decided to develop more complex, multicellular eye models.

### 3.2. 3D Models

Three-dimensional models better replicate the organism-environment compared to two-dimensional cultures. Cells grow in every dimension and closely replicate tissue in vitro, which complements 2D cell culture [65]. Although 3D models provide us with more information than 2D models, they are more challenging to handle. Multilayer models respond to more and more questions about corneal damage and disfigurement, but they are still far from the complex equipment that the eye is. For example, they lack the lacrimal apparatus responsible for cleansing and supporting regeneration [66]. Few 3D cornea models have been developed to this point.

EpiOcular™, developed in 2010, was obtained from cultured human epithelial cells. The cells showed a morphology and expression of biomarkers similar to the intact human cornea and maintained its thickness and permeability [67]. The EpiOcular model was used to assess the eye irritation potential of surfactant and surfactant-based formulations. Based on the protocol, the compound is considered to be an irritant when more than 50% of the cell die as compared to the negative control [68,69]. This test is validated and under review by the European Center for the Validation of Alternative Methods (ECVAM).

Clonetics (cHCEC) was developed in 2011 and was obtained from human corneal epithelial cells. Research using the model provides information on the assessment of corneal penetration by various chemical compounds (e.g., ophthalmic drugs) [68]. cHCEC was examined by RT-PCR for the expression profile of drug-metabolizing enzymes (e.g., CYP P450s and UGT1A1) and transporters in cHCE in comparison to the human cornea [70].

The SkinEthic (HCE) 3D cornea model comprises immortalized human mucosa cells; cells are grown at the air-liquid interface using a polycarbonate membrane. Under appropriate conditions, the cells differentiate and form the three-dimensional (3D) stratified epithelium and have non-keratin structures [69]. The advantage of the model is the possibility of administering dissolved substances in organic and inorganic solvents at any concentration (a very concentrated solution or minimal drug application concentrations can be applied) and the preservation of conditions similar to the eye mucosa of the human eye. Immortalized cells are grown in a dedicated medium and form a histologically multilayered construct with a thickness of 60 μm. The HCE secretes the same mucins found in the human cornea in vivo and expresses CD44 and keratin. This model is used to study phototoxicity, irritation, corrosivity, and the transport of substances [71,72].

The LabCyte CORNEA-MODEL is produced from normal human cornea epithelial cells [73]. It was developed by differentiating and stratifying cornea epithelial cells and is meant to be used to identify irritant chemicals in eye irritation tests. The corneal epithelial cells are cultivated on an inert filter substrate for 13 days with a medium containing 5% FBS. Proliferating cells build up in a multilayer structure consisting of a fully differentiated epithelium with features of the average human corneal epithelial tissue [74].

The limited source of corneal tissue to form a 3D-model, the short-lived life cycle of the corneal cells themselves, and the time-consuming culture contribute to problems with the industrialization of culture. The use of many commercially available 3D models was limited by the rapid differentiation of cells leading to problems with maintaining cell culture [75]. Many attempts have been made to increase the in vitro culture cycle of corneal epithelial cells concerning telomerase reverse transcription gene transfection, viral transfection, and the induction of spontaneous mutations. Nevertheless, the abnormal phenotype of these cells, which can lead to the potential risk of tumorigenesis, is not desired in the construction of new cornea models. Therefore, Li et al. enriched cornea cells with limbal stem cells providing additional expansion and development stimulation [75]. The addition of limbal stem cells promoted development and cell expansion. Moreover, it enabled the large-scale production of a new 3D model. Use of the corneal stromal layer of the animal to stimulate a specific microenvironment for limbal stem cells resulted in their differentiation into cornea epithelial cells.

Zuguo et al. proposed a new in vitro xeropthalmia model by dissecting the conjunctival epithelium and subconjunctival matrix, culturing it on a collagen I coated dish submerged in a culture medium. The in vitro dry eye model is obtained after 4–20 days. The invention can be used to research dry eye squamous metaplasia, ocular surface epithelial barrier damage, epithelial mucin change, to test new drugs, or to find new methods for dry eye treatment [76].

The model created by Minami et al. consists of bovine epithelial, stromal, and endothelial cells in a collagen gel matrix. The epithelium consists of five to six layers, and the epithelial cells produce keratin, which is a fundamental multilayer model for the cornea [77]. In addition, some corneal models use cell lines from different animals. In these models, individual layers come from mice, rabbits, bovines, and pigs [65,78,79].

### 3.3. 2.5D Models

2.5D models seem to be an alternative approach compared to 2D and 3D cultures. In 2D, cells are grown on a flat surface, while 3D models are based on cells embedded in an extracellular matrix (ECM) and/or scaffolds that provide a proper three-dimensional environment. In 2.5D cultures, cells are grown in an extracellular matrix (ECM) layer which often is not flat but unregular with projections and grooves, thus, providing an intermediate between 2D and 3D conditions [80].

### 3.4. Ex Vivo Models

One alternative to the Draize test is harvesting organs for examination from animals used for meat (ex vivo model). Eyeballs are isolated from bovines (BCOP), rabbits (IRE), pigs (PCOP), and chickens (ICE). This test was accepted internationally in 2009 and is used to research if significant tissue damage can occur [81]. Tests on the models mentioned above are based mainly on histological and light transmittance through cornea analysis. The pigs' cornea provides the highest degree of similarity to the human cornea, especially in tests involving substances dissolved in water [82]. Unfortunately, all the models mentioned above have serious drawbacks, primarily resulting from anatomical differences. In addition, these models can only be used to study individual eye structures. Therefore, they do not allow for general-purpose research. It is vital to create cell microenvironments that support tissue differentiation and changes, tissue-tissue communication, and spatiotemporal chemical and mechanical gradients of the microenvironment of living organs [83].

Yu F. proposed an ex vivo mammalian cornea culture system used for chemical tests of consumer products [84]. This system closely resembles in vivo testing by maintaining the corneal structure, architecture, and epithelial cell interaction. The cornea or the whole eye is excised and placed on an agar or collagen scaffold. It is then submerged in a culture medium until the medium covers the limbus. The upper part of the cornea is not submerged in medium. The tested reagent is administered directly to the cornea. The inventor claims that the system may be used to replace the use of Draize's test in many situations. This system allows drug testing without using live animals. The corneas or eyeballs may be, for example, easily acquired when dissecting rabbits for meat or fur industry purposes.

### *3.5. Spheroids, Organoids, and Organ-on-Chips*

New techniques and technologies in cell culture allow the development of more proper and scientific-useful models for ocular research. Here we described three types of it: spheroids, organoids, and organ-on-chips, a summary of which is presented in Table 2.

#### 3.5.1. Spheroids

Spheroids are self-assembly aggregated cells that spontaneously organize themselves into spherical-shaped structures. This phenomenon occurs naturally during embryogenesis, morphogenesis, or organogenesis. In in vitro culture, single cells may constitute multicellular spheroids after applying appropriate cell culture techniques (i.e., pellet culture, the hanging drop method, culture in the extracellular matrix, or others) [85]. Spheroids may have a different biological response to various factors due to the presence of a concentration gradient of nutrients, oxygen, or metabolites between cells from the outside and the inside part of the spheroid. Spheroids are mainly used in cancer research [86,87]. However, the technique of 3D multicellular culture with spheroids is also used in cellular research. Lu et al., using air-lifting 3D spheroid formation techniques, developed an in vitro model for research on the ocular surface and tear film systems. The model was composed of rabbit conjunctival epithelium and lacrimal gland cell spheroids [88]. The model allowed for the creation of the aqueous and mucin layers of the tear film, which may facilitate research on dry eye. A Japanese-German research group generated multicellular spheroids from human-donor RPE cells cultured in a methylcellulose matrix [89,90]. The model mimics the in vitro drusen model, which might help understand the pathogenesis of drusen-related diseases such as AMD. Sherwin's group from New Zeeland developed methods for isolation and propagation of spheroid human peripheral cornea using a clear cornea component of the rim isolated from a donor [91,92]. They found that generated spheroids implanted into frozen-stored corneoscleral tissue worked as limbal stem cell centers and proliferated to reproduce limbal cells. Spheroids are also used in ocular cancer research. There are several spheroid models of retinoblastoma (cells isolated from human intraocular tumors) which are used to develop new cancer treatments [93] or to understand retinoblastoma pathophysiology [94,95].

#### 3.5.2. Organoids

Organoids are stem cell derived 3D structures with organ-level functions. They are composed of self-organizing organ-specific cells derived from embryonic stem cells, induced pluripotent stem cells, or organ-restricted adult stem cells [96,97].

One of the most well-known ocular organoid models is a model described by Eiraku et al. [98]. The authors used mouse embryonic stem cells and show that ESCs in differentiation medium are self-organizing into optic-cups in 3D culture. Susaimanickam et al. developed an organoid model based on human embryonic stem cells (ESCs) or human induced pluripotent stem cells (iPSCs) cultured in a retinal differentiation medium supplemented with noggin [99]. The addition of noggin is crucial because of the protein's (a BMP inhibitor) involvement in the retinal differentiation of pluripotent stem cells during embryonic and organoid development. After two weeks, the culture gave rise to retinal and corneal primordia, and after six to eight weeks, primordia developed into minicorneas with specific morphological and marker similarities to the human cornea. This model may be used in basic research and regenerative applications. In addition, the use of organoid models with different ranges of time culture could provide us with data regarding drug toxicity in different stages of eye development. A congruous model or cornea organoids was developed by Foster et al. [100]. In this model of the cornea, three distinct cell types with the expression of key epithelial, stromal and endothelial cell markers were obtained. Mellough et al. in 2012 showed that ESC and iPSC cultured in ventral neural induction media (VNIM) supplemented with noggin, Dickkopf-1, Insulin-like growth factor 1, Lefty A, Human Sonic Hedgehog, and 3, 30, 5-triiodo-L-thyronine may develop retinal photoreceptor cells [101]. Later, they showed that VNIM can differentiate both EPS and iPSC cells, but the presence of IGF-1 is essential for the development of 3D ocular-like structures containing retinal pigmented epithelium, neural retina, primitive lens, and corneal-like structures [102]. In the latest work, Mellough et al. found that different embryoid bodies' (EBs) generation protocols affect the method and maintenance conditions that determine the later differentiation and maturation of retinal organoids [103]. The generation of more advanced in vitro multiocular organoids from human iPSCs cells was proposed by Isla-Magrané et al. [104]. In this protocol, organoids are differentiated in three different media, which leads to obtaining multicellular organoids after 150 days. Firstly, 75% confluent hiPSCs were cultured on Matrigel in an induction medium (DMEM/F12, 5% fetal bovine serum, nonessential amino acids, GlutaMax, N2, B27, β-glycerol phosphate, nicotinamide, Noggin, DKK1, bFGF) for 30 days. Next, all-trans retinoic acid (ATRA) was added for the next 60 days. Finally, cells maintained in a medium with ATRA for the next 60 days develop multiocular and corneal organoids, and cells cultured without ATRA and with triiodothyronine develop retinal organoids, RPE organoids, and multiocular organoids.

Recently, the National Centre for the Replacement Refinement & Reduction of Animals in Research (NC3Rs) and The National Eye Institute established a relationship that will result in the construction of organoids for drug screening, disease modeling, and regenerative medicine [105]. Therefore, a retinal 3D model is constantly being developed to fulfill those criteria [105–107]. The retinal 3D model contains bioprinted Müller cells, microglia, neurons, and RPE cells [108].

### 3.5.3. Organ-on-Chips

Organ-on-chips (OoC) are structures created by combining microfluidic technology, biomaterials, and cell culture methods [97]. Many organ-on-chips were used to research the permeability of the epithelium. Puleo et al. created a microfluidic device consisting of a bilayer structure of a corneal epithelial layer, a layer of stromal cells, and collagen vitrigel substrate [109]. Bennet et al. invented a cornea organ chip including epithelial layers, Bowman's membrane, basement membrane, and a device simulating tear flow dynamics. The measurement of epithelium permeability underflow showed results similar to in vivo measurements [110]. Cornea and retina chips are powerful and promising in vitro tools to study drug effects and therapeutic approaches, yet the chips are still minimal and straightforward 60 [97].

Recently, Seo and Huh proposed a cornea-on-chip "human blinking eye model" [111]. The system mimicked spontaneous eye blinking in humans with keratinocytes cultured to mimic the epithelial cells and form a corneal structure. Blinking imitation was performed by integrating a tear chamber in a 3D-printed eyelid [112].

DynaMiTES' Dynamic Micro Tissue Engineering System was developed from cornea immortalized cells. The system allowed for the measuring of transepithelial electrical resistance in real-time by implementing two electrodes into the system, providing a non-invasive way to monitor cell conditions [113].

Although organoids and organ-on-chips carry indisputable benefits, their potential in drug testing has yet to be closely examined. The main issues concerning drug assays relate to permeation and accessibility of the ocular surface of the tested models [97].

### 3.6. In Silico Analysis

In silico analysis is often used to meet the 3Rs regulations (replacement, reduction, and refinement) [114]. Many in silico models have been proposed up to this point in time. One of them is a quantitative structure-property relationship (QSPR) model proposed by Vincze et al. to study corneal permeability. The model is based on corneal-PAMPA (Parallel artificial membrane permeability assay) experimental data and different in silico drug transport parameters (Caco-2 and jejunal permeability) [115]. The test provided good predictions and is suitable for efficiently shortening the examined drugs list, provided we have comparable experimental data at our disposal. However, although promising, in silico studies currently do not provide us with enough data to regard drugs as safe. Therefore, experimental testing should be carried out to confirm the result of the studies.

**Table 2.** Spheroids, organoids, and microphysiological models for in vitro ocular research.

| | | **Spheroids** | | |
|---|---|---|---|---|
| Initial cells | Targeted cells | Culture techniques | Research possibilities | References |
| primary rabbit conjunctival epithelial cells (CECs) and lacrimal gland (LG) cell | cells that produce the aqueous and mucin layers of the tear film | an orbital shaker than Matrigel® matrix | dry eye disease | [88] |
| human RPE cells (hRPECs) obtained from donors | retinal pigment epithelium | culture in methylcellulose | Drusen-associated degeneration in the retina | [89,90] |
| human peripheral cornea | transplantable elements for limbal stem cell repopulation and limbal reconstruction | clear cornea component of the rim from a human donor | Regenerative medicine | [91,92] |
| human retoinoblastoma | human retoinoblastoma | different techniques | retinoblastoma | [93–95] |
| | | **Organoids** | | |
| Initial cells | Targeted cells | Culture techniques | Research possibilities | References |
| mouse embryonic stem cell (ESCs) | mechanically rigid pigment epithelium, embryonic optic cup, stratified neural retinal tissue | G-MEM supplemented with knockout serum replacement, nonessential amino acids, pyruvate, mercaptoethanol. | development of eye, eye disorders, disease modeling, | [98] |
| human embryonic stem cells (ESCs) or human induced pluripotent stem cells (iPSCs) | after two weeks: retinal and corneal primordia, after six to eight weeks: primordia developed into minicorneas with specific morphological and marker similarities to the human cornea. | culture in a retinal differentiation medium supplemented with noggin. | neurodevelopmental disorders, disease modeling, | [99] |
| human induced pluripotent stem cells | cornea, harboring three distinct cell types with the expression of key epithelial, stromal and endothelial cell markers. | multistep procotol | investigating corneal developmental processes and their disruptions in diseased condition | [100] |
| human embryonic stem cells (ESCs), human induced pluripotent stem cells (iPSCs) | 3D ocular-like structures contains: retinal pigmented epithelium, neural retina, primitive lens and corneal-like structures. | differentiation: ventral neural induction media (VNIM) with IGF-1 | development of eye, eye disorders, disease modeling, | [102] |
| human induced pluripotent stem cells (iPSCs) | 3D multiocular organoids contans: retinal pigment epithelium, retina, and cornea. | multistep procotol | model the crosstalk between different cell types in eye development and disease | [104] |

**Table 2.** *Cont.*

| Organs on Chip | | | | |
|---|---|---|---|---|
| Initial cells | Targeted cells | Culture techniques | Research possibilities | References |
| epithelium/stromal cells and keartnocytes isolated from rabbit eyes | microfluidic device consisting of a bilayer structure of corneal epithelial layer, a layer of stromal cells, and collagen vitrigel substrate | microfluidic devices containing collagen vitrigel (CV) | miniaturizing the standard transepithelial permeability (TEP) assay in order to measure the integrity of an array of corneal tissue micropatches. | [109] |
| immortalized human corneal epithelial cells | microengineered corneal epithelium-on-a-chip | porous membrane embedded microfluidic platform separated a chip into an apical and basal side | preclinical evaluations of potential therapeutic drugs and to mimic the environment of the human cornea. | [110] |
| human cells derived from the cornea and conjunctiva | mimic spontaneous eye blinking in humans | dome-shaped three-dimensional (3D) scaffolds in in vivo-like spatial arrangements | used for disease modeling and drug testing | [111] |
| human corneal epithelial (HCE–T) cells | dynamic cell cultivation and dynamic drug absorption testing on physiological barriers | DynaMiTES | improvement of common in vitro drug testing procedures | [112] |

## 4. Conclusions

Tissue engineering is one of the most rapidly developing scientific disciplines. It allows an easy and more reliable study of the effects of various factors and substances (including drugs). The development of this field will contribute to the invention of more advanced methods of combating diseases, repairing damaged tissues as a result of trauma, and to the ability to change and improve the function of given structures. At the same time, it will limit the number of animals used for experiments, which are now often indispensable research models. Currently, scientists are trying to fine-tune in vitro models and combine as many elements as possible to create a fully functional organ. One of the paths leading to this goal is the development of bioreactors. Bioreactors extend the time of in vitro culturing through specific, periodic exchanges of the culture medium. Physical factors are strictly controlled, e.g., temperature, pH, oxygen, and carbon dioxide. Additionally, they enable the precise delivery of nutrients and the removal of unnecessary metabolites from the nutrient solution [116,117].

All of the research models mentioned above have their limitations and advantages. Different Draize test alternative models provide more extensive flexibility in our research. Currently, 2D cultures are the most common research models. The reason is that 2D cultures are relatively inexpensive, more modulable, and easy to maintain [118]. Because of reproducible results obtained in controlled conditions [119], big-scale screening assays should be performed on these models. The main weakness of 2D models is their low ability to recreate the complexity of different cell classes and matrices interaction [118]. On the other hand, 3D multilayer models seem to be sufficient for small-scale drug toxicity and irritation assays. These models more closely resemble the eye microenvironment and consider cell-to-cell interactions, providing more relevant results. Moreover, 2D and 3D models seem to be limited in immunological disorders, such as allergy or sensitivity, because of their low of complexity. This problem could be addressed with organoids, which generate remarkable research outcomes, but only after long and arduous steps of standardization and testing [118].

Both in vitro and ex vivo models share one major limitation: the lack of vascularization [120]. The immune cells and vascularization should be introduced to these models to address this problem more appropriately. Organ-on-chip technology may be applied to facilitate the manipulation of more complex research models [120,121]. For example, including blood vessels in the model is possible by applying a forced flow supplied by on-chip technology. All that remains is to hope that the current development of in vitro models in ocular research allows for the complete elimination of the need to conduct tests on living organisms in the near future.

**Author Contributions:** Conceptualization: S.L. and Ł.S.; investigation: K.L., R.S. and A.L.; writing—original draft: R.S., K.L., M.S., E.K. and A.L.; writing—review and editing: A.Z., Ł.S. and S.L.; visualization: R.S. and Ł.S.; supervision: S.L. All authors have read and agreed to the published version of the manuscript.

**Funding:** The study was supported by Polish National Centre for Research and Development project no: DOB-1-6/1/PS/2014.

**Institutional Review Board Statement:** The study did not require ethical approval.

**Informed Consent Statement:** Not applicable.

**Data Availability Statement:** The data reviewed in this study are available on request from the corresponding author. The data are not publicly available due to founding agreement limitations.

**Conflicts of Interest:** The authors declare that they have no conflict of interest.

## References

1. Levkovitch-Verbin, H. Animal models of optic nerve diseases. *Eye* **2004**, *18*, 1066–1074. [CrossRef] [PubMed]
2. Pennesi, M.E.; Neuringer, M.; Courtney, R.J. Animal models of age related macular degeneration. *Mol. Asp. Med.* **2012**, *33*, 487–509. [CrossRef] [PubMed]
3. Wilhelmus, K.R. The Draize Eye Test. *Surv. Ophthalmol.* **2001**, *45*, 493–515. [CrossRef]
4. Liebsch, M.; Grune, B.; Seiler, A.; Butzke, D.; Oelgeschläger, M.; Pirow, R.; Adler, S.; Riebeling, C.; Luch, A. Alternatives to animal testing: Current status and future perspectives. *Arch. Toxicol.* **2011**, *85*, 841–858. [CrossRef]
5. DelMonte, D.W.; Kim, T. Anatomy and physiology of the cornea. *J. Cataract Refract. Surg.* **2011**, *37*, 588–598. [CrossRef]
6. Danilova, I.; Medvedeva, S.; Shmakova, S.; Chereshneva, M.; Sarapultsev, A.; Sarapultsev, P. Pathological changes in the cellular structures of retina and choroidea in the early stages of alloxan-induced diabetes. *World J. Diabetes* **2018**, *9*, 239–251. [CrossRef]
7. Pfeffer, B.A.; Philp, N.J. Cell culture of retinal pigment epithelium: Special Issue. *Exp. Eye Res.* **2014**, *126*, 1–4. [CrossRef]
8. Strauss, O. The Retinal Pigment Epithelium in Visual Function. *Physiol. Rev.* **2005**, *85*, 845–881. [CrossRef]
9. Salazar, J.J.; Ramírez, A.I.; De Hoz, R.; Salobrar-Garcia, E.; Rojas, P.; Fernández-Albarral, J.A.; López-Cuenca, I.; Rojas, B.; Triviño, A.; Ramírez, J.M. *Anatomy of the Human Optic Nerve: Structure and Function*; Intech: London, UK, 2019. [CrossRef]
10. Komai, Y.; Ushiki, T. The three-dimensional organization of collagen fibrils in the human cornea and sclera. *Investig. Ophthalmol. Vis. Sci.* **1991**, *32*, 2244–2258.
11. Nabi, I.R.; Mathews, A.P.; Cohen-Gould, L.; Gundersen, D.; Rodriguez-Boulan, E. Immortalization of polarized rat retinal pigment epithelium. *J. Cell Sci.* **1993**, *104*, 37–49. [CrossRef]
12. Badyal, D.K.; Desai, C. Animal use in pharmacology education and research: The changing scenario. *Indian J. Pharmacol.* **2014**, *46*, 257–265. [CrossRef] [PubMed]
13. Hunt, R.C.; Davis, A.A. Altered expression of keratin and vimentin in human retinal pigment epithelial cells in vivo and in vitro. *J. Cell. Physiol.* **1990**, *145*, 187–199. [CrossRef] [PubMed]
14. Davis, A.A.; Bernstein, P.S.; Bok, D.; Turner, J.; Nachtigal, M.; Hunt, R.C. A human retinal pigment epithelial cell line that retains epithelial characteristics after prolonged culture. *Investig. Ophthalmol. Vis. Sci.* **1995**, *36*, 955–964.
15. Catanuto, P.; Espinosa-Heidmann, D.; Pereira-Simon, S.; Sanchez, P.; Salas, P.; Hernandez, E.; Cousins, S.W.; Elliot, S.J. Mouse retinal pigmented epithelial cell lines retain their phenotypic characteristics after transfection with human papilloma virus: A new tool to further the study of RPE biology. *Exp. Eye Res.* **2009**, *88*, 99–105. [CrossRef] [PubMed]
16. Zech, J.C.; Pouvreau, I.; Cotinet, A.; Goureau, O.; Le Varlet, B.; De Kozak, Y. Effect of cytokines and nitric oxide on tight junctions in cultured rat retinal pigment epithelium. *Investig. Ophthalmol. Vis. Sci.* **1998**, *39*, 1600–1608.
17. Chang, C.W.; Ye, L.; Defoe, D.M.; Caldwell, R.B. Serum inhibits tight junction formation in cultured pigment epithelial cells. *Investig. Ophthalmol. Vis. Sci.* **1997**, *38*, 1082–1093.
18. Israel, P.; Masterson, E.; Goldman, A.I.; Wiggert, B.; Chader, G.J. Retinal pigment epithelial cell differentiation in vitro. Influence of culture medium. *Investig. Ophthalmol. Vis. Sci.* **1980**, *19*, 720–727.
19. Oka, M.S.; Landers, R.A.; Bridges, C.D. A serum-free defined medium for retinal pigment epithelial cells. *Exp. Cell Res.* **1984**, *154*, 537–547. [CrossRef]
20. Defoe, D.M.; Ahmad, A.; Chen, W.; Hughes, B.A. Membrane Polarity of the $Na^+$-$K^+$ Pump in Primary Cultures of Xenopus Retinal Pigment Epithelium. *Exp. Eye Res.* **1994**, *59*, 587–596. [CrossRef]

21. Engelmann, K.; Valtink, M. RPE cell cultivation. *Graefe's Arch. Clin. Exp. Ophthalmol.* **2004**, *242*, 65–67. [CrossRef]
22. Dunn, K.C.; Aotaki-Keen, A.E.; Putkey, F.R.; Hjelmeland, L.M. ARPE-19, A Human Retinal Pigment Epithelial Cell Line with Differentiated Properties. *Exp. Eye Res.* **1996**, *62*, 155–169. [CrossRef] [PubMed]
23. Holtkamp, G.M.; Van Rossem, M.; de Vos, A.F.; Willekens, B.; Peek, R.; Kijlstra, A. Polarized secretion of IL-6 and IL-8 by human retinal pigment epithelial cells. *Clin. Exp. Immunol.* **1998**, *112*, 34–43. [CrossRef] [PubMed]
24. Matsunaga, H.; Handa, J.T.; Aotaki-Keen, A.; Sherwood, S.W.; West, M.D.; Hjelmeland, L.M. Beta-galactosidase histochemistry and telomere loss in senescent retinal pigment epithelial cells. *Investig. Ophthalmol. Vis. Sci.* **1999**, *40*, 197–202.
25. Rambhatla, L.; Chiu, C.-P.; Glickman, R.D.; Rowe-Rendleman, C. In vitro differentiation capacity of telomerase immortalized human RPE cells. *Investig. Ophthalmol. Vis. Sci.* **2002**, *43*, 1622–1630.
26. Bodnar, A.G.; Ouellette, M.; Frolkis, M.; Holt, S.E.; Chiu, C.P.; Morin, G.B.; Harley, C.B.; Shay, J.W.; Lichtsteiner, S.; Wright, W.E. Extension of Life-Span by Introduction of Telomerase into Normal Human Cells. *Science* **1998**, *279*, 349–352. [CrossRef]
27. Kuznetsova, A.V.; Kurinov, A.M.; Aleksandrova, M.A. Cell Models to Study Regulation of Cell Transformation in Pathologies of Retinal Pigment Epithelium. *J. Ophthalmol.* **2014**, *2014*, e801787. [CrossRef]
28. Chadwick, B.P. Variation in Xi chromatin organization and correlation of the H3K27me3 chromatin territories to transcribed sequences by microarray analysis. *Chromosoma* **2007**, *116*, 147–157. [CrossRef]
29. Culver-Cochran, A.E.; Chadwick, B.P. The WSTF-ISWI Chromatin Remodeling Complex Transiently Associates with the Human Inactive X Chromosome during Late S-Phase Prior to BRCA1 and γ-H2AX. *PLoS ONE* **2012**, *7*, e50023. [CrossRef]
30. Kawazu, K.; Yamada, K.; Nakamura, M.; Ota, A. Characterization of cyclosporin A transport in cultured rabbit corneal epithelial cells: P-glycoprotein transport activity and binding to cyclophilin. *Investig. Ophthalmol. Vis. Sci.* **1999**, *40*, 1738–1744.
31. Kawazu, K.; Shiono, H.; Tanioka, H.; Ota, A.; Ikuse, T.; Takashina, H.; Kawashima, Y. Beta adrenergic antagonist permeation across cultured rabbit corneal epithelial cells grown on permeable supports. *Curr. Eye Res.* **1998**, *17*, 125–131. [CrossRef]
32. Seigel, G.M. Review: R28 retinal precursor cells: The first 20 years. *Mol. Vis.* **2014**, *20*, 301–306. [PubMed]
33. Nakanishi, Y.; Nakamura, M.; Mukuno, H.; Kanamori, A.; Seigel, G.M.; Negi, A. Latanoprost rescues retinal neuro-glial cells from apoptosis by inhibiting caspase-3, which is mediated by p44/p42 mitogen-activated protein kinase. *Exp. Eye Res.* **2006**, *83*, 1108–1117. [CrossRef] [PubMed]
34. Adamus, G.; Machnicki, M.; Seigel, G.M. Apoptotic retinal cell death induced by antirecoverin autoantibodies of cancer-associated retinopathy. *Investig. Ophthalmol. Vis. Sci.* **1997**, *38*, 283–291.
35. Adamus, G.; Amundson, D.; Seigel, G.M.; Machnicki, M. Anti-Enolase-α Autoantibodies in Cancer-Associated Retinopathy: Epitope Mapping and Cytotoxicity on Retinal Cells. *J. Autoimmun.* **1998**, *11*, 671–677. [CrossRef] [PubMed]
36. Luthra, S.; Narayanan, R.; Marques, L.E.A.; Chwa, M.; Kim, D.W.; Dong, J.; Seigel, G.M.; Neekhra, A.; Gramajo, A.L.; Brown, D.J.; et al. Evaluation of in vitro effects of bevacizumab (avastin) on retinal pigment epithelial, neurosensory retinal, and microvascular endothelial cells. *Retina* **2006**, *26*, 512–518. [CrossRef] [PubMed]
37. Narayanan, R.; Kenney, M.C.; Kamjoo, S.; Trinh, T.-H.T.; Seigel, G.M.; Resende, G.P.; Kuppermann, B.D. Toxicity of Indocyanine Green (ICG) in Combination with Light on Retinal Pigment Epithelial Cells and Neurosensory Retinal Cells. *Curr. Eye Res.* **2005**, *30*, 471–478. [CrossRef]
38. Wu, X.; Reiter, C.E.N.; Antonetti, D.A.; Kimball, S.R.; Jefferson, L.S.; Gardner, T.W. Insulin Promotes Rat Retinal Neuronal Cell Survival in a p70S6K-dependent Manner. *J. Biol. Chem.* **2004**, *279*, 9167–9175. [CrossRef]
39. Shen, J.; Wu, Y.; Xu, J.-Y.; Zhang, J.; Sinclair, S.H.; Yanoff, M.; Xu, G.; Li, W.; Xu, G.-T. ERK- and Akt-Dependent Neuroprotection by Erythropoietin (EPO) against Glyoxal-AGEs via Modulation of Bcl-xL, Bax, and BAD. *Investig. Opthalmol. Vis. Sci.* **2010**, *51*, 35–46. [CrossRef]
40. Mukuno, H.; Nakamura, M.; Kanamori, A.; Nagai, A.; Negi, A.; Seigel, G.M. Unoprostone isopropyl rescues retinal progenitor cells from apoptosis in Vitro. *Curr. Eye Res.* **2004**, *29*, 457–464. [CrossRef]
41. Van Bergen, N.J.; Wood, J.P.M.; Chidlow, G.; Trounce, I.A.; Casson, R.J.; Ju, W.-K.; Weinreb, R.N.; Crowston, J.G. Recharacterization of the RCG-5 Retinal Ganglion Cell Line. *Investig. Opthalmol. Vis. Sci.* **2009**, *50*, 4267–4272. [CrossRef]
42. Krishnamoorthy, R.R.; Clark, A.F.; Daudt, D.; Vishwanatha, J.K.; Yorio, T. A Forensic Path to RGC-5 Cell Line Identification: Lessons Learned. *Investig. Opthalmol. Vis. Sci.* **2013**, *54*, 5712–5719. [CrossRef] [PubMed]
43. Wheway, G.; Nazlamova, L.; Turner, D.; Cross, S. 661W Photoreceptor Cell Line as a Cell Model for Studying Retinal Ciliopathies. *Front. Genet.* **2019**, *10*, 308. [CrossRef] [PubMed]
44. Sayyad, Z.; Sirohi, K.; Radha, V.; Swarup, G. 661W is a retinal ganglion precursor-like cell line in which glaucoma-associated optineurin mutants induce cell death selectively. *Sci. Rep.* **2017**, *7*, 16855. [CrossRef]
45. Kim, Y.J.; Kim, S.; Jung, Y.; Jung, E.; Kwon, H.J.; Kim, J. Eupatilin rescues ciliary transition zone defects to ameliorate ciliopathy-related phenotypes. *J. Clin. Investig.* **2018**, *128*, 3642–3648. [CrossRef] [PubMed]
46. Uematsu, M.; Mohamed, Y.H.; Onizuka, N.; Ueki, R.; Inoue, D.; Fujikawa, A.; Kitaoka, T. A novel in vivo corneal trans-epithelial electrical resistance measurement device. *J. Pharmacol. Toxicol. Methods* **2015**, *76*, 65–71. [CrossRef] [PubMed]
47. Ramaesh, K.; Dhillon, B. Ex Vivo Expansion of Corneal Limbal Epithelial/Stem Cells for Corneal Surface Reconstruction. *Eur. J. Ophthalmol.* **2003**, *13*, 515–524. [CrossRef]
48. Ward, S.L.; Walker, T.L.; Dimitrijevich, S.D. Evaluation of chemically induced toxicity using an in vitro model of human corneal epithelium. *Toxicol. In Vitro* **1997**, *11*, 121–139. [CrossRef]

49. Sie, N.M.; Yam, G.H.-F.; Soh, Y.Q.; Lovatt, M.; Dhaliwal, D.; Kocaba, V.; Mehta, J.S. Regenerative capacity of the corneal transition zone for endothelial cell therapy. *Stem Cell Res. Ther.* **2020**, *11*, 523. [CrossRef]
50. Pistsov, M.Y.; Sadovnikova, E.Y.; Danilov, S.M. Human corneal endothelial cells: Isolation, characterization and long-term cultivation. *Exp. Eye Res.* **1988**, *47*, 403–414. [CrossRef]
51. Parekh, M.; Romano, V.; Hassanin, K.; Testa, V.; Wongvisavavit, R.; Ferrari, S.; Haneef, A.; Willoughby, C.; Ponzin, D.; Jhanji, V.; et al. Biomaterials for corneal endothelial cell culture and tissue engineering. *J. Tissue Eng.* **2021**, *12*, 2041731421990536. [CrossRef]
52. Götze, T.; Valtink, M.; Nitschke, M.; Gramm, S.; Hanke, T.; Engelmann, K.; Werner, C. Cultivation of an immortalized human corneal endothelial cell population and two distinct clonal subpopulations on thermo-responsive carriers. *Graefes Arch. Clin. Exp. Ophthalmol.* **2008**, *246*, 1575–1583. [CrossRef] [PubMed]
53. Frausto, R.F.; Le, D.J.; Aldave, A.J. Transcriptomic Analysis of Cultured Corneal Endothelial Cells as a Validation for Their Use in Cell Replacement Therapy. *Cell Transplant.* **2016**, *25*, 1159–1176. [CrossRef] [PubMed]
54. Valtink, M.; Gruschwitz, R.; Funk, R.H.W.; Engelmann, K. Two Clonal Cell Lines of Immortalized Human Corneal Endothelial Cells Show either Differentiated or Precursor Cell Characteristics. *Cells Tissues Organs* **2008**, *187*, 286–294. [CrossRef] [PubMed]
55. McCabe, K.L.; Kunzevitzky, N.J.; Chiswell, B.P.; Xia, X.; Goldberg, J.L.; Lanza, R. Efficient Generation of Human Embryonic Stem Cell-Derived Corneal Endothelial Cells by Directed Differentiation. *PLoS ONE* **2015**, *10*, e0145266. [CrossRef] [PubMed]
56. Ali, M.; Khan, S.Y.; Gottsch, J.D.; Hutchinson, E.K.; Khan, A.; Riazuddin, S.A. Pluripotent stem cell–derived corneal endothelial cells as an alternative to donor corneal endothelium in keratoplasty. *Stem Cell Rep.* **2021**, *16*, 2320–2335. [CrossRef]
57. Saha, P. A Primary Culture Model of Rabbit Conjunctival Epithelial Cells Exhibiting Tight Barrier Properties. *Curr. Eye Res.* **1996**, *15*, 1163–1169. [CrossRef]
58. Yang, J.J.; Ueda, H.; Kim, K.-J.; Lee, V.H. Meeting future challenges in topical ocular drug delivery:: Development of an air-interfaced primary culture of rabbit conjunctival epithelial cells on a permeable support for drug transport studies. *J. Control. Release* **2000**, *65*, 1–11. [CrossRef]
59. Civiale, C.; Paladino, G.; Marino, C.; Trombetta, F.; Pulvirenti, T.; Enea, V. Multilayer Primary Epithelial Cell Culture from Bovine Conjunctiva as a Model for in vitro Toxicity Tests. *Ophthalmic Res.* **2003**, *35*, 126–136. [CrossRef]
60. Reichl, S.; Müller-Goymann, C.C. The use of a porcine organotypic cornea construct for permeation studies from formulations containing befunolol hydrochloride. *Int. J. Pharm.* **2003**, *250*, 191–201. [CrossRef]
61. Scuderi, N.; Alfano, C.; Paolini, G.; Marchese, C.; Scuderi, G. Transplantation of autologous cultivated conjunctival epithelium for the restoration of defects in the ocular surface. *Scand. J. Plast. Reconstr. Surg. Hand Surg.* **2002**, *36*, 340–348. [CrossRef]
62. Gipson, I.K.; Spurr-Michaud, S.; Argueso, P.; Tisdale, A.; Ng, T.F.; Russo, C.L. Mucin Gene Expression in Immortalized Human Corneal–Limbal and Conjunctival Epithelial Cell Lines. *Investig. Opthalmol. Vis. Sci.* **2003**, *44*, 2496–2506. [CrossRef] [PubMed]
63. Diebold, Y.; Calonge, M.; Enríquez de Salamanca, A.; de Salamanca, A.E.; Callejo, S.; Corrales, R.M.; Sa'ez, V.; Siemasko, K.F.; Stern, M.E. Characterization of a spontaneously immortalized cell line (IOBA-NHC) from normal human conjunctiva. *Investig. Opthalmol. Vis. Sci.* **2003**, *44*, 4263–4274. [CrossRef] [PubMed]
64. Barile, F.A. Validating and troubleshooting ocular invitro toxicology tests. *J. Pharmacol. Toxicol. Methods* **2010**, *61*, 136–145. [CrossRef] [PubMed]
65. Shafaie, S.; Hutter, V.; Cook, M.T.; Brown, M.B.; Chau, D.Y.S. In Vitro Cell Models for Ophthalmic Drug Development Applications. *BioRes. Open Access* **2016**, *5*, 94–108. [CrossRef] [PubMed]
66. Tegtmeyer, S.; Papantoniou, I.; Müller-Goymann, C.C. Reconstruction of an in vitro cornea and its use for drug permeation studies from different formulations containing pilocarpine hydrochloride. *Eur. J. Pharm. Biopharm.* **2001**, *51*, 119–125. [CrossRef]
67. Jung, K.-M.; Lee, S.-H.; Ryu, Y.-H.; Jang, W.-H.; Jung, H.-S.; Han, J.-H.; Seok, S.-H.; Park, J.-H.; Son, Y.; Park, Y.-H.; et al. A new 3D reconstituted human corneal epithelium model as an alternative method for the eye irritation test. *Toxicol. In Vitro* **2011**, *25*, 403–410. [CrossRef]
68. Kulkarni, A.; Chang, W.W.; Shen, J.; Welty, D. Use of Clonetics® Human Corneal Epithelial Cell Model for Evaluating Corneal Penetration and Hydrolysis of Ophthalmic Drug Candidates. *Investig. Opthalmol. Vis. Sci.* **2011**, *52*, 3259.
69. Cotovio, J.; Grandidier, M.-H.; Lelièvre, D.; Bremond, C.; Amsellem, C.; Maloug, S.; Ovigne, J.-M.; Loisel-Joubert, S.; Van Der Lee, A.; Minondo, A.-M.; et al. In vitro assessment of eye irritancy using the Reconstructed Human Corneal Epithelial SkinEthic™ HCE model: Application to 435 substances from consumer products industry. *Toxicol. In Vitro* **2010**, *24*, 523–537. [CrossRef]
70. Xiang, C.D.; Batugo, M.; Gale, D.C.; Zhang, T.; Ye, J.; Li, C.; Zhou, S.; Wu, E.Y.; Zhang, E.Y. Characterization of Human Corneal Epithelial Cell Model As a Surrogate for Corneal Permeability Assessment: Metabolism and Transport. *Drug Metab. Dispos.* **2009**, *37*, 992–998. [CrossRef]
71. van Goethem, F.; Adriaens, E.; Alépée, N.; Straube, F.; De Wever, B.; Cappadoro, M.; Catoire, S.; Hansen, E.; Wolf, A.; Vanparys, P. Prevalidation of a new in vitro reconstituted human cornea model to assess the eye irritating potential of chemicals. *Toxicol. In Vitro* **2006**, *20*, 1–17. [CrossRef]
72. de Brugerolle, A. SkinEthic Laboratories, a Company Devoted to Develop and Produce In Vitro Alternative Methods to Animal Use. *ALTEX-Altern. Anim. Exp.* **2007**, *24*, 167–171. [CrossRef] [PubMed]

73. Katoh, M.; Hamajima, F.; Ogasawara, T.; Hata, K.-I. Establishment of a new in vitro test method for evaluation of eye irritancy using a reconstructed human corneal epithelial model, LabCyte CORNEA-MODEL. *Toxicol. In Vitro* **2013**, *27*, 2184–2192. [CrossRef] [PubMed]
74. Katoh, M.; Uemura, N.; Hamajima, F.; Ogasawara, T.; Hata, K. ichiro Morphological Characterization of a Reconstructed Human Corneal Epithelial Model (LabCyte CORNEA-MODEL) as an Alternative to the Draize Eye Test for the Assessment of Eye Irritation. *Altern. Anim. Test. Exp.* **2012**, *17*, 2–8. [CrossRef]
75. Li, X.; Lu, Y.; Deng, Z. Recombinant Cornea Model for Ocular Surface Irritation Evaluation and Preparation Method Thereof. Patent No: CN110066847A, 2019. Available online: https://worldwide.espacenet.com/patent/search/family/067364436/publication/CN110066847A?q=%22Recombinant%20Cornea%20Model%20for%20Ocular%20Surface%20Irritation%20Evaluation%20and%20Preparation%22 (accessed on 5 April 2022).
76. Zuguo, L.; Hui, L.; Wei, L.; Yangluowa, Q. Construction Method for Dry Eye Model In Vitro. Patent No: CN101508974B, 2009. Available online: https://worldwide.espacenet.com/patent/search/family/041001507/publication/CN101508974B?q=%22Construction%20Method%20for%20Dry%20Eye%20Model%20in%20Vitro%22 (accessed on 5 April 2022).
77. Minami, Y.; Sugihara, H.; Oono, S. Reconstruction of cornea in three-dimensional collagen gel matrix culture. *Investig. Ophthalmol. Vis. Sci.* **1993**, *34*, 2316–2324.
78. Schneider, A.I.; Maier-Reif, K.; Graeve, T. Constructing an in vitro cornea from cultures of the three specific corneal cell types. *In Vitro Cell. Dev. Biol. Anim.* **1999**, *35*, 515–526. [CrossRef]
79. Zieske, J.D.; Mason, V.S.; Wasson, M.E.; Meunier, S.F.; Nolte, C.J.; Fukai, N.; Olsen, B.R.; Parenteau, N.L. Basement Membrane Assembly and Differentiation of Cultured Corneal Cells: Importance of Culture Environment and Endothelial Cell Interaction. *Exp. Cell Res.* **1994**, *214*, 621–633. [CrossRef]
80. Shamir, E.R.; Ewald, A.J. Three-dimensional organotypic culture: Experimental models of mammalian biology and disease. *Nat. Rev. Mol. Cell Biol.* **2014**, *15*, 647–664. [CrossRef]
81. Tchao, R. Trans-Epithelial Permeability of Fluorescein in Vitro as an Assay to Determine Eye Irritants. *Altern. Methods Toxicol.* **1988**, *6*, 271–283.
82. van den Berghe, C.; Guillet, M.C.; Compan, D. Performance of porcine corneal opacity and permeability assay to predict eye irritation for water-soluble cosmetic ingredients. *Toxicol. In Vitro* **2005**, *19*, 823–830. [CrossRef]
83. Huh, D.; Hamilton, G.A.; Ingber, D.E. From 3D cell culture to organs-on-chips. *Trends Cell Biol.* **2011**, *21*, 745–754. [CrossRef]
84. Yu, F.-S.X. Corneal Organ Culture System for Chemical Toxicity Tests of Consumer Products. Patent No: WO 99/51763, 1999. Available online: https://patentimages.storage.googleapis.com/c8/4a/0d/d3b3a03378b3e4/WO1999051763A1.pdf (accessed on 17 May 2022).
85. Ryu, N.-E.; Lee, S.-H.; Park, H. Spheroid Culture System Methods and Applications for Mesenchymal Stem Cells. *Cells* **2019**, *8*, 1620. [CrossRef] [PubMed]
86. Matak, D.; Brodaczewska, K.K.; Lipiec, M.; Szymanski, Ł.; Szczylik, C.; Czarnecka, A.M. Colony, hanging drop, and methylcellulose three dimensional hypoxic growth optimization of renal cell carcinoma cell lines. *Cytotechnology* **2017**, *69*, 565–578. [CrossRef] [PubMed]
87. Hofmann, S.; Cohen-Harazi, R.; Maizels, Y.; Koman, I. Patient-derived tumor spheroid cultures as a promising tool to assist personalized therapeutic decisions in breast cancer. *Transl. Cancer Res.* **2022**, *11*, 134–147. [CrossRef] [PubMed]
88. Lu, Q.; Yin, H.; Grant, M.P.; Elisseeff, J.H. An In Vitro Model for the Ocular Surface and Tear Film System. *Sci. Rep.* **2017**, *7*, 6163. [CrossRef]
89. Sato, R.; Yasukawa, T.; Kacza, J.; Eichler, W.; Nishiwaki, A.; Iandiev, I.; Ohbayashi, M.; Kato, A.; Yafai, Y.; Bringmann, A.; et al. Three-Dimensional Spheroidal Culture Visualization of Membranogenesis of Bruch's Membrane and Basolateral Functions of the Retinal Pigment Epithelium. *Investig. Opthalmol. Vis. Sci.* **2013**, *54*, 1740–1749. [CrossRef]
90. Usui, H.; Nishiwaki, A.; Landiev, L.; Kacza, J.; Eichler, W.; Wako, R.; Kato, A.; Takase, N.; Kuwayama, S.; Ohashi, K.; et al. In vitro drusen model: Three-dimensional spheroid culture of retinal pigment epithelial cells. *J. Cell Sci.* **2018**, *132*, jcs.215798. [CrossRef]
91. Yoon, J.J.; Wang, E.F.; Ismail, S.; McGhee, J.J.; Sherwin, T. Sphere-forming cells from peripheral cornea demonstrate polarity and directed cell migration. *Cell Biol. Int.* **2013**, *37*, 949–960. [CrossRef]
92. Mathan, J.J.; Ismail, S.; McGhee, J.J.; McGhee, C.N.J.; Sherwin, T. Sphere-forming cells from peripheral cornea demonstrate the ability to repopulate the ocular surface. *Stem Cell Res. Ther.* **2016**, *7*, 81. [CrossRef]
93. Winter, U.; Aschero, R.; Fuentes, F.; Buontempo, F.; Zugbi, S.; Sgroi, M.; Sampor, C.; Abramson, D.H.; Carcaboso, A.M.; Schaiquevich, P. Tridimensional Retinoblastoma Cultures as Vitreous Seeds Models for Live-Cell Imaging of Chemotherapy Penetration. *Int. J. Mol. Sci.* **2019**, *20*, 1077. [CrossRef]
94. Bond, W.S.; Akinfenwa, P.Y.; Perlaky, L.; Hurwitz, M.Y.; Hurwitz, R.L.; Chévez-Barrios, P. Tumorspheres but Not Adherent Cells Derived from Retinoblastoma Tumors Are of Malignant Origin. *PLoS ONE* **2013**, *8*, e63519. [CrossRef]
95. Tang, Z.; Ma, H.; Mao, Y.; Ai, S.; Zhang, P.; Nie, C.; Gao, Y.; Lu, R. Identification of stemness in primary retinoblastoma cells by analysis of stem-cell phenotypes and tumorigenicity with culture and xenograft models. *Exp. Cell Res.* **2019**, *379*, 110–118. [CrossRef] [PubMed]
96. Clevers, H. Modeling Development and Disease with Organoids. *Cell* **2016**, *165*, 1586–1597. [CrossRef] [PubMed]
97. Manafi, N.; Shokri, F.; Achberger, K.; Hirayama, M.; Mohammadi, M.H.; Noorizadeh, F.; Hong, J.; Liebau, S.; Tsuji, T.; Quinn, P.M.J.; et al. Organoids and organ chips in ophthalmology. *Ocul. Surf.* **2020**, *19*, 1–15. [CrossRef] [PubMed]

98. Eiraku, M.; Takata, N.; Ishibashi, H.; Kawada, M.; Sakakura, E.; Okuda, S.; Sekiguchi, K.; Adachi, T.; Sasai, Y. Self-organizing optic-cup morphogenesis in three-dimensional culture. *Nature* **2011**, *472*, 51–56. [CrossRef]
99. Susaimanickam, P.J.; Kiral, F.R.; Park, I.-H. Region Specific Brain Organoids to Study Neurodevelopmental Disorders. *Int. J. Stem Cells* **2022**, *15*, 26–40. [CrossRef]
100. Foster, J.W.; Wahlin, K.; Adams, S.M.; Birk, D.E.; Zack, D.J.; Chakravarti, S. Cornea organoids from human induced pluripotent stem cells. *Sci. Rep.* **2017**, *7*, 41286. [CrossRef]
101. Mellough, C.B.; Sernagor, E.; Moreno-Gimeno, I.; Steel, D.H.W.; Lako, M. Efficient Stage-Specific Differentiation of Human Pluripotent Stem Cells Toward Retinal Photoreceptor Cells. *Stem Cells* **2012**, *30*, 673–686. [CrossRef]
102. Mellough, C.B.; Collin, J.; Khazim, M.; White, K.; Sernagor, E.; Steel, D.H.W.; Lako, M. IGF-1 Signaling Plays an Important Role in the Formation of Three-Dimensional Laminated Neural Retina and Other Ocular Structures From Human Embryonic Stem Cells. *Stem Cells* **2015**, *33*, 2416–2430. [CrossRef]
103. Mellough, C.B.; Collin, J.; Queen, R.; Hilgen, G.; Dorgau, B.; Zerti, D.; Felemban, M.; White, K.; Sernagor, E.; Lako, M. Systematic Comparison of Retinal Organoid Differentiation from Human Pluripotent Stem Cells Reveals Stage Specific, Cell Line, and Methodological Differences. *Stem Cells Transl. Med.* **2019**, *8*, 694–706. [CrossRef]
104. Isla-Magrané, H.; Veiga, A.; García-Arumí, J.; Duarri, A. Multiocular organoids from human induced pluripotent stem cells displayed retinal, corneal, and retinal pigment epithelium lineages. *Stem Cell Res. Ther.* **2021**, *12*, 581. [CrossRef]
105. Mazerik, J.N.; Becker, S.; Sieving, P.A. 3-D retina organoids. *Cell Med.* **2018**, *10*, 2155179018773758. [CrossRef] [PubMed]
106. Georgiou, M.; Chichagova, V.; Hilgen, G.; Dorgau, B.; Sernagor, E.; Armstrong, L.; Lako, M. Room temperature shipment does not affect the biological activity of pluripotent stem cell-derived retinal organoids. *PLoS ONE* **2020**, *15*, e0233860. [CrossRef] [PubMed]
107. Chichagova, V.; Hallam, D.; Nicholds, M.; Thomas, R.; Lako, M.; Armstrong, L. Using Statistical Models to Identify Key Variavles Critical for Generation of Retinal Organoids from IPSC. *Investig. Opthalmol. Vis. Sci.* **2018**, *59*, 565.
108. National Centre for Replacement Refinement & Reduction of Animals in Research Retinal 3D. Available online: https://nc3rs.org.uk/crackit/retinal-3d (accessed on 21 May 2022).
109. Puleo, C.M.; Ambrose, W.M.; Takezawa, T.; Elisseeff, J.; Wang, T.-H. Integration and application of vitrified collagen in multilayered microfluidic devices for corneal microtissue culture. *Lab Chip* **2009**, *9*, 3221–3227. [CrossRef] [PubMed]
110. Bennet, D.; Estlack, Z.; Reid, T.; Kim, J. A microengineered human corneal epithelium-on-a-chip for eye drops mass transport evaluation. *Lab Chip* **2018**, *18*, 1539–1551. [CrossRef] [PubMed]
111. Seo, J.; Byun, W.Y.; Frank, A.; Massaro-Giordano, M.; Lee, V.; Bunya, V.Y.; Huh, D. A Human Blinking "Eye-on-a-Chip". *Investig. Opthalmol. Vis. Sci.* **2016**, *57*, 3872.
112. Seo, J.; Byun, W.Y.; Alisafaei, F.; Georgescu, A.; Yi, Y.-S.; Massaro-Giordano, M.; Shenoy, V.B.; Lee, V.; Bunya, V.Y.; Huh, D. Multiscale reverse engineering of the human ocular surface. *Nat. Med.* **2019**, *25*, 1310–1318. [CrossRef]
113. Beiβner, N.; Mattern, K.; Dietzel, A.; Reichl, S. DynaMiTES—A dynamic cell culture platform for in vitro drug testing PART 2—Ocular DynaMiTES for drug absorption studies of the anterior eye. *Eur. J. Pharm. Biopharm.* **2018**, *126*, 166–176. [CrossRef] [PubMed]
114. In Silico Cancer Research towards 3R-PMC. Available online: https://www.ncbi.nlm.nih.gov/pmc/articles/PMC5897933/ (accessed on 22 April 2022).
115. Vincze, A.; Dargó, G.; Rácz, A.; Balogh, G.T. A corneal-PAMPA-based in silico model for predicting corneal permeability. *J. Pharm. Biomed. Anal.* **2021**, *203*, 114218. [CrossRef]
116. Trone, M.-C.; Campolmi, N.; Gauthier, A.-S.; Nefzaoui, C.; Bernard, A.; Naigeon, N.; Nangoum-Fosso, T.; Piselli, S.; Garraud, O.; Acquart, S.; et al. Conception and optimization of a corneal bioreactor. *Acta Ophthalmol.* **2013**, *91*. [CrossRef]
117. Martin, I.; Wendt, D.; Heberer, M. The role of bioreactors in tissue engineering. *Trends Biotechnol.* **2004**, *22*, 80–86. [CrossRef] [PubMed]
118. Alfonsetti, M.; Castelli, V.; D'Angelo, M.; Benedetti, E.; Allegretti, M.; Barboni, B.; Cimini, A. Looking for In Vitro Models for Retinal Diseases. *Int. J. Mol. Sci.* **2021**, *22*, 10334. [CrossRef] [PubMed]
119. Hornof, M.; Toropainen, E.; Urtti, A. Cell culture models of the ocular barriers. *Eur. J. Pharm. Biopharm.* **2005**, *60*, 207–225. [CrossRef]
120. Achberger, K.; Probst, C.; Haderspeck, J.; Bolz, S.; Rogal, J.; Chuchuy, J.; Nikolova, M.; Cora, V.; Antkowiak, L.; Haq, W.; et al. Merging organoid and organ-on-a-chip technology to generate complex multi-layer tissue models in a human retina-on-a-chip platform. *eLife* **2019**, *8*, e46188. [CrossRef] [PubMed]
121. O'Hara-Wright, M.; Gonzalez-Cordero, A. Retinal organoids: A window into human retinal development. *Development* **2020**, *147*, dev189746. [CrossRef]

International Journal of
*Molecular Sciences*

MDPI

*Review*

# A Review of Defatting Strategies for Non-Alcoholic Fatty Liver Disease

**Erin Nicole Young, Murat Dogan, Christine Watkins, Amandeep Bajwa, James D. Eason, Canan Kuscu and Cem Kuscu ***

Transplant Research Institute, James D. Eason Transplant Institute, Department of Surgery, College of Medicine, The University of Tennessee Health Science Center, Memphis, TN 38163, USA
* Correspondence: ckuscu1@uthsc.edu

**Abstract:** Non-alcoholic fatty liver disease is a huge cause of chronic liver failure around the world. This condition has become more prevalent as rates of metabolic syndrome, type 2 diabetes, and obesity have also escalated. The unfortunate outcome for many people is liver cirrhosis that warrants transplantation or being unable to receive a transplant since many livers are discarded due to high levels of steatosis. Over the past several years, however, a great deal of work has gone into understanding the pathophysiology of this disease as well as possible treatment options. This review summarizes various defatting strategies including in vitro use of pharmacologic agents, machine perfusion of extracted livers, and genomic approaches targeting specific proteins. The goal of the field is to reduce the number of necessary transplants and expand the pool of organs available for use.

**Keywords:** liver; steatosis; defatting; molecular biology

**Citation:** Young, E.N.; Dogan, M.; Watkins, C.; Bajwa, A.; Eason, J.D.; Kuscu, C.; Kuscu, C. A Review of Defatting Strategies for Non-Alcoholic Fatty Liver Disease. *Int. J. Mol. Sci.* **2022**, *23*, 11805. https://doi.org/10.3390/ijms231911805

Academic Editors: Patrick C. Baer and Ralf Schubert

Received: 10 September 2022
Accepted: 28 September 2022
Published: 5 October 2022



## 1. Introduction

The liver is a complicated organ, and one of its main functions is to regulate lipid homeostasis through the interplay of various hormones, nuclear receptors, and transcription factors [1]. This includes conducting synthesis of new lipids, coordinating their transport to other parts of the body, and utilizing them as energy substrates [2]. Any imbalance between these pathways can result in the accumulation of fat. In the absence of alcohol consumption, non-alcoholic fatty liver disease (NAFLD) is characterized by the build-up of ectopic fat in the liver, also known as steatosis [3]. NAFLD is the current liver pandemic sweeping the globe. It is estimated that about 25% of the world's population is affected by NAFLD [4]. There is a very close association with dyslipidemia, type 2 diabetes, central obesity, and metabolic syndrome, each with an occurrence of 69%, 23%, 51%, and 43%, respectively [4]. Consequently, the rate of NAFLD diagnoses has been steadily rising with the ever-increasing obesity burden. Although the prevalence of this disease is so high, public knowledge of its existence and effects is still lacking.

More information is constantly being revealed on the pathophysiology of NAFLD, however, the general concept is that more lipids are being retained than the hepatocyte is able to expel. The four major pathways regulating lipid acquisition and disposal include uptake of free circulating fatty acids, de novo lipogenesis, fatty acid oxidation and transportation of lipid to the outside of liver as very low-density lipoproteins (VLDL) [5] (Figure 1). Focusing on the uptake of circulating lipids, it is directly proportional to the concentration of plasma free fatty acids (FFAs) that are mainly derived from the body's supply of adipose tissue [6]. This mechanism will be explored more in terms of genetic modifications that can be made to lipid transport proteins. Lipid accumulation can be characterized as either microsteatosis or macrosteatosis. Microsteatosis represents small lipid droplets that do not displace the cell's nucleus, and these livers generally are not associated with many complications. Macrosteatosis, which involves displacement of the nucleus, is a condition that can range from simple lipid build-up with minimal effects to non-alcoholic steatohepatitis (NASH) that could result in fibrosis, cirrhosis, and eventual end-stage liver disease [3]. With these complications in mind, one can imagine how NAFLD has become a leading cause for liver transplantation as well as the reason for many livers not being suitable for donation. Donor liver steatosis is a significant risk factor for post-transplant complications due to its increased susceptibility to ischemia reperfusion injury (IRI) which could result in post-operative graft dysfunction, graft loss and requirement for re-transplantation [7].

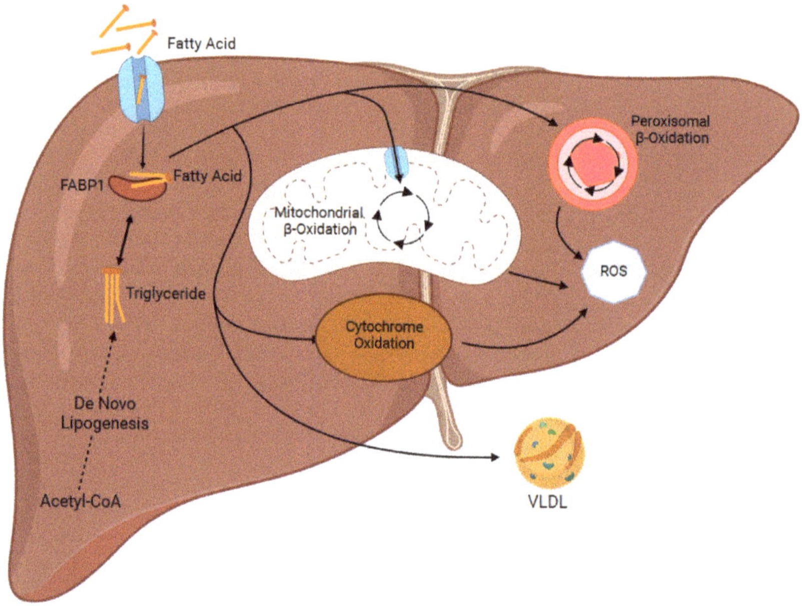


**Figure 1.** Major pathways regulating lipid acquisition and disposal inside the liver. Uptake of free circulating lipids and de novo lipogenesis increase the amount of fat in liver cells. Several fatty acid oxidation mechanism and transport of fat as low density lipoproteins (VLDL) decrease the amount of fat inside the liver cells (Created with BioRender.com).

Regardless of how fat accumulates in the hepatocytes, its presence can result in endoplasmic reticulum, oxidative, and mitochondrial stress along with impaired autophagy [3]. The two-hit hypothesis was proposed in 1998 by Day and James to explain the pathogenesis of NAFLD [8]. According to their model, the "first hit" occurs when triglycerides (TGs) build-up in the hepatocytes. Inflammation and necrosis follow in the lipid-filled hepatocytes when the TGs undergo peroxidation, thus signifying the "second hit". With the progression of more cellular damage, NAFLD can evolve into NASH that is characterized by increased amounts of inflammation. Of the patients with NASH, approximately 5–18% develop cirrhosis, and when combined with fibrosis, about 38% develop cirrhosis [9]. Patients with NASH and fibrosis or cirrhosis are at an increased risk of hepatocellular carcinoma.

A large amount of research has gone into understanding the pathophysiology of NAFLD in order to find ways to prevent and possibly reverse fat accumulation in hepatocytes. This paper will review current in vitro defatting techniques and machine perfusion therapy along with exploring possible genomic approaches that could affect fatty acid transport into hepatocytes as well as de novo lipogenesis. Targeting specific genes and proteins involved in the mechanism of steatosis could be a potential avenue of future treatment for both donors and those on the transplant list.

## 2. In Vitro Defatting Techniques

Over the years, there have been several studies conducted to determine the effects of in vitro defatting strategies. In 2013, Nativ et al. explored the effect of macrosteatosis reduction approaches on lipid droplet size as well as hepatocyte viability and functions specific to the liver [10]. In this experiment, primary hepatocytes from lean Zucker rats were cultured in a medium with equal amounts of both oleic and linoleic acids to induce steatosis [10–12]. Then, forskolin (a glucagon mimetic), the PPAR$\alpha$ agonists (GW7647, GW501516), scoparone, hypericin (a pregnane X receptor [PXR] ligand), visfatin (an adipokine), and amino acids were used as a cocktail for defatting purposes [9]. This cocktail had been shown to reduce macrosteatosis by activating hepatocellular TG metabolism [12]. This study showed that macrosteatosis can be inducible in primary hepatocytes. They also confirmed that macrosteatosis can be reversable with SRS (steatosis reduction supplements). They also concluded that accelerated macrosteatotic reduction led to a faster recovery of urea secretion and bile canalicular formation with the same viability as seen in lean rat hepatocytes. These results indicate that hepatocyte functional recovery is dependent on either macrosteatosis reduction time or direct effects of the SRS.

Nativ et al. conducted another study to determine the sensitivity of hepatocytes to hypoxia/reoxygenation (H/R) stress after exposure to defatting protocol [13]. They used the same protocol including SRS enriched with L-carnitine. They showed that the amount of steatosis is correlated with vulnerability to H/R injury, and lean or microsteatotic hepatocytes are resistant to injury. These macrosteatotic hepatocytes have less ability to produce ATP and that is related to increased production of reactive oxygen species. Lean and microsteatotic hepatocytes also showed improved activity in liver function like urea secretion and bile canalicular transport compared to those with macrosteatosis. They recommended that the lipid lowering defatting agents and combinations could be beneficial to overcome H/R stress and provide a possible recovery mechanism for discarded macrosteatotic liver grafts [13].

Another experiment conducted in 2016 by Yarmush et al. examined the use of a defatting cocktail, as described by Nativ et al. [10,13], on HepG2 cells with induced steatosis [14]. They performed their experiment under normoxic and hyperoxic conditions. They showed that TG storage levels decreased along with an increase in beta-oxidation, the tricarboxylic acid cycle, and the urea cycle. All these parameters were augmented by SRS and hyperoxic conditions. They also found that the rate of extracellular glucose uptake was miniscule compared to the amount supplied by glycogenolysis within the cell which has also active glycolysis. In conclusion, both glycolysis and beta-oxidation were occurring at the same time, which is not expected under normal conditions. Typically, glycolysis occurs when the body is in the fed state while beta-oxidation occurs during fasting. However, the combination of defatting agents does not mimic any known metabolic conditions, so the unusual result could be anticipated.

In 2018, Boteon et al. focused to determine the effect of defatting agents on primary human liver cells [15]. This was the first study to use primary human hepatocytes (PHH), which were isolated from discarded donor livers in this type of experiment. PHH cells were incubated with standard medium supplemented with FFAs, consisting of palmitic, linoleic, and oleic acids. Forskolin, GW7647, hypericin, scoparone, GW501516, visfatin, and L-carnitine were added and incubated for 48 h. They successfully reduced the intracellular lipid accumulation by 54% and TG levels by 35%. Furthermore, production of ketone bodies was increased, indicating that beta-oxidation was occurring at a higher rate. In cytotoxicity experiments, they used human intra-hepatic endothelial cells (HIEC) and human cholangiocytes, and the viability was measured for all cultures, including PHH. They demonstrated an 11% increase in viability of PHH treated with defatting drugs compared to the fatty control group. Moreover, there was no difference in viability between the treated and control groups of HIEC while the viability of human cholangiocytes improved, although this result was not statistically significant. It was the first study to prove that defatting drug cocktails have efficacy in reducing lipid content of PHH while causing no harm to non-parenchymal cells.

Recently, Aoudjehane and colleagues used a novel defatting cocktail in 3 different human culture models: PHH with induced steatosis, PHH isolated from a steatotic liver, and precision-cut liver slices (PCLS) from a steatotic liver [16]. They used a similar defatting cocktail (including forskolin, L-carnitine, and a PPARα agonist GW7647) in addition to two new agents, rapamycin and necrosulfonamide. Rapamycin is an immunosuppressant that can reduce steatosis by inhibiting mammalian target of rapamycin (mTOR). This action promotes lipogenesis, TG secretion, and macro-autophagy [17–19]. Necrosulfonamide (NSA) is an inhibitor of an effector in the necroptosis pathway that recently was revealed as a regulator of TG storage in the liver [20]. The new cocktail showed a significant decrease in lipid droplets and TG levels in steatosis-induced PHH, PHH isolated from fatty livers and also in PCLSs. Additionally, they reported a reduction in endoplasmic reticulum stress and reactive oxygen species production. By using PCLS, this study was the first to demonstrate defatting agents are successful in a model that maintains the 3D structure of liver tissue and the interactions between hepatocytes and other liver cell types. Table 1 summarizes chemicals that have been used in in vitro experiments.

**Table 1.** Summary of In Vitro Defatting Techniques.

| First Author | Year | In Vitro Model | Defatting Strategy | Effects of Agents |
|---|---|---|---|---|
| Mao et al. [10] | 2013 | Rat hepatocytes | Forskolin, PPARα and δ agonist, scoparone, hypericin, visfatin, amino acids | Faster steatosis reduction; recovery of urea secretion and bile canalicular formation |
| Nativ et al. [13] | 2014 | Rat hepatocytes | Forskolin, PPARα and δ agonist, scoparone, hypericin, visfatin, amino acids ± L-carnitine ± 90% $O_2$ | Higher reduction in TGs, increase in β-oxidation and ATP levels with L-carnitine and hyperoxia |
| Yarmush et al. [14] | 2016 | Human hepatoma cells | Forskolin, PPARα and δ agonist, scoparone, hypericin, visfatin, amino acids ± 90% $O_2$ | Decreased TGs, increased β-oxidation, TCA cycle and urea cycle, especially with hyperoxia |
| Boteon et al. [15] | 2018 | PHH, HIEC, human cholangiocytes | Forskolin, PPARα and δ agonist, scoparone, hypericin, visfatin, L-carnitine | PHH-decrease in lipids and TGs; increased viability of PHH and cholangiocytes; no cytotoxic effects on HIEC |
| Aoudjehane et al. [16] | 2020 | PHH, PHH from human fatty liver, human PCLS | Forskolin, L-carnitine, PPAR α and δ agonist, rapamycin, necrosulfonamide | Decrease in lipids and TGs and endoplasmic reticulum stress and production of reactive oxygen species |

## 3. Machine Perfusion Defatting Techniques

### *3.1. Preclinical Studies*

The detrimental effects of static cold storage on transplantable livers have produced a need for a more efficient strategy to store organs for future use, such as ex vivo machine perfusion. Ex vivo machine perfusion has been shown to not only reduce storage injury, but also provides an opportunity to treat damaged livers before transplantation. Much research has gone into this field over recent years leading it to become an alternate strategy to reduce preservation injury [21].

Bessems et al. conducted a study to compare cold storage versus machine perfusion [21]. They induced macrosteatosis in a rat model by feeding it a methionine and choline-deficient diet before using either hypothermic cold storage or machine perfusion (4 °C) for 24 h. They reported that machine perfused livers had significantly less damage as well as higher bile production, ammonia clearance, urea production, oxygen consumption, and ATP levels compared to static cold storage samples. Additionally, Kron et al. tested similar experiment to determine the viability of machine perfused grafts after transplantation [22]. After inducing macrosteatosis, livers were transplanted after either <1 h of cold storage, 12 h of cold storage, 12 h of cold storage followed by 1 h of hypothermic oxygenated perfusion (HOPE) or 12 h of cold storage followed by 1 h of hypothermic nonoxygenated perfusion (HNPE). They reported that HOPE therapy before transplantation resulted in significantly decreased reperfusion injury evidenced by less oxidative stress, nuclear injury, macrophage activation and fibrosis after one week. However, these protective effects were lost with the absence of oxygen in the perfusate, and this study did not find any reduction in the level of steatosis after HOPE therapy. Early trials of normothermic (37 °C) machine perfusion with an oxygenated blood-based perfusion system showed that a reduction in steatosis could be achieved in porcine models [23]. Even without the addition of a pharmacologic defatting cocktail, Jamieson et al. observed a 13% reduction in diet-induced liver steatosis after 48 h of normothermic ex vivo machine perfusion. However, other studies have since shown that the use of defatting cocktails further enhances the effect of machine perfusion.

Nagrath et al. showed results describing a 65% reduction in hepatocyte TG content after normothermic perfusion with a defatting cocktail (PPARα ligand (GW7647), a PPARδ ligand (GW501516), hypericin, scoparone, forskolin and visfatin) for three hours [12]. They concluded that more oxygen availability led to an increase in beta-oxidation and subsequent reduction in steatosis. Although, normothermic perfusion alone still reduced TG content by 30%, demonstrating the inherent defatting capabilities of this technique. They further tested if subnormothermic (20 °C) perfusion with the same defatting cocktail [12] would have an equivalent effect while being easier to maintain [24]. The results revealed no significant reduction in intracellular lipid content after six hours of perfusion in livers from obese Zucker rats. This indicates that higher temperatures may be required for cellular metabolism, specifically for beta oxidation, to take place at a high rate.

Another novel mechanism for liver defatting was described by Vakili et al. and involves the use of glial cell line-derived neurotrophic factor (GDNF) [25]. They had previously explained the mechanism behind the protective effect of GDNF against high-fat diet-induced steatosis in mice by reducing PPARγ expression [26] and wanted to test the potential of it before transplantation. Steatotic and lean donor livers were both perfused with either the vehicle, GDNF, or the same defatting cocktail described previously. They found that GDNF was equally effective as the defatting agents [12] at reducing TG content in hepatocytes (>40% reduction); however, GDNF induced less liver damage than the defatting cocktail, indicated by a significant rise in lactate dehydrogenase activity. Moreover, GDNF may prove to be a more suitable option for treating livers ready for transplant.

In a recent study, Raigani et al. demonstrated how a defatting cocktail in normothermic perfusion improves markers of cell viability in mouse model [27]. They implemented the same defatting cocktail supplemented with L-carnitine and amino acids [12]. They showed a reduction in perfusate lactate and better bile quality along with a decrease in inflammatory markers such as tumor necrosis factor-α (TNFα), NF-κB, and apoptosis markers, specifically caspase-3 and Fas cell surface death receptor. The results also showed an increase in gene expression of mitochondrial beta-oxidation markers; however, there was not a significant reduction in hepatocyte steatosis. They concluded that perhaps a clinical improvement in liver function is more important than decreasing lipid content alone. Table 2 summarize the preclinical studies.

### 3.2. Clinical Trials

Although much work has been done dealing with fatty livers in animal models, researchers are still working out the kinks when conducting studies on human livers. In 2010, Guarrera and colleagues first experimented with human livers to see if hypothermic machine perfusion (HMP) would preserve the organs better than cold storage, and they found that HMP is promising to use before transplantation and may improve graft function depending on early biochemical markers [28]. Other studies have looked at using hypothermic machine perfusion to assess the quality of liver grafts [29,30]. Both groups used human livers, some potentially transplantable and some discarded mainly due to steatosis. The studies found that damaged livers released higher levels of injury markers such as AST, ALT, and lactate dehydrogenase (LDH). Discarded donor livers also had a lower ATP recovery rate compared to the potentially transplantable group. At hypothermic temperatures, both groups also found that the morphology was preserved with or without oxygen supplementation.

**Table 2.** Summary of Machine Perfusion Techniques–Preclinical Studies.

| First Author | Year | Temperature of Perfusion | Additional Agents | Effects of Perfusion |
|---|---|---|---|---|
| Bessems et al. [21] | 2007 | Hypothermic | None | Less cell damage; increased bile production, ammonia clearance, urea production, $O_2$ consumption, and ATP levels |
| Kron et al. [22] | 2017 | HOPE<br>HNPE | None | HOPE: less oxidative stress, nuclear injury, macrophage activation and fibrosis; no decrease in steatosis<br>HNPE: loss of protective effects seen with HOPE therapy |
| Jamieson et al. [23] | 2011 | Normothermic | None | 13% reduction in steatosis |
| Nagrath et al. [12] | 2009 | Normothermic | PPARα and δ ligands, hypericin, scoparone, forskolin and visfatin | 65% reduction in TG content |
| Liu et al. [24] | 2013 | Subnormothermic | PPARα and δ ligands, hypericin, scoparone, forskolin and visfatin | No significant reduction in steatosis |
| Vakili et al. [25] | 2016 | Normothermic | GDNF or PPARα and δ ligands, hypericin, scoparone, forskolin and visfatin | GDNF: equally effective as defatting agents at lowering TGs, and caused less liver damage (rise in LDH activity) |
| Raigani et al. [27] | 2020 | Normothermic | PPARα and δ ligands, hypericin, scoparone, forskolin, visfatin, L-carnitine and amino acids | Decreased perfusate lactate, better bile quality, and decreased inflammatory markers; increased β-oxidation markers; no significant reduction in steatosis |

Clinical trials using HOPE to preserve livers destined for transplant have also been underway in recent years. These studies evaluated liver function in transplant patients after their organ was preserved using HOPE or static cold storage. One study specifically looked at the risk of non-anastomotic biliary strictures (NAS) and found that livers preserved with HOPE had significantly less occurrences of biliary strictures, post reperfusion syndrome, and early allograft dysfunction [31]. Two other studies have also used HOPE to preserve grafts but observed differences in liver enzymes and function after transplantation [32,33]. Both groups found that patients who received livers preserved using HOPE had significantly lower levels of liver injury enzymes, graft dysfunction, 90-day complications and hospital stay. These results are very promising for the future of liver transplantation.

A great deal of research has also gone into the effects of normothermic machine perfusion on liver graft quality. There have been several studies specifically using normothermic perfusion to assess graft viability by evaluating markers like injury enzymes, lactate clearance, and bile production [34–36]. These studies were performed on previously discarded livers and determined that they were suitable for transplant after normothermic perfusion. Grafts perfused before transplant typically resulted in better patient outcomes and less post-surgery complications. There has also been more interest in normothermic perfusion for treating hepatic steatosis. For example, Liu et al. investigated changes in the lipid profiles on ten perfused livers [37]. They found that perfusate TG levels significantly increased from 1-h to 24-h of treatment; however, there was a decrease in perfusate levels of total cholesterol, high-density lipoproteins, and low-density lipoproteins. Additionally, there was no significant decrease in steatosis histologically. They believed their findings to be due to differences in hepatic morphology compared to animal models along with the chronic accumulation of fat in humans compared to diet-induced steatosis in animal experiments. Although the increase in perfusate TGs indicates that active metabolism occurred in the grafts. Moreover, normothermic machine perfusion may be a potential starting point for liver defatting if combined with pharmacological cocktails.

Boteon and colleagues conducted an experiment using normothermic machine perfusion supplemented with a previously described defatting cocktail [12] to assess intracellular lipid reduction in ten discarded livers [38]. Compared to a control group perfused with vehicle only, the five livers perfused with defatting agents for six hours had a reduction in tissue TGs and macrosteatosis by 38% and 40%, respectively. The team also saw increased beta-oxidation with higher ATP production along with enhanced viability markers such as urea production, bile production, and lowered injury enzymes. With such promising results, one could get excited about the future of pre-transplant liver treatment. Even though the same group has previously demonstrated minimal cytotoxicity of these drugs directly on cholangiocytes and intrahepatic endothelial cells [16], the systemic effects have not yet been investigated in humans. More extensive trials would need to take place to establish drug safety before this defatting strategy could be implemented in patients. In Table 3, perfusion methods in clinical trials have been summarized.

**Table 3.** Summary of Machine Perfusion Techniques–Clinical Trials.

| First Author | Year | Temperature of Perfusion | Additional Agents | Effects of Perfusion |
|---|---|---|---|---|
| Guarrera et al. [28] | 2010 | Hypothermic | None | May improve graft function |
| Monbaliu et al. [29]<br>Abudhaise et al. [30] | 2012<br>2018 | Hypothermic | None | Discarded livers had higher levels of injury markers in the perfusate |
| van Rijn et al. [31] | 2021 | HOPE | None | Less occurrence of biliary strictures, post reperfusion syndrome, and allograft dysfunction |
| Czigany et al. [32]<br>Ravaioli et al. [33] | 2021<br>2020 | HOPE | None | Lower levels of liver injury enzymes, graft dysfunction, 90-day complications, and hospital stay |
| Watson et al. [34]<br>Mergental et al. [35]<br>Quintini et al. [36] | 2018<br>2020<br>2022 | Normothermic | None | Discarded livers were suitable for transplantation after perfusion |
| Liu et al. [37] | 2018 | Normothermic | None | Increased perfusate TG levels during treatment; no significant decrease in steatosis histologically |
| Boteon et al. [38] | 2019 | Normothermic | PPARα and δ ligands, hypericin, scoparone, forskolin and visfatin | Reduction in TGs and macrosteatosis, increased β-oxidation, higher ATP levels, and enhanced viability |

## 4. Genomic Approaches for Defatting Strategies

Hepatic steatosis is the result of dysfunction between the four major pathways regulating lipid metabolism in the liver. These pathways include the uptake of fatty acids (FAs), de novo lipogenesis, beta-oxidation, and transport out of hepatocytes [5]. Abnormal levels of proteins associated with both the uptake and transport of lipids and de novo lipogenesis have been documented in patients with NAFLD, and these provide a potential target for treatment. Specifically, this section will focus on recent studies that have incorporated genomic approaches to influence the processes involved in intracellular lipid accumulation.

To begin with hepatocyte lipid transport, FAs are primarily moved into the cells by transporters with diffusion playing a much smaller role [39]. The main proteins involved in this process include fatty acid transport proteins (FATP), cluster of differentiation 36 (CD36), and caveolins, all located in the plasma membrane. Out of six possible FATPs, FATP2 and FATP5 are primarily found in the liver and have been implicated in the pathogenesis of NAFLD [40]. Studies have shown that both FATP2 and FATP5 are expressed at higher levels in patients with NAFLD that progresses to NASH [5].

Falcon et al. conducted a study to characterize the role of FATP2 in lipid transport [41]. By using an adeno-associated virus (AAV)-based knockdown strategy, they were able to inhibit the function of FATP2 in vivo using mouse models. One week after injection with the viruses, a 40% decrease in intracellular lipid uptake was observed in the FATP2 knockdown mice compared to control. The team also found that knockdown of FATP2 resulted in lowered liver TGs but did not affect hepatic free and esterified cholesterol levels. Liver injury enzymes, liver histology, and feeding behavior were also not affected by inhibition of FATP2, indicating this as a promising route for further investigation. Additionally, early studies also using AAV-based strategies showed that knockout of FATP5 also significantly reduces FA uptake in hepatocytes, TG content, and reverses steatosis [42,43].

CD36 is also found at higher levels in patients with NAFLD. CD36 is a translocase protein that aids in the transport of long-chain FAs and is regulated by PPAR-γ, pregnane X receptor, and liver X receptor [44]. Wilson et al. determined the role of CD36 in the pathogenesis of NAFLD [45]. They found that deletion of the *Cd36* gene in mice fed with a high-fat diet resulted in reduced liver lipid content and hepatocyte FA uptake. Additionally, the mice had improved whole-body insulin sensitivity and reduced liver inflammatory markers, making this protein an excellent target for NAFLD gene therapy.

The caveolin protein family consists of three members, termed caveolins-1, 2, and 3, found in the plasma membrane and whose function is to facilitate protein trafficking and lipid droplet formation [40]. Early studies found that caveolin-1 levels were increased in mice fed a steatosis-inducing high-fat diet for 14 weeks, indicating that caveolins may be implicated in the pathogenesis of NAFLD [46]. In contrast, recently, Li et al. demonstrated that wild-type mice fed a high-fat diet with NAFLD had markedly reduced expression of the caveolin-1 gene [47]. Similarly, mice with caveolin-1 knockdown had augmented steatosis, increased plasma cholesterol, and elevated liver injury enzymes, whereas overexpression of the gene resulted in significantly attenuated lipid accumulation in hepatocytes. Another study aimed to determine the protective mechanism behind caveolins in NAFLD through both in vitro and in vivo methods [48]. Taken together, this study revealed decreased levels of caveolins and autophagy-related proteins when exposed to high levels of FAs for an extended period of time. They further concluded that the inhibition of Akt/mTOR pathway was involved in the protective role of caveolin-1 in autophagy and lipid metabolism in NAFLD.

Fatty acid binding protein (FABP) is another lipid transport protein that is found intracellularly. Following passage through the plasma membrane, lipids are not allowed to travel freely through the cytosol; instead, FABPs shuttle them between different organelles where various metabolic processes take place. Specifically, FABP1 is the predominant isoform found in the liver [40]. It is thought that FABP1 is cytoprotective due to its ability to facilitate the oxidation or TG incorporation of potentially lipotoxic FAs that accumulate in the cytosol [49]. An early study demonstrated that patients with NAFLD had much higher mRNA levels of FABP1 compared to controls [50]. The authors concluded that this was a compensatory mechanism where the cell attempted to store or release excess lipid. However, the enhanced FA trafficking may lead to storage of harmful lipid levels, promoting steatosis. Knocking out FABP1 in mice resulted in decreased hepatic TGs and lipid disposal pathways, as expected [51]. Additionally, a more recent study of FABP1 knockout found that expression of inflammatory and oxidative stress markers, as well as a marker of lipid peroxidation, were also significantly decreased [52]. Findings like these indicate that attenuation of FABP1 production could be a potential route of treatment for NAFLD.

De novo lipogenesis (DNL) enables the liver to synthesize new FAs from acetyl-CoA [5]. The two main enzymes involved are acetyl-CoA carboxylase (ACC) and fatty acid synthase (FASN). Once the new FAs have been synthesized, they must undergo a myriad of modifications before being stored as TGs or exported as VLDL particles [5]. Thus, an increased rate of DNL could easily cause accumulation of lipids inside hepatocytes, leading to steatosis and eventually NAFLD.

Two key transcription factors involved in the regulation of DNL include sterol regulatory element-binding protein1c (SREBP1c) and carbohydrate regulatory element-binding protein (ChREBP) [5]. Beginning with SREBP1c, it is activated by insulin and liver X receptor α [53]. Studies have found elevated levels of SREBP1c in patients with NAFLD [54] along with an expected rise in hepatic TG levels in mice genetically engineered to overexpress the protein [55]. Moreover, SREBP1c knockout mice displayed decreased levels of mRNAs encoding enzymes like ACC and FASN, both critical for DNL [56]. ChREBP, on the other hand, is only activated by carbohydrates and does not facilitate fat-induced lipogenesis, whereas high-fat diets may reduce the activity of ChREBP [5]. Lizuka et al. reported that ChREBP knockout in mice resulted in reduced hepatic TG synthesis by 65%, but insulin resistance, delayed glucose clearance, and intolerance to simple sugars was also observed [57]. This goes to show the importance of ChREBP in both DNL and glucose metabolism. Another experiment demonstrated that ChREBP knockout protected against fructose-induced steatosis in mice while enhancing hepatic damage through increased cholesterol synthesis and resultant cytotoxicity [58]. This finding leads the authors to believe that ChREBP may have a cytoprotective effect by limiting cholesterol toxicity. Thus, increased levels of ChREBP in patients with NAFLD may be due to a defense mechanism to prevent progression to NASH. It has also been postulated that lipogenesis may cause steatosis but prevents disease conversion to NASH [59]. Finally, hepatic overexpression of ChREBP in mice produced steatosis from upregulated DNL, but also maintained insulin sensitivity and glucose tolerance [60]. As detailed, both transcription factors of DNL play important roles in the pathogenesis of NAFLD, however, it was concluded that SREBP1c is the predominant regulator [60]. Patients with NAFLD had lowered levels of ChREBP while SREBP1c was upregulated, causing increased activity of ACC and FASN.

As previously stated, ACC is a crucial enzyme that regulates the rate-limiting step of DNL and is elevated in response to SREBP1c in NAFLD patients. Early studies were conducted with knockout of different isomers of the enzyme. Mao et al. generated liver-specific ACC1 knockout mice where generation of malonyl-CoA, the product of ACC1, was 75% lower compared to control mice, and the livers accumulated 40–70% less TGs after feeding a fat-free diet for 10 days [61]. However, the synthesis of lipogenic enzymes was increased in ACC1 knockout livers, possibly due to overexpression of ACC2. Another study demonstrated that inhibition of both ACC1 and ACC2 achieved reversal of hepatic steatosis, reduced malonyl-CoA levels, and improved insulin sensitivity by increasing the rate of beta-oxidation [62]. Very recently, a great deal of work has been done to create small molecule inhibitors of ACC1 and 2. Matsumoto et al. used GS-0976 (fircosostat) in mice with diet-induced steatosis to inhibit both isoforms which resulted in significantly reduced TG hepatocyte content, histologically [63]. The small molecule also reduced the areas of hepatic fibrosis and treated high levels of liver injury markers, indicating a potential use in NASH as well as NAFLD. Another novel small molecule inhibitor specific for ACC1, called compound 1, has been investigated for its potential role in the treatment of NAFLD [64]. It has been shown that dual inhibition of ACC1 and 2 causes increase in plasma TG levels, which could be harmful to some patients [63]. However, selective ACC1 inhibition in mice did not affect plasma TG levels compared to controls while still achieving a significant reduction in hepatic steatosis and fibrosis. The study concluded that this effect is due to preserved activity of ACC2 that compensates for the loss of malonyl-CoA production from ACC1. Overall, these drugs show great promise in the future of NAFLD/NASH treatment.

FASN is another key enzyme involved in DNL whose primary role is converting malonyl-CoA to palmitate [5]. Similar to ACC, studies have also been done to show to effects of modifying the activity of FASN in hepatocytes. A study was conducted to evaluate the effects of FASN knockout in mice fed a zero-fat diet [65]. After prolonged fasting, the mice unexpectedly displayed hypoglycemia, fatty liver, and defects in PPARα expression which is a transcription factor integral to beta oxidation of FAs. Overexpression of FASN did not induce any histological changes in liver; however, complete genetic ablation of FASN resulted in a decline in cell proliferation and a rise in apoptosis [65]. Although inhibition of FASN may improve liver viability by decreasing DNL, it seems to have more important regulatory properties that are disastrous for the cell if interrupted.

Another enzyme involved in liver lipid metabolism is stearoyl-CoA desaturase-1 (SCD1), and it is responsible for conversion of saturated Fas to monounsaturated Fas [5]. This effect is thought to be protective against NAFLD [60]. To corroborate this thought, Li et al. discovered that incubation of hepatocytes with saturated Fas lowered cell viability while incubation with monounsaturated Fas did not affect viability even though there was enhanced lipid accumulation [66]. Within the same study, SCD1 knockout was performed in mouse hepatocytes with diet-induced NASH and resulted in increased fibrosis and cellular apoptosis. Therefore, inhibition of SCD1 could exacerbate NAFLD or NASH due to excessive build-up of cytotoxic lipid species when monounsaturated Fas are not created to be safely stored. Overall, SCD1 proves to be a potential effective target for the treatment of NAFLD progression.

In a recent study, it has been shown that the transcription factor zinc fingers and homeoboxes 2 (ZHX2) also has a role in NASH models [67]. ZHX2 suppressed NASH progression in steatotic hepatic cells and downgraded inflammation and fibrosis in liver. Conversely, knocking out ZHX2 exacerbated NASH progression in animal models which was confirmed by increased lipid accumulation, aggravated inflammation, and increased fibrosis scores in liver. This protective effect is predominantly through activation of the phosphatase and tensin homolog (PTEN) gene by ZHX2.

In addition to the characterized genes and their functions in lipid accumulation in hepatocytes, functional genomics will help us find novel genes for defatting purposes. Recently, Hilgendor et al. used genome wide CRISPR screening for adipocytes and sorted the cells according to their lipid content. They identified candidate adipogenic regulators [68]. Similar experiments will shed light on hepatocyte lipid accumulation and will give us some novel targets to prevent fat accumulation in hepatocytes.

Improvement of techniques and a more affordable cost in next generation sequencing have allowed researchers to accumulate more genetic information from patients. In this regard, genome-wide association studies (GWAS) will identify genes and their associated alterations in the genome (called SNPs) with a trait or disease like NAFLD. Two independent groups used large cohorts (10k and 20k) to find a novel association between NAFLD and target genes in European ancestry [69,70]. PNPLA3 has been characterized as a risk factor for the susceptibility of fat accumulation. Further studies in different geographic locations and with more diversity in race are necessary to conclude the gene-trait interaction. Table 4 summarizes the gene manipulations for liver defatting.

**Table 4.** Summary of Genomic Approaches for Liver Defatting.

| Gene | First Author | Year | Type of Modification | Effect of Genomic Modification |
|---|---|---|---|---|
| *FATP2* | Falcon et al. [41] | 2010 | Knockdown | Decreased lipid uptake and lower liver TGs |
| *FATP5* | Doege et al. [42]<br>Doege et al. [43] | 2006<br>2008 | Knockout | Decreased lipid uptake, lower TGs, and reverses steatosis |
| *CD36* | Wilson et al. [45] | 2016 | Knockout | Decreased lipid uptake, lower TGs, improved insulin sensitivity and reduced inflammatory markers |
| *Caveolin-1* | Li et al. [47]<br>Li et al. [47] | 2017<br>2017 | Knockdown<br>Overexpression | Increased steatosis, plasma cholesterol, and liver injury enzymes<br>Decreased lipid accumulation |
| *FABP1* | Martin et al. [51]<br>Mukai et al. [52] | 2009<br>2017 | Knockout<br>Knockout | Lower TGs and decreased lipid disposal pathways<br>Decreased expression of inflammatory markers |
| *SREBP1c* | Shimano et al. [55]<br>Liang et al. [56] | 1997<br>2002 | Overexpression<br>Knockout | Higher TG levels<br>Decreased ACC and FASN mRNA levels (needed for DNL) |
| *ChREBP* | Lizuka et al. [57]<br>Zhang et al. [58]<br>Benhamed et al. [60] | 2004<br>2017<br>2012 | Knockout<br>Knockout<br>Overexpression | Lower TGs, but higher insulin resistance, delayed glucose clearance and simple sugar intolerance<br>Protection against steatosis but enhanced hepatic damage<br>Produced steatosis but maintained insulin sensitivity and glucose tolerance |

**Table 4.** *Cont.*

| Gene | First Author | Year | Type of Modification | Effect of Genomic Modification |
|---|---|---|---|---|
| *ACC1/ACC2* | Mao et al. [61] | 2006 | Knockout | Less production of malonyl-CoA, less TG accumulation; increased synthesis of lipogenic enzymes |
| | Savage et al. [62] | 2006 | Knockdown | Reversed steatosis, reduced malonyl-CoA, improved insulin sensitivity, and increased β-oxidation |
| | Matsumoto et al. [63] | 2020 | Small molecule inhibitors (ACC1/2) | Lower TGs, reduced fibrosis and lowered liver injury markers; higher plasma TGs |
| | Tamura et al. [64] | 2021 | Small molecule inhibitors (ACC1) | Reduction in steatosis and fibrosis; no change in plasma TGs |
| *FASN* | Li et al. [65] | 2016 | Knockout | Hypoglycemia, liver steatosis, and decreased β-oxidation; decline in cell proliferation and rise in apoptosis |
| *SCD1* | Li et al. [66] | 2009 | Knockout | Increased fibrosis and cellular apoptosis |
| *ZHX2* | Zhao et al. [67] | 2022 | Overexpression | Stopped progression of steatosis and reduced liver inflammation |
| | | | Knockout | Increased lipid accumulation, increased fibrosis and enhanced liver inflammation |

## 5. Additional Interventions for Liver Defatting

Both surgical and medical interventions have been shown to have an effect on people diagnosed with NAFLD. For example, bariatric surgery is an option for patients with obesity-related diseases who are able to commit to long-term medical follow-up. It has been shown that after bariatric surgery, patients have significant improvement in steatosis (resolution of NASH in 80% of patients in first year following surgery), which is related to the improvement in insulin resistance [71,72]. In addition to the resolution of NASH, decreasing fibrosis in the liver could be beneficial for the health of the liver [73]. However, some studies have controversial results regarding improvement in liver fibrosis in NASH patients [72]. In a recent long term follow-up study, NASH resolution in liver samples was observed in 84% of patients 5 years after bariatric surgery, and the reduction in liver fibrosis was most prominent in the first year with continued progression through 5 years [74].

Another option to possibly treat people with NAFLD includes the use of currently available medicines. One of the major drivers of NASH is insulin resistance as well as the dominant characteristics of type 2 diabetes and obesity. In addition to insulin resistance, dysfunction and dysregulation of adipose tissue leads to increased circulating fatty acids, carbohydrates and causes generation of hepatic lipid accumulation, cell injury on hepatocytes, increased inflammation, and liver fibrosis in the long run [75–77]. Glucagon-like peptide 1 (GLP-1) is an intestinal hormone that has a role in regulation of glucose metabolism. It stimulates insulin secretion, proinsulin gene expression and β-cell proliferative and anti-apoptotic pathways, as well as inhibiting glucagon release, gastric emptying, and food intake [78]. GLP-1 receptor agonists are a class of drugs that are beneficial to break the insulin resistance pattern of NAFLD. Liraglutide (one of the FDA approved GLP-1 receptor agonists) has been shown to improve liver function levels and reduce liver fat progression [79]. It has histologically been proved that it is beneficial for NASH resolution at liver tissue [80]. Another GLP-1 receptor agonist, semaglutide, is approved for the treatment of type 2 diabetes [81] and is being studied for use in weight management [82]. Semaglutide has a similar mechanism of action as liraglutide, but the metabolic effects of semaglutide are more robust [83–85]. Besides its effect on control of diabetes and weight loss, it reduces the AST levels and inflammation markers in liver [86]. In a recent randomized, placebo-controlled, phase 2 trial, it has been showed that semaglutide increased the NASH resolution compared to placebo in patients with biopsy-confirmed NASH and liver fibrosis. However, they did not show improvement in liver fibrosis [87].

## 6. Conclusions

NAFLD is an ever-growing problem affecting millions of people each day, but there is not currently a recommended treatment to prevent or cure this formidable disease. The increasing number of liver transplants due to NAFLD has led to a huge need for suitable donor livers to be available. In addition, many potential grafts are discarded because they are also too steatotic for successful transplantation. Much research has focused into defatting strategies to treat those living individuals with NAFLD and techniques to generate suitable donor grafts for transplantation. Medical and surgical interventions, targeting specific proteins or identification of novel target genes for defatting purposes, defatting using pharmacologic agents, and machine perfusion of extracted livers are all exciting avenues to explore for future treatment of this disease (Figure 2). However, issues are likely to arise with any method due to the extremely interconnected physiology of the liver where one change could have many effects far from the intended result. Therefore, all possibilities must be taken into careful consideration when developing treatments for NAFLD.

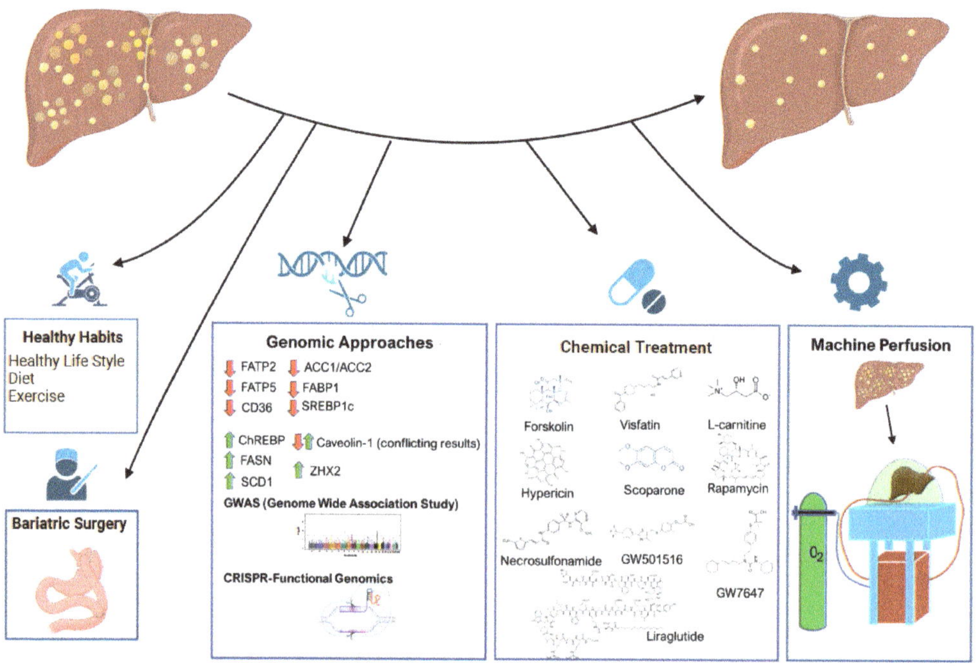


**Figure 2.** Overview of the different approaches for defatting purposes. Red arrow shows the beneficial effect of down regulation of target genes while green arrow demonstrates the positive effect of upregulation on defatting in third panel. Fourth panel summarizes the structure of known chemicals from in vitro defatting experiment. Perfusion of liver on normo- and hypothermic machine has great premise for future therapeutic approaches (Created with BioRender.com).

**Author Contributions:** Conceptualization, E.N.Y. and C.K. (Cem Kuscu); methodology, E.N.Y., M.D. and C.K. (Cem Kuscu); investigation, E.N.Y., M.D.; resources, J.D.E., C.K. (Cem Kuscu); data curation, E.N.Y.; writing—original draft preparation, E.N.Y., C.K. (Cem Kuscu); writing—review and editing; E.N.Y., M.D., C.W., A.B., J.D.E., C.K. (Canan Kuscu), C.K. (Cem Kuscu); visualization, E.N.Y., M.D., and C.K. (Cem Kuscu); supervision, A.B., J.D.E., C.K. (Canan Kuscu), C.K. (Cem Kuscu); project administration, A.B., J.D.E., C.K. (Canan Kuscu), C.K. (Cem Kuscu); funding acquisition, A.B., J.D.E., C.K. (Canan Kuscu) and C.K. (Cem Kuscu). All authors have read and agreed to the published version of the manuscript.

**Funding:** Individuals were supported by the National Institute of Diabetes and Digestive and Kidney Diseases of the National Institutes of Health (NIH) under award number R01DK117183 and DK132230 (AB), institutional startup by the James D. Eason Transplant Institute. The content is solely the responsibility of the authors and does not necessarily represent the official views of the NIH.

**Institutional Review Board Statement:** Not applicable.

**Informed Consent Statement:** Not applicable.

**Data Availability Statement:** Not applicable.

**Conflicts of Interest:** The authors declare no conflict of interest.

## References


1. Bechmann, L.P.; Hannivoort, R.A.; Gerken, G.; Hotamisligil, G.S.; Trauner, M.; Canbay, A. The interaction of hepatic lipid and glucose metabolism in liver diseases. *J. Hepatol.* **2012**, *56*, 952–964. [CrossRef] [PubMed]
2. Nguyen, P.; Leray, V.; Diez, M.; Serisier, S.; Le Bloc'H, J.; Siliart, B.; Dumon, H. Liver lipid metabolism. *J. Anim. Physiol. Anim. Nutr.* **2008**, *92*, 272–283. [CrossRef] [PubMed]
3. Machado, M.V.; Cortez-Pinto, H. Non-alcoholic fatty liver disease: What the clinician needs to know. *World J. Gastroenterol.* **2014**, *20*, 12956–12980. [CrossRef] [PubMed]
4. Maurice, J.; Manousou, P. Non-alcoholic fatty liver disease. *Clin. Med.* **2018**, *18*, 245–250. [CrossRef] [PubMed]
5. Ipsen, D.H.; Lykkesfeldt, J.; Tveden-Nyborg, P. Molecular mechanisms of hepatic lipid accumulation in non-alcoholic fatty liver disease. *Cell Mol. Life Sci.* **2018**, *75*, 3313–3327. [CrossRef] [PubMed]
6. Tamura, S.; Shimomura, I. Contribution of adipose tissue and de novo lipogenesis to nonalcoholic fatty liver disease. *J. Clin. Investig.* **2005**, *115*, 1139–1142. [CrossRef]
7. Goldaracena, N.; Barbas, A.S.; Selzner, M. Normothermic and subnormothermic ex-vivo liver perfusion in liver transplantation. *Curr. Opin. Organ Transpl.* **2016**, *21*, 315–321. [CrossRef]
8. Day, C.P.; James, O.F. Steatohepatitis: A tale of two "hits"? *Gastroenterology* **1998**, *114*, 842–845. [CrossRef]
9. Mazilescu, L.I.; Selzner, M.; Selzner, N. Defatting strategies in the current era of liver steatosis. *JHEP Rep.* **2021**, *3*, 100265. [CrossRef]
10. Nativ, N.I.; Yarmush, G.; Chen, A.; Dong, D.; Henry, S.D.; Guarrera, J.V.; Klein, K.M.; Maguire, T.; Schloss, R.; Berthiaume, F.; et al. Rat hepatocyte culture model of macrosteatosis: Effect of macrosteatosis induction and reversal on viability and liver-specific function. *J. Hepatol.* **2013**, *59*, 1307–1314. [CrossRef]
11. Berthiaume, F.; Barbe, L.; Mokuno, Y.; MacDonald, A.D.; Jindal, R.; Yarmush, M.L. Steatosis reversibly increases hepatocyte sensitivity to hypoxia-reoxygenation injury. *J. Surg. Res.* **2009**, *152*, 54–60. [CrossRef] [PubMed]
12. Nagrath, D.; Xu, H.; Tanimura, Y.; Zuo, R.; Berthiaume, F.; Avila, M.; Yarmush, R.; Yarmush, M.L. Metabolic preconditioning of donor organs: Defatting fatty livers by normothermic perfusion ex vivo. *Metab. Eng.* **2009**, *11*, 274–283. [CrossRef] [PubMed]
13. Nativ, N.I.; Yarmush, G.; So, A.; Barminko, J.; Maguire, T.J.; Schloss, R.; Berthiaume, F.; Yarmush, M.L. Elevated sensitivity of macrosteatotic hepatocytes to hypoxia/reoxygenation stress is reversed by a novel defatting protocol. *Liver Transpl.* **2014**, *20*, 1000–1011. [CrossRef] [PubMed]
14. Yarmush, G.; Santos, L.; Yarmush, J.; Koundinyan, S.; Saleem, M.; Nativ, N.I.; Schloss, R.S.; Yarmush, M.L.; Maguire, T.J.; Berthiaume, F. Metabolic Flux Distribution during Defatting of Steatotic Human Hepatoma (HepG2) Cells. *Metabolites* **2016**, *6*, 1. [CrossRef]
15. Boteon, Y.; Wallace, L.; Boteon, A.P.C.S.; Mirza, D.F.; Mergental, H.; Bhogal, R.H.; Afford, S. An effective protocol for pharmacological defatting of primary human hepatocytes which is non-toxic to cholangiocytes or intrahepatic endothelial cells. *PLoS ONE* **2018**, *13*, e0201419. [CrossRef]
16. Aoudjehane, L.; Gautheron, J.; Le Goff, W.; Goumard, C.; Gilaizeau, J.; Nget, C.S.; Savier, E.; Atif, M.; Lesnik, P.; Morichon, R.; et al. Novel defatting strategies reduce lipid accumulation in primary human culture models of liver steatosis. *Dis. Model. Mech.* **2020**, *13*, dmm042663. [CrossRef]
17. Lin, C.-W.; Zhang, H.; Li, M.; Xiong, X.; Chen, X.; Chen, X.; Dong, X.C.; Yin, X.-M. Pharmacological promotion of autophagy alleviates steatosis and injury in alcoholic and non-alcoholic fatty liver conditions in mice. *J. Hepatol.* **2013**, *58*, 993–999. [CrossRef]
18. Waskowicz, L.R.; Zhou, J.; Landau, D.J.; Brooks, E.D.; Lim, A.; A Yavarow, Z.; Kudo, T.; Zhang, H.; Wu, Y.; Grant, S.; et al. Bezafibrate induces autophagy and improves hepatic lipid metabolism in glycogen storage disease type Ia. *Hum. Mol. Genet.* **2019**, *28*, 143–154. [CrossRef]
19. Zhou, W.; Ye, S. Rapamycin improves insulin resistance and hepatic steatosis in type 2 diabetes rats through activation of autophagy. *Cell Biol. Int.* **2018**, *42*, 1282–1291. [CrossRef]
20. Xu, H.; Du, X.; Liu, G.; Huang, S.; Du, W.; Zou, S.; Tang, D.; Fan, C.; Xie, Y.; Wei, Y.; et al. The pseudokinase MLKL regulates hepatic insulin sensitivity independently of inflammation. *Mol. Metab.* **2019**, *23*, 14–23. [CrossRef]
21. Bessems, M.; Doorschodt, B.M.; Kolkert, J.L.; Vetelainen, R.L.; van Vliet, A.K.; Vreeling, H.; van Marle, J.; van Gulik, T.M. Preservation of steatotic livers: A comparison between cold storage and machine perfusion preservation. *Liver Transpl.* **2007**, *13*, 497–504. [CrossRef] [PubMed]
22. Kron, P.; Schlegel, A.; Mancina, L.; Clavien, P.-A.; Dutkowski, P. Hypothermic oxygenated perfusion (HOPE) for fatty liver grafts in rats and humans. *J. Hepatol.* **2017**. [CrossRef] [PubMed]
23. Jamieson, R.W.; Zilvetti, M.; Roy, D.; Hughes, D.; Morovat, A.; Coussios, C.C.; Friend, P.J. Hepatic steatosis and normothermic perfusion-preliminary experiments in a porcine model. *Transplantation* **2011**, *92*, 289–295. [CrossRef] [PubMed]
24. Liu, Q.; Berendsen, T.; Izamis, M.-L.; Uygun, B.; Yarmush, M.; Uygun, K. Perfusion defatting at subnormothermic temperatures in steatotic rat livers. *Transpl. Proc.* **2013**, *45*, 3209–3213. [CrossRef]

25. Vakili, S.T.T.; Kailar, R.; Rahman, K.; Nezami, B.G.; Mwangi, S.M.; Anania, F.A.; Srinivasan, S. Glial cell line-derived neurotrophic factor-induced mice liver defatting: A novel strategy to enable transplantation of steatotic livers. *Liver Transpl.* **2016**, *22*, 459–467. [CrossRef]
26. Mwangi, S.M.; Peng, S.; Nezami, B.G.; Thorn, N.; Farris, A.B.; Jain, S.; Laroui, H.; Merlin, D.; Anania, F.; Srinivasan, S. Glial cell line-derived neurotrophic factor protects against high-fat diet-induced hepatic steatosis by suppressing hepatic PPAR-$\gamma$ expression. *Am. J. Physiol. Liver Physiol.* **2016**, *310*, G103–G116. [CrossRef]
27. Raigani, S.; Carroll, C.; Griffith, S.; Pendexter, C.; Rosales, I.; Deirawan, H.; Beydoun, R.; Yarmush, M.; Uygun, K.; Yeh, H. Improvement of steatotic rat liver function with a defatting cocktail during ex situ normothermic machine perfusion is not directly related to liver fat content. *PLoS ONE* **2020**, *15*, e0232886. [CrossRef]
28. Guarrera, J.V.; Henry, S.D.; Samstein, B.; Odeh-Ramadan, R.; Kinkhabwala, M.; Goldstein, M.J.; Ratner, L.E.; Renz, J.F.; Lee, H.T.; Brown, J.R.S.; et al. Hypothermic machine preservation in human liver transplantation: The first clinical series. *Am. J. Transpl.* **2010**, *10*, 372–381. [CrossRef]
29. Monbaliu, D.; Liu, Q.; Libbrecht, L.; De Vos, R.; Vekemans, K.; Debbaut, C.; Detry, O.; Roskams, T.; Van Pelt, J.; Pirenne, J. Preserving the morphology and evaluating the quality of liver grafts by hypothermic machine perfusion: A proof-of-concept study using discarded human livers. *Liver Transpl.* **2012**, *18*, 1495–1507. [CrossRef]
30. Abudhaise, H.; Davidson, B.R.; Demuylder, P.; Luong, T.V.; Fuller, B. Evolution of dynamic, biochemical, and morphological parameters in hypothermic machine perfusion of human livers: A proof-of-concept study. *PLoS ONE* **2018**, *13*, e0203803. [CrossRef]
31. van Rijn, R.; Schurink, I.J.; de Vries, Y.; Berg, A.P.V.D.; Cerisuelo, M.C.; Murad, S.D.; Erdmann, J.I.; Gilbo, N.; de Haas, R.J.; Heaton, N.; et al. Hypothermic Machine Perfusion in Liver Transplantation—A Randomized Trial. *N. Engl. J. Med.* **2021**, *384*, 1391–1401. [CrossRef] [PubMed]
32. Czigany, Z.; Pratschke, J.; Froněk, J.; Guba, M.; Schöning, W.; Raptis, D.A.; Andrassy, J.; Kramer, M.; Strnad, P.; Tolba, R.H.; et al. Hypothermic Oxygenated Machine Perfusion Reduces Early Allograft Injury and Improves Post-transplant Outcomes in Extended Criteria Donation Liver Transplantation from Donation After Brain Death: Results from a Multicenter Randomized Controlled Trial (HOPE ECD-DBD). *Ann. Surg.* **2021**, *274*, 705–712. [CrossRef] [PubMed]
33. Ravaioli, M.; De Pace, V.; Angeletti, A.; Comai, G.; Vasuri, F.; Baldassarre, M.; Maroni, L.; Odaldi, F.; Fallani, G.; Caraceni, P.; et al. Hypothermic Oxygenated New Machine Perfusion System in Liver and Kidney Transplantation of Extended Criteria Donors:First Italian Clinical Trial. *Sci Rep.* **2020**, *10*, 6063. [CrossRef]
34. Watson, C.J.E.; Kosmoliaptsis, V.; Pley, C.; Randle, L.; Fear, C.; Crick, K.; Gimson, A.E.; Allison, M.; Upponi, S.; Brais, R.; et al. Observations on the ex situ perfusion of livers for transplantation. *Am. J. Transpl.* **2018**, *18*, 2005–2020. [CrossRef]
35. Mergental, H.; Laing, R.W.; Kirkham, A.J.; Perera, M.T.P.R.; Boteon, Y.L.; Attard, J.; Barton, D.; Curbishley, S.; Wilkhu, M.; Neil, D.A.H.; et al. Transplantation of discarded livers following viability testing with normothermic machine perfusion. *Nat. Commun.* **2020**, *11*, 2939. [CrossRef] [PubMed]
36. Quintini, C.; Del Prete, L.; Simioni, A.; Del Angel, L.; Uso, T.D.; D'Amico, G.; Hashimoto, K.; Aucejo, F.; Fujiki, M.; Eghtesad, B.; et al. Transplantation of declined livers after normothermic perfusion. *Surgery* **2022**, *171*, 747–756. [CrossRef] [PubMed]
37. Liu, Q.; Nassar, A.; Buccini, L.; Iuppa, G.; Soliman, B.; Pezzati, D.; Hassan, A.; Blum, M.; Baldwin, W.; Bennett, A.; et al. Lipid metabolism and functional assessment of discarded human livers with steatosis undergoing 24 hours of normothermic machine perfusion. *Liver Transpl.* **2018**, *24*, 233–245. [CrossRef] [PubMed]
38. Boteon, Y.L.; Attard, J.; Boteon, A.P.C.S.; Wallace, L.; Reynolds, G.; Hubscher, S.; Mirza, D.F.; Mergental, H.; Bhogal, R.H.; Afford, S.C. Manipulation of Lipid Metabolism During Normothermic Machine Perfusion: Effect of Defatting Therapies on Donor Liver Functional Recovery. *Liver Transpl.* **2019**, *25*, 1007–1022. [CrossRef]
39. Mashek, D.G. Hepatic fatty acid trafficking: Multiple forks in the road. *Adv. Nutr.* **2013**, *4*, 697–710. [CrossRef]
40. Koo, S.H. Nonalcoholic fatty liver disease: Molecular mechanisms for the hepatic steatosis. *Clin. Mol. Hepatol.* **2013**, *19*, 210–215. [CrossRef]
41. Falcon, A.; Doege, H.; Fluitt, A.; Tsang, B.; Watson, N.; Kay, M.A.; Stahl, A. FATP2 is a hepatic fatty acid transporter and peroxisomal very long-chain acyl-CoA synthetase. *Am. J. Physiol. Endocrinol. Metab.* **2010**, *299*, E384–E393. [CrossRef]
42. Doege, H.; Baillie, R.A.; Ortegon, A.M.; Tsang, B.; Wu, Q.; Punreddy, S.; Hirsch, D.; Watson, N.; Gimeno, R.E.; Stahl, A. Targeted deletion of FATP5 reveals multiple functions in liver metabolism: Alterations in hepatic lipid homeostasis. *Gastroenterology* **2006**, *130*, 1245–1258. [CrossRef] [PubMed]
43. Doege, H.; Grimm, D.; Falcon, A.; Tsang, B.; Storm, T.A.; Xu, H.; Ortegon, A.M.; Kazantzis, M.; Kay, M.A.; Stahl, A. Silencing of hepatic fatty acid transporter protein 5 in vivo reverses diet-induced non-alcoholic fatty liver disease and improves hyperglycemia. *J. Biol. Chem.* **2008**, *283*, 22186–22192. [CrossRef] [PubMed]
44. Silverstein, R.L.; Febbraio, M. CD36, a scavenger receptor involved in immunity, metabolism, angiogenesis, and behavior. *Sci. Signal.* **2009**, *2*, re3. [CrossRef] [PubMed]
45. Wilson, C.G.; Tran, J.L.; Erion, D.M.; Vera, N.B.; Febbraio, M.; Weiss, E.J. Hepatocyte-Specific Disruption of CD36 Attenuates Fatty Liver and Improves Insulin Sensitivity in HFD-Fed Mice. *Endocrinology* **2016**, *157*, 570–585. [CrossRef] [PubMed]
46. Qiu, Y.; Liu, S.; Chen, H.-T.; Yu, C.-H.; Teng, X.-D.; Yao, H.-T.; Xu, G.-Q. Upregulation of caveolin-1 and SR-B1 in mice with non-alcoholic fatty liver disease. *Hepatobiliary Pancreat. Dis. Int.* **2013**, *12*, 630–636. [CrossRef]

47. Li, M.; Chen, D.; Huang, H.; Wang, J.; Wan, X.; Xu, C.; Li, C.; Ma, H.; Yu, C.; Li, Y. Caveolin1 protects against diet induced hepatic lipid accumulation in mice. *PLoS ONE* **2017**, *12*, e0178748. [CrossRef]
48. Xue, W.; Wang, J.; Jiang, W.; Shi, C.; Wang, X.; Huang, Y.; Hu, C. Caveolin-1 alleviates lipid accumulation in NAFLD associated with promoting autophagy by inhibiting the Akt/mTOR pathway. *Eur. J. Pharmacol.* **2020**, *871*, 172910. [CrossRef]
49. Wang, G.; Bonkovsky, H.L.; de Lemos, A.; Burczynski, F.J. Recent insights into the biological functions of liver fatty acid binding protein 1. *J. Lipid Res.* **2015**, *56*, 2238–2247. [CrossRef]
50. Higuchi, N.; Kato, M.; Tanaka, M.; Miyazaki, M.; Takao, S.; Kohjima, M.; Kotoh, K.; Enjoji, M.; Nakamuta, M.; Takayanagi, R. Effects of insulin resistance and hepatic lipid accumulation on hepatic mRNA expression levels of apoB, MTP and L-FABP in non-alcoholic fatty liver disease. *Exp. Ther. Med.* **2011**, *2*, 1077–1081. [CrossRef]
51. Martin, G.G.; Atshaves, B.P.; Huang, H.; McIntosh, A.L.; Williams, B.J.; Pai, P.-J.; Russell, D.H.; Kier, A.B.; Schroeder, F. Hepatic phenotype of liver fatty acid binding protein gene-ablated mice. *Am. J. Physiol. Gastrointest. Liver Physiol.* **2009**, *297*, G1053–G1065. [CrossRef] [PubMed]
52. Mukai, T.; Egawa, M.; Takeuchi, T.; Yamashita, H.; Kusudo, T. Silencing of FABP1 ameliorates hepatic steatosis, inflammation, and oxidative stress in mice with nonalcoholic fatty liver disease. *FEBS Open Biol.* **2017**, *7*, 1009–1016. [CrossRef] [PubMed]
53. Sanders, F.W.; Griffin, J.L. De novo lipogenesis in the liver in health and disease: More than just a shunting yard for glucose. *Biol. Rev. Camb. Philos. Soc.* **2016**, *91*, 452–468. [CrossRef] [PubMed]
54. Kohjima, M.; Enjoji, M.; Higuchi, N.; Kato, M.; Kotoh, K.; Yoshimoto, T.; Fujino, T.; Yada, M.; Yada, R.; Harada, N.; et al. Re-evaluation of fatty acid metabolism-related gene expression in nonalcoholic fatty liver disease. *Int. J. Mol. Med.* **2007**, *20*, 351–358. [CrossRef]
55. Shimano, H.; Horton, J.D.; Shimomura, I.; Hammer, R.E.; Brown, M.S.; Goldstein, J.L. Isoform 1c of sterol regulatory element binding protein is less active than isoform 1a in livers of transgenic mice and in cultured cells. *J. Clin. Investig.* **1997**, *99*, 846–854. [CrossRef]
56. Liang, G.; Yang, J.; Horton, J.D.; Hammer, R.E.; Goldstein, J.L.; Brown, M.S. Diminished hepatic response to fasting/refeeding and liver X receptor agonists in mice with selective deficiency of sterol regulatory element-binding protein-1c. *J. Biol. Chem.* **2002**, *277*, 9520–9528. [CrossRef]
57. Iizuka, K.; Bruick, R.K.; Liang, G.; Horton, J.D.; Uyeda, K. Deficiency of carbohydrate response element-binding protein (ChREBP) reduces lipogenesis as well as glycolysis. *Proc. Natl. Acad. Sci. USA* **2004**, *101*, 7281–7286. [CrossRef]
58. Zhang, D.; Tong, X.; Vandommelen, K.; Gupta, N.; Stamper, K.; Brady, G.F.; Meng, Z.; Lin, J.; Rui, L.; Omary, B.; et al. Lipogenic transcription factor ChREBP mediates fructose-induced metabolic adaptations to prevent hepatotoxicity. *J. Clin. Investig.* **2017**, *127*, 2855–2867. [CrossRef]
59. Ducheix, S.; Vegliante, M.C.; Villani, G.; Napoli, N.; Sabbà, C.; Moschetta, A. Is hepatic lipogenesis fundamental for NAFLD/NASH? A focus on the nuclear receptor coactivator PGC-1β. *Cell Mol. Life Sci.* **2016**, *73*, 3809–3822. [CrossRef]
60. Benhamed, F.; Denechaud, P.-D.; Lemoine, M.; Robichon, C.; Moldes, M.; Bertrand-Michel, J.; Ratziu, V.; Serfaty, L.; Housset, C.; Capeau, J.; et al. The lipogenic transcription factor ChREBP dissociates hepatic steatosis from insulin resistance in mice and humans. *J. Clin. Investig.* **2012**, *122*, 2176–2194. [CrossRef]
61. Mao, J.; DeMayo, F.J.; Li, H.; Abu-Elheiga, L.; Gu, Z.; Shaikenov, T.E.; Kordari, P.; Chirala, S.S.; Heird, W.C.; Wakil, S.J. Liver-specific deletion of acetyl-CoA carboxylase 1 reduces hepatic triglyceride accumulation without affecting glucose homeostasis. *Proc. Natl. Acad. Sci. USA* **2006**, *103*, 8552–8557. [CrossRef] [PubMed]
62. Savage, D.B.; Choi, C.S.; Samuel, V.T.; Liu, Z.-X.; Zhang, D.; Wang, A.; Zhang, X.-M.; Cline, G.W.; Yu, X.X.; Geisler, J.G.; et al. Reversal of diet-induced hepatic steatosis and hepatic insulin resistance by antisense oligonucleotide inhibitors of acetyl-CoA carboxylases 1 and 2. *J. Clin. Investig.* **2006**, *116*, 817–824. [CrossRef] [PubMed]
63. Matsumoto, M.; Yashiro, H.; Ogino, H.; Aoyama, K.; Nambu, T.; Nakamura, S.; Nishida, M.; Wang, X.; Erion, D.M.; Kaneko, M. Acetyl-CoA carboxylase 1 and 2 inhibition ameliorates steatosis and hepatic fibrosis in a MC4R knockout murine model of nonalcoholic steatohepatitis. *PLoS ONE* **2020**, *15*, e0228212. [CrossRef] [PubMed]
64. Tamura, Y.O.; Sugama, J.; Iwasaki, S.; Sasaki, M.; Yasuno, H.; Aoyama, K.; Watanabe, M.; Erion, D.M.; Yashiro, H. Selective Acetyl-CoA Carboxylase 1 Inhibitor Improves Hepatic Steatosis and Hepatic Fibrosis in a Preclinical Nonalcoholic Steatohepatitis Model. *J. Pharmacol. Exp. Ther.* **2021**, *379*, 280–289. [CrossRef]
65. Li, L.; Pilo, G.M.; Li, X.; Cigliano, A.; Latte, G.; Che, L.; Joseph, C.; Mela, M.; Wang, C.; Jiang, L.; et al. Inactivation of fatty acid synthase impairs hepatocarcinogenesis driven by AKT in mice and humans. *J. Hepatol.* **2016**, *64*, 333–341. [CrossRef]
66. Li, Z.Z.; Berk, M.; McIntyre, T.M.; Feldstein, A.E. Hepatic lipid partitioning and liver damage in nonalcoholic fatty liver disease: Role of stearoyl-CoA desaturase. *J. Biol. Chem.* **2009**, *284*, 5637–5644. [CrossRef]
67. Zhao, Y.; Gao, L.; Jiang, C.; Chen, J.; Qin, Z.; Zhong, F.; Yan, Y.; Tong, R.; Zhou, M.; Yuan, A.; et al. The transcription factor zinc fingers and homeoboxes 2 alleviates NASH by transcriptional activation of phosphatase and tensin homolog. *Hepatology* **2022**, *75*, 939–954. [CrossRef]
68. Hilgendorf, K.I.; Johnson, C.T.; Han, K.; Rabiee, A.; Demeter, J.; Cheng, R.; Zhu, Y.; Jiang, Z.; Svensson, K.J.; Bassik, M.C.; et al. A CRISPR-based genome-wide screen for adipogenesis reveals new insights into mitotic expansion and lipogenesis. *bioRxiv* **2020**. [CrossRef]

69. Anstee, Q.M.; Darlay, R.; Cockell, S.; Meroni, M.; Govaere, O.; Tiniakos, D.; Burt, A.D.; Bedossa, P.; Palmer, J.; Liu, Y.-L.; et al. Genome-wide association study of non-alcoholic fatty liver and steatohepatitis in a histologically characterised cohort☆. *J. Hepatol.* **2020**, *73*, 505–515. [CrossRef]
70. Namjou, B.; Lingren, T.; Huang, Y.; Parameswaran, S.; Cobb1, B.L.; Stanaway, I.B.; Connolly, J.J.; Mentch, F.D.; Benoit, B.; Niu, X.; et al. eMERGE Network, Xanthakos SA, Harley JB. GWAS and enrichment analyses of non-alcoholic fatty liver disease identify new trait-associated genes and pathways across eMERGE Network. *BMC Med.* **2019**, *17*, 135. [CrossRef]
71. Mathurin, P.; Gonzalez, F.; Kerdraon, O.; Leteurtre, E.; Arnalsteen, L.; Hollebecque, A.; Louvet, A.; Dharancy, S.; Cocq, P.; Jany, P.; et al. The evolution of severe steatosis after bariatric surgery is related to insulin resistance. *Gastroenterology* **2006**, *130*, 1617–1624. [CrossRef] [PubMed]
72. Lassailly, G.; Caiazzo, R.; Buob, D.; Pigeyre, M.; Verkindt, H.; Labreuche, J.; Raverdy, V.; Leteurtre, E.; Dharancy, S.; Louvet, A.; et al. Bariatric surgery reduces features of nonalcoholic steatohepatitis in morbidly obese patients. *Gastroenterology* **2015**, *149*, 379–388. [CrossRef] [PubMed]
73. Sanyal, A.J.; Brunt, E.M.; Kleiner, D.E.; Kowdley, K.V.; Chalasani, N.; Lavine, J.E.; Ratziu, V.; McCullough, A. Endpoints and clinical trial design for nonalcoholic steatohepatitis. *Hepatology* **2011**, *54*, 344–353. [CrossRef]
74. Lassailly, G.; Caiazzo, R.; Ntandja-Wandji, L.-C.; Gnemmi, V.; Baud, G.; Verkindt, H.; Ningarhari, M.; Louvet, A.; Leteurtre, E.; Raverdy, V.; et al. Bariatric Surgery Provides Long-term Resolution of Nonalcoholic Steatohepatitis and Regression of Fibrosis. *Gastroenterology* **2020**, *159*, 1290–1301. [CrossRef] [PubMed]
75. Gastaldelli, A.; Cusi, K. From NASH to diabetes and from diabetes to NASH: Mechanisms and treatment options. *JHEP Rep.* **2019**, *1*, 312–328. [CrossRef] [PubMed]
76. Armstrong, M.J. BASL and the Dame Sheila Sherlock Award 2016. Glucagon-like peptide-1 analogues in nonalcoholic steatohepatitis: From bench to bedside. *Clin. Liver Dis.* **2017**, *10*, 32–35. [CrossRef]
77. Sanyal, A.J. Past, present and future perspectives in nonalcoholic fatty liver disease. *Nat. Rev. Gastroenterol. Hepatol.* **2019**, *16*, 377–386. [CrossRef]
78. Drucker, D.J. The biology of incretin hormones. *Cell Metab.* **2006**, *3*, 153–165. [CrossRef]
79. Cusi, K. Incretin-Based Therapies for the Management of Nonalcoholic Fatty Liver Disease in Patients with Type 2 Diabetes. *Hepatology* **2019**, *69*, 2318–2322. [CrossRef]
80. Armstrong, M.J.; Gaunt, P.; Aithal, G.P.; Barton, D.; Hull, D.; Parker, R.; Hazlehurst, J.M.; Guo, K.; Abouda, G.; A Aldersley, M.; et al. Liraglutide safety and efficacy in patients with non-alcoholic steatohepatitis (LEAN): A multicentre, double-blind, randomised, placebo-controlled phase 2 study. *Lancet* **2016**, *387*, 679–690. [CrossRef]
81. Ozempic (Semaglutide) Prescribing Information. January 2020. Available online: https://www.accessdata.fda.gov/drugsatfda_docs/label/2020/209637s003lbl.pdf (accessed on 24 August 2022).
82. Kushner, R.F.; Calanna, S.; Davies, M.; Dicker, D.; Garvey, W.T.; Goldman, B.; Lingvay, I.; Thomsen, M.; Wadden, T.A.; Wharton, S.; et al. Semaglutide 2.4 mg for the Treatment of Obesity: Key Elements of the STEP Trials 1 to 5. *Obesity* **2020**, *28*, 1050–1061. [CrossRef] [PubMed]
83. Capehorn, M.; Catarig, A.-M.; Furberg, J.; Janez, A.; Price, H.; Tadayon, S.; Vergès, B.; Marre, M. Efficacy and safety of once-weekly semaglutide 1.0mg vs once-daily liraglutide 1.2mg as add-on to 1-3 oral antidiabetic drugs in subjects with type 2 diabetes (SUSTAIN 10). *Diabetes Metab.* **2020**, *46*, 100–109. [CrossRef] [PubMed]
84. O'Neil, P.M.; Birkenfeld, A.L.; McGowan, B.; Mosenzon, O.; Pedersen, S.D.; Wharton, S.; Carson, C.G.; Jepsen, C.H.; Kabisch, M.; Wilding, J.P.H. Efficacy and safety of semaglutide compared with liraglutide and placebo for weight loss in patients with obesity: A randomised, double-blind, placebo and active controlled, dose-ranging, phase 2 trial. *Lancet* **2018**, *392*, 637–649. [CrossRef]
85. Nauck, M.A.; Meier, J.J. Management of endocrine Disease: Are all GLP-1 agonists equal in the treatment of type 2 diabetes? *Eur. J. Endocrinol.* **2019**, *181*, R211–R234. [CrossRef] [PubMed]
86. Newsome, P.; Francque, S.; Harrison, S.; Ratziu, V.; Van Gaal, L.; Calanna, S.; Hansen, M.; Linder, M.; Sanyal, A. Effect of semaglutide on liver enzymes and markers of inflammation in subjects with type 2 diabetes and/or obesity. *Aliment. Pharmacol. Ther.* **2019**, *50*, 193–203. [CrossRef] [PubMed]
87. Newsome, P.N.; Buchholtz, K.; Cusi, K.; Linder, M.; Okanoue, T.; Ratziu, V.; Sanyal, A.J.; Sejling, A.-S.; Harrison, S.A. A Placebo-Controlled Trial of Subcutaneous Semaglutide in Nonalcoholic Steatohepatitis. *N. Engl. J. Med.* **2021**, *384*, 1113–1124. [CrossRef]

International Journal of *Molecular Sciences*

MDPI

*Article*

# Convolutional Neural Network Model Based on 2D Fingerprint for Bioactivity Prediction

**Hamza Hentabli [1,2,*], Billel Bengherbia [1], Faisal Saeed [2,3,*], Naomie Salim [2], Ibtehal Nafea [4], Abdelmoughni Toubal [1] and Maged Nasser [5]**

1 Laboratory of Advanced Electronics Systems (LSEA), University of Medea, Medea 26000, Algeria
2 UTM Big Data Centre, Ibnu Sina Institute for Scientific and Industrial Research, Universiti Teknologi Malaysia, Johor Bahru 81310, Johor, Malaysia
3 DAAI Research Group, Department of Computing and Data Science, School of Computing and Digital Technology, Birmingham City University, Birmingham B4 7XG, UK
4 College of Computer Science and Engineering, Taibah University, Medina 41477, Saudi Arabia
5 School of Computer Sciences, Universiti Sains Malaysia, Gelugor 11800, Penang, Malaysia
* Correspondence: hentabli_hamza@yahoo.fr (H.H.); faisal.saeed@bcu.ac.uk (F.S.)

**Abstract:** Determining and modeling the possible behaviour and actions of molecules requires investigating the basic structural features and physicochemical properties that determine their behaviour during chemical, physical, biological, and environmental processes. Computational approaches such as machine learning methods are alternatives to predicting the physiochemical properties of molecules based on their structures. However, the limited accuracy and high error rates of such predictions restrict their use. In this paper, a novel technique based on a deep learning convolutional neural network (CNN) for the prediction of chemical compounds' bioactivity is proposed and developed. The molecules are represented in the new matrix format Mol2mat, a molecular matrix representation adapted from the well-known 2D-fingerprint descriptors. To evaluate the performance of the proposed methods, a series of experiments were conducted using two standard datasets, namely the MDL Drug Data Report (MDDR) and Sutherland, datasets comprising 10 homogeneous and 14 heterogeneous activity classes. After analysing the eight fingerprints, all the probable combinations were investigated using the five best descriptors. The results showed that a combination of three fingerprints, ECFP4, EPFP4, and ECFC4, along with a CNN activity prediction process, achieved the highest performance of 98% AUC when compared to the state-of-the-art ML algorithms NaiveB, LSVM, and RBFN.

**Keywords:** activity prediction model; biological activities; bioactive molecules; convolutional neural network; deep learning

**Citation:** Hentabli, H.; Bengherbia, B.; Saeed, F.; Salim, N.; Nafea, I.; Toubal, A.; Nasser, M. Convolutional Neural Network Model Based on 2D Fingerprint for Bioactivity Prediction. *Int. J. Mol. Sci.* **2022**, *23*, 13230. https://doi.org/10.3390/ijms232113230

Academic Editor: Alexandre G. de Brevern

Received: 14 September 2022
Accepted: 27 October 2022
Published: 30 October 2022



## 1. Introduction

Extraction of the structural activity relationship (SAR) [1,2] information from chemical datasets relies on the pairwise structural comparison of all toxicophore features and small molecules, which highlights the degree of the structural relationship between the compounds [3–6]. The Quantitative Structure–Activity Relationship (QSAR) can correlate the compound's chemical and structural features with its physicochemical or biological properties. The molecular descriptors are applied for encoding the features, while the QSAR model identifies the mathematical relationship between the descriptors and the biological features or other relevant properties of the known ligands for predicting the unknown ligands. These QSAR studies are able to reduce the failure costs of potential drug molecules, as they easily identify the promising lead molecules and reduce the number of expensive experiments. These are considered important tools in the pharmaceutical industry since they have identified many high-quality leads during the early stages of drug discovery.

A great deal of information is contained in the molecular structure of a compound: For example, it indicates the number of elements or describes its shape and electrostatic field [7,8]. The collection of atoms that constitute a molecule can be symbolically represented in many ways. It is not easy to determine the optimum approach that represents the molecular structure that is suited for all applications [9–11].

Generally, molecules are represented using their molecular or structural formulae and line drawings, which indicate the number of atoms for various elements present in the single molecule of a compound, for example, $H_2O$ indicates the presence of two hydrogens and one oxygen atom in a water molecule. In many cases, the molecular formula alone cannot represent the chemical structure. For instance, in isomers, molecules with a similar molecular formula show a different atomic arrangement. The structural formula depicts the molecular structure and represents the individual bonds between all atoms as lines.

Many chemoinformatics methods are based on numerical descriptors that include a description of the molecular structure and properties. These descriptors are used as input data for various statistical and data mining techniques. The other types of property descriptors are generally used in the diversity analysis, selection of the representative compound subsets, combinatorial library design, and QSAR studies. Thus, the fingerprint $\mathcal{X}$ of molecule A is represented using a sequence of numbers:

$$\mathcal{X}_{\mathrm{A}} = \{x_1, x_2, x_3, \ldots, x_{\mathrm{n}}\}$$

where $x_{\mathrm{i}}$ refers to the i-th structural unit in molecule A, i.e., bonds, atoms, or fragments. The value n represents the length or size of all fingerprints, i.e., the number of molecular properties.

The 2D fingerprint descriptors are also used to provide a rapid screening step during substructure and similarity searches [1,10]. These 2D fingerprints are categorised based on the methods used, for example, the fragment dictionary and hashed methods illustrated in Figure 1. The fingerprints are generated using a fingerprinting process that converts a chemical structure into a binary form (i.e., a string of 0 s and 1 s). The binary form depicts the chemical shorthand, which indicates the presence/absence of the structural features in a molecule.

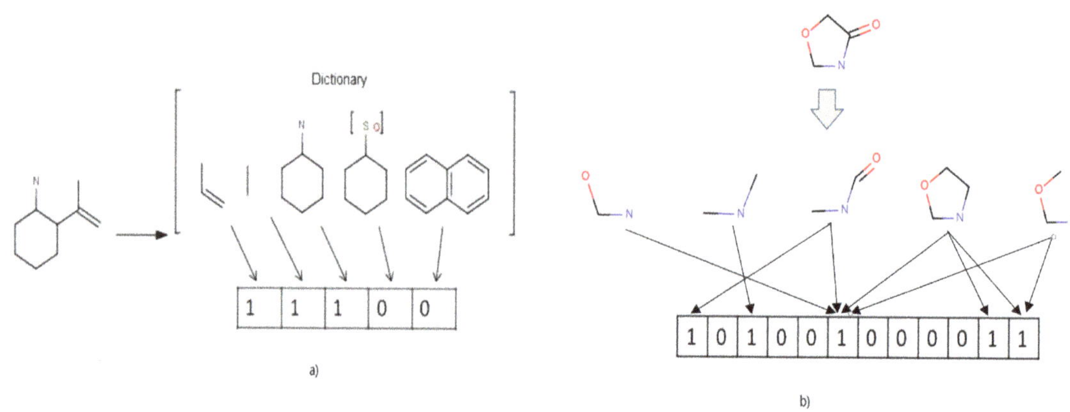


**Figure 1.** Two examples showing the generation of a molecular fingerprint: (**a**) Dictionary-based fingerprint and (**b**) hashed-based fingerprint.

The molecule-based fingerprints are represented by dividing the molecules into fragments of specific substructures and structural features. In this kind of representation, the fingerprint length is based on the number of fragments present in the dictionary, where every bit position in the binary string is assigned to one particular sub-structural feature in the dictionary. Thus, the bits can individually or in combination represent the presence or absence of the features [10,12].

The state-of-the-art 2D fingerprint technique used in the present study was based on QSAR, which can predict and measure all biological activities of the compounds. In this study, eight different 2D fingerprints were investigated for bioactivity prediction, which was generated using the PaDEL descriptor software. Here, the 2D fingerprint descriptors were used with the CNN model for predicting the biological activities and studying the combination and the integration of various fingerprints in the CNN architecture. The next sections describe the background and design of the novel technique. The performance of the proposed technique was evaluated after conducting several experiments based on the structure or bioactivity prediction.

## 2. Results

The proposed code was implemented in public DL software, Keras [13], based on Theano [14]. The experiments were conducted using the Dell Precision T1700 CPU system with 16 GB memory and the professional-grade NVIDIA GeForce GTX 1060 6 GB graphics.

The proposed novel CNN model for predicting the molecular bioactivities was a ligand-based activity prediction or target-fishing technique that could be used for unknown chemical compounds. It was a deep learning system consisting of an adapted molecular matrix representation, "Mol2mat", which incorporated all the substructural data on the molecules based on their fingerprint features for predicting their activities. This proposed CNN method was then compared to three different ML algorithms described in the WEKA-Workbench, NaiveB, LSVM, and RBFN, using optimal parameters obtained from previous work using the same datasets [15], as previously explained in Section 4.4.

We also determined the computing prediction accuracy of this deep learning system by applying the technique described in Section 4.2, using eight fingerprint representatives. The results derived from these fingerprints were then compared using the Analysis of variance (ANOVA) technique as a significance test and a violin-plot with boxplot charts. The five fingerprint representatives that showed the best CNN configuration were further chosen as the best representatives. This encompassed Stage 1 of the analysis and is described in detail below. In Stage 2, these five representatives were assessed using all probable combinations, such as 2, 3, 4, or 5. The results acquired from Stage 2 were further compared using their violin-plot charts, and the best fingerprint combination was noted. Stage 2 is described in more detail below. In Stage 3, all results were compared for the best combination derived from the previous stages with three known ML algorithms, NaiveB, LSVM, and RBFN. The proposed CNN model in this paper will be henceforth referred to as CNNfp.

### 2.1. Benchmarking

The proposed technique was evaluated by comparing it with three other machine learning methods using WEKA-Workbench [16] methods, including a Naive Bayesian classifier (NaiveB) [17], LibSVM [18], and a Radial basis function network (RBFN) [19]. Finding the best values for the classifier's parameters is a difficult task. However, the best probable setup for the LSVM classifier was identified by the WEKA-Workbench. In this paper, the linear kernel was used for SVM, and the values of 0.1, 1.0, and 0.001 were used for the Gamma, Cost, and Epsilon parameters, respectively. For the NaiveB classifier, a supervised discretisation technique was used to convert the numeric attributes to the nominal attributes, while the minimal standard deviation limit for the RBFN classifier was 0.01. All the remaining parameters of the classifiers used the default values in the WEKA-Workbench.

### 2.2. Stage 1

In this stage, the prediction accuracies of the 24 activity classes present in an MDDR1, MDDR2, and Sutherland datasets were determined and compared using eight fingerprint representatives. Figure 2 summarises the CNN configuration, which used the Mol2mat molecular representation.

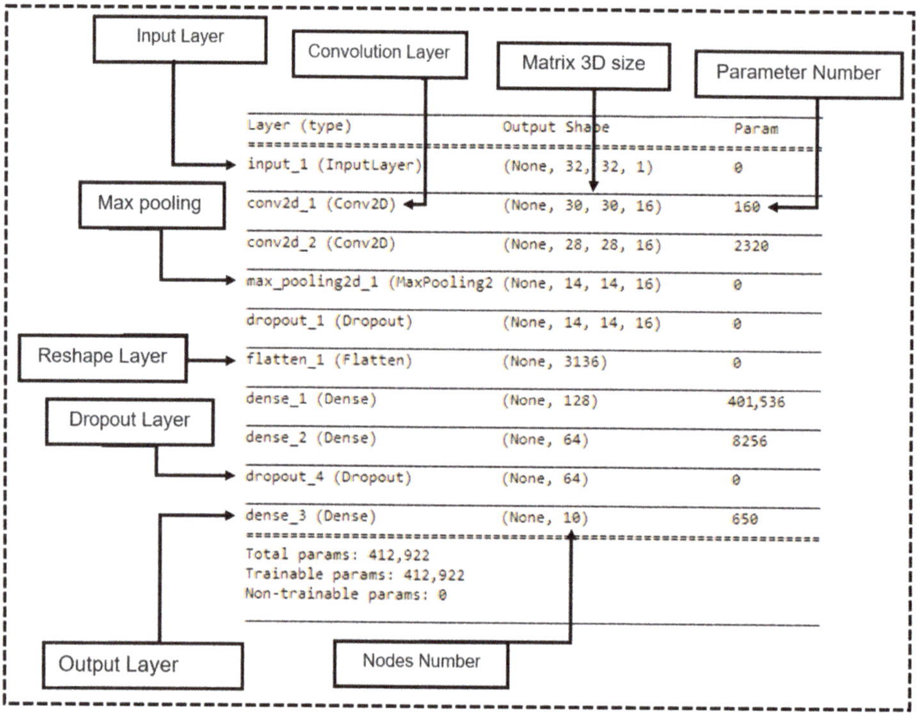


**Figure 2.** A summary of the proposed CNN configuration that uses the Mol2Mat representation.

In Stage 1, the eight fingerprints described above were studied based on two parameters. The first parameter included the accuracy response vs. the number of iterations, while the second parameter included the MSE response vs. the number of epochs. These were studied in a 2D graph consisting of the training data results.

Figure 3a presents a graphical result for the number of iterations vs. the accuracy. It also presents eight lines of the different fingerprints. The ECFC4 fingerprint displayed a speed augmentation in their prediction accuracy from the third epoch, whereas the EPFP4 fingerprint showed better accuracy in 17 epochs. However, the AlogP and the MDL fingerprints displayed the lowest prediction accuracy values. The mean squared error or loss value showed similar results to the accuracy performance, as shown in Figure 3b. The novel CNN model could accurately predict biological activities with an average MSE value of 0.0054 for ECFC4 and 0.002 for the ECFP4 fingerprints.

Figure 4 shows the comparison of the prediction accuracy values for Stage 1 experiments that were conducted using the CNN model for eight fingerprint representatives using the violin-plot charts. The construction of violin-plot charts is shown on the right-hand side of this figure.

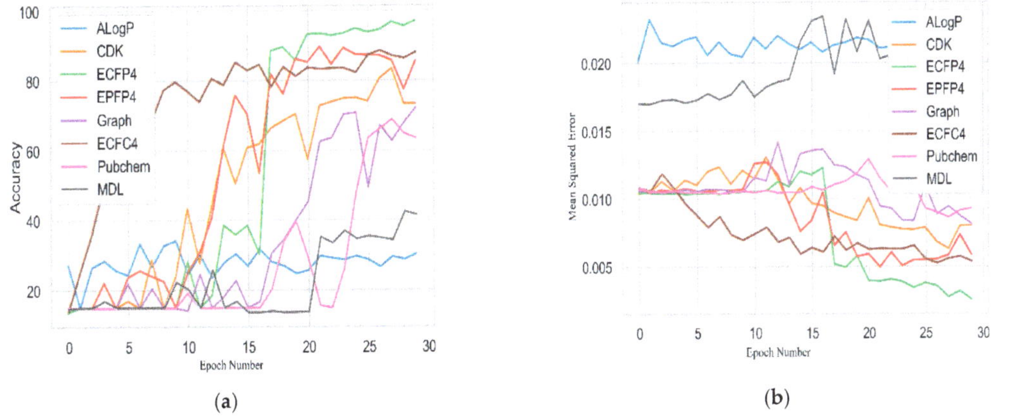


**Figure 3.** Evaluation of eight fingerprints based on their (**a**) accuracy and (**b**) MSE performance.

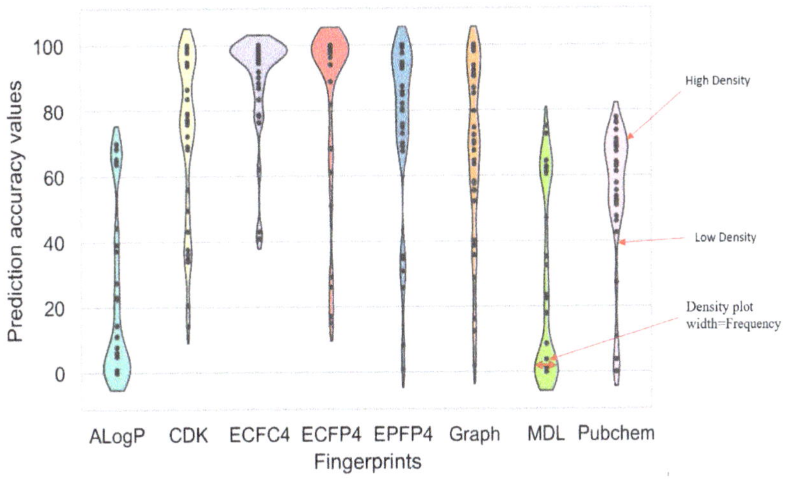


**Figure 4.** Prediction accuracy values of the CNN model for the eight fingerprint representatives using the violin-plot charts.

The violin-plot charts are able to remove the conventional boxplot elements and plot each activity class as a single point. Figure 2 indicates that the eight fingerprint representatives showed a clear difference in their average prediction accuracy values. The ECFC4 showed the best average accuracy of 90.17. The graph fingerprint came next with a value of 74.84, closely followed by the CDKFp and ECFP4 fingerprints, which showed similar average accuracy values of 72.28 and 71.97, respectively. The worst average accuracy values were displayed by PubChem (53.88), MDL 26.25, and AlogP, with an accuracy value of only 22.45. Using these results, and based on the ANOVA significant test results, a small *p*-value of 0.04 was noted, which highlighted the difference between all the fingerprints.

Furthermore, the AlogP, MDL, and PubChem fingerprints were regarded as the worst contenders as they showed a higher variance between all the biological activity classes. Thus, CDK, ECFP4, ECFC4, EPFP4, and graph were some of the best fingerprints and could be forwarded to Stage 2 to improve all the results based on the probable combination cases of two, three, four, or five of the best fingerprints. The combinations were based on the fusion of the extracted feature levels.

In this stage, we used better techniques to combine the various sources of knowledge available in the area of deep learning [20–22]. Firstly, we proposed a feature extraction step to present each selected molecular fingerprint. This combination significantly improved the models, since they could benefit from every molecular fingerprint and combine all the extracted features from various sources after a flattened layer, which followed the max-pooling layer. This helped them convert the 2D matrix data into the vector. As a result, they could process the output data using the fully connected layers, known as the dense layers. This section described the CNN architecture utilised in this study and how many CNN architectures can be combined into a single model. The next section will describe the performance evaluation.

### 2.3. Stage 2

In this stage, the prediction accuracies for the different combination cases of the five fingerprint representatives were determined. Table 1 presents 26 possible combinations for these five fingerprints, including combinations of two, three, four, and five combinations of the CDK, ECFP4, ECFC4, EPFP4, and graph fingerprints. Henceforth, each combination case will be based on its name (A–Z), and each row will represent one combination case. Case A consists of two combinations, while Case Z consists of five fingerprint combinations.

**Table 1.** Probable combination cases for the five best fingerprints.

| Labels | Combination | CDK | ECFP4 | EPFP4 | Graph | ECFC4 |
|---|---|---|---|---|---|---|
| A | 2 | √ | √ | | | |
| B | 2 | √ | | √ | | |
| C | 2 | √ | | | √ | |
| D | 2 | √ | | | | √ |
| E | 2 | | √ | √ | | |
| F | 2 | | √ | | √ | |
| G | 2 | | √ | | | √ |
| H | 2 | | | √ | √ | |
| I | 2 | | | √ | | √ |
| J | 2 | | | | √ | √ |
| K | 3 | √ | √ | √ | | |
| L | 3 | √ | √ | | √ | |
| M | 3 | √ | √ | | | √ |
| N | 3 | √ | | √ | √ | |
| O | 3 | √ | | √ | | √ |
| P | 3 | √ | | | √ | √ |
| Q | 3 | | √ | √ | √ | |
| R | 3 | | √ | √ | | √ |
| S | 3 | | √ | | √ | √ |
| T | 3 | | | √ | √ | √ |
| U | 4 | √ | √ | √ | √ | |
| V | 4 | √ | √ | √ | | √ |
| W | 4 | √ | √ | | √ | √ |
| X | 4 | √ | | √ | √ | √ |
| Y | 4 | | √ | √ | √ | √ |
| Z | 5 | √ | √ | √ | √ | √ |

The colors are used to differentiate between each level. Combination of 2 blue; Combination of 3 orange; Combination of 4 yellow; Combination of 2 green.

The 26 combinations of the five fingerprints were investigated, as shown in Table 1. Figures 5 and 6 summarise the CNN configuration for the combination case between the CDK, ECFP4, and EPFP4 fingerprints, referred to as "K", as an example using the Mol2mat molecular representation. As seen in both figures, the model has three branches, with a matrix (32 × 32) as the input and two Conv. layers and max-pooling for each branch concatenate layer to merge all extracted features into one array. Finally, there are two hidden layers with 256 and 128 neurons and an output layer with 10 outputs. Rectified linear activation functions are used in each hidden layer, and a SoftMax activation function is used in the output layer.

| Layer (type) | Output Shape | Param | Connected to |
|---|---|---|---|
| input_1 (InputLayer) | (None, 32, 32, 1) | 0 | |
| input_2 (InputLayer) | (None, 32, 32, 1) | 0 | |
| input_3 (InputLayer) | (None, 32, 32, 1) | 0 | |
| conv2d_1 (Conv2D) | (None, 30, 30, 16) | 160 | input_1[0][0] |
| conv2d_3 (Conv2D) | (None, 30, 30, 16) | 160 | input_2[0][0] |
| conv2d_5 (Conv2D) | (None, 30, 30, 16) | 160 | input_3[0][0] |
| conv2d_2 (Conv2D) | (None, 28, 28, 16) | 2320 | conv2d_1[0][0] |
| conv2d_4 (Conv2D) | (None, 28, 28, 16) | 2320 | conv2d_3[0][0] |
| conv2d_6 (Conv2D) | (None, 28, 28, 16) | 2320 | conv2d_5[0][0] |
| max_pooling2d_1 (MaxPooling2D) | (None, 14, 14, 16) | 0 | conv2d_2[0][0] |
| max_pooling2d_2 (MaxPooling2D) | (None, 14, 14, 16) | 0 | conv2d_4[0][0] |
| max_pooling2d_3 (MaxPooling2D) | (None, 14, 14, 16) | 0 | conv2d_6[0][0] |
| dropout_1 (Dropout) | (None, 14, 14, 16) | 0 | max_pooling2d_1[0][0] |
| dropout_2 (Dropout) | (None, 14, 14, 16) | 0 | max_pooling2d_2[0][0] |
| dropout_3 (Dropout) | (None, 14, 14, 16) | 0 | max_pooling2d_3[0][0] |
| flatten_1 (Flatten) | (None, 3136) | 0 | dropout_1[0][0] |
| flatten_2 (Flatten) | (None, 3136) | 0 | dropout_2[0][0] |
| flatten_3 (Flatten) | (None, 3136) | 0 | dropout_3[0][0] |
| concatenate_1 (Concatenate) | (None, 9408) | 0 | flatten_1[0][0]<br>flatten_2[0][0]<br>flatten_3[0][0] |
| dense_1 (Dense) | (None, 256) | 2,408,704 | concatenate_1[0][0] |
| dense_2 (Dense) | (None, 128) | 32,896 | dense_1[0][0] |
| dropout_4 (Dropout) | (None, 128) | 0 | dense_2[0][0] |
| dense_3 (Dense) | (None, 10) | 1290 | dropout_4[0][0] |

Total params: 2,450,330
Trainable params: 2,450,330
Non-trainable params: 0

**Figure 5.** A summary of the CNN configuration for a combination case named "K" using a Mol2mat representation.

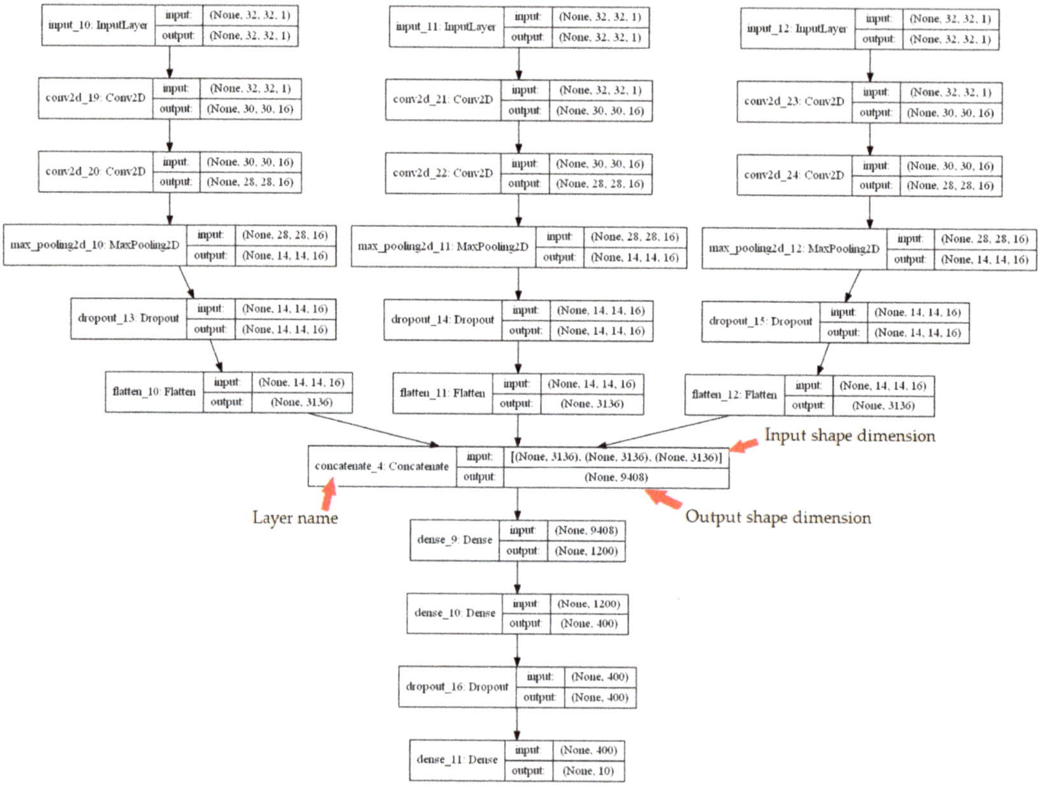


**Figure 6.** A CNN Model configuration for a combination case named "K" using the Mol2mat representation.

Figure 7 compares the prediction accuracy values for the Stage 2 experiments for all 26 combination cases with the help of violin-plot charts.

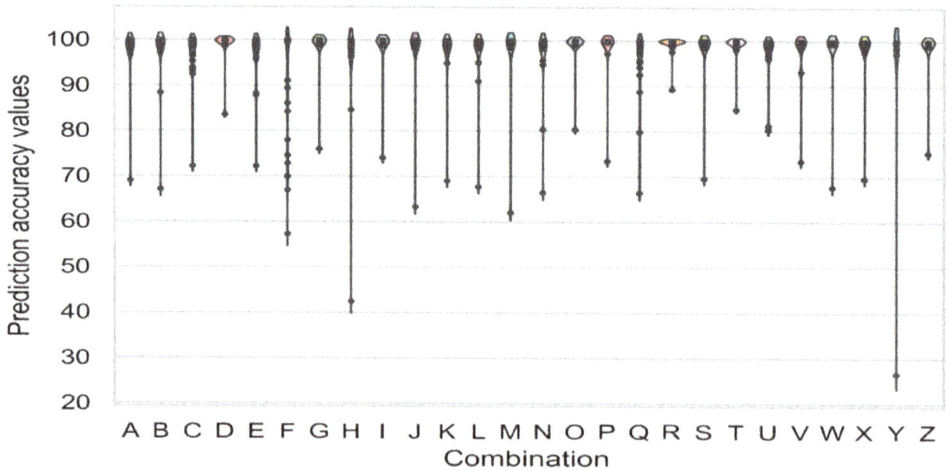


**Figure 7.** Prediction accuracy values for the CNN model were applied to the 26 combination cases of the five best fingerprints with the help of violin-plot charts.

The results in Figure 7 show a $p$-value of 0.031 based on the ANOVA significance test results, indicating that the difference between all the combination cases is significant. The violin-plot charts plotted each activity class as the point. It was seen that the D, O, R, and T combination cases displayed the highest prediction accuracy, >80%, and a low variance amongst all the activity classes. The combination cases were plotted in different boxplot charts to determine the distribution of the activity classes based on the low- and high-diversity values noted for each activity class. Figure 8 compares the prediction accuracies for all experiments in Stage 2 for the D, O, R, and T combination cases, which were plotted using the Boxplot charts.

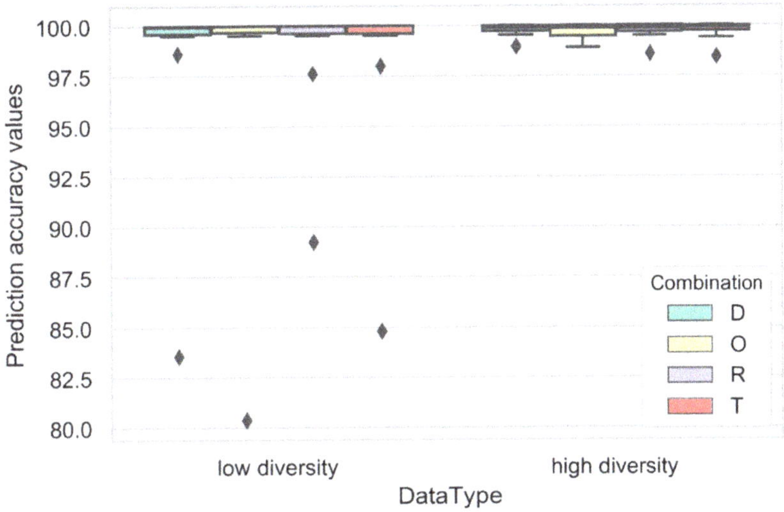


**Figure 8.** A comparison of the prediction accuracies for the D, O, R, and T combination cases, plotted using the boxplot charts.

Based on the violin-plot charts presented in Figure 7 and the Boxplot chart shown in Figure 8, a $p$-value of 0.048 was calculated based on the ANOVA significance test results. This indicated the significance of the difference between all the models. The R combination displayed the best average prediction accuracy of 99.17, indicating that a combination of the three fingerprints, ECFP4, EPFP4, and ECFC4, showed good performance compared to the other combinations.

The R combination also showed a lower variance of 5.52 compared to the other cases. Furthermore, this combination showed higher stability even when placed in a high- or low-diversity class. Meanwhile, the D, O, and T combinations displayed a mean prediction accuracy of 97.45, 97.03, and 97.72, respectively. They also displayed higher variance than the R combination. These combination cases showed a variance prediction accuracy of 12.62, 17.97, and 10.81, respectively, indicating that R was the best fingerprint combination seen in Stage 2.

### 2.4. *Stage 3*

In Stage 3, the authors compared the results for the best combination of ECFP4, EPFP4, and ECFC4, as established in Stage 2, with those obtained from the standard ML algorithms existing in a WEKA-Workbench: NaiveB, LSVM, and RBFN.

Tables 2–4 show the sensitivity, specificity, and AUC values for all the datasets used here. A visual inspection of all tables could be used to compare the performance of the prediction accuracies of all four algorithms. However, the authors applied a quantitative boxplot chart to compare these algorithms. This process quantifies the agreement level between all the multiple sets and ranks the different objects.

**Table 2.** Sensitivity, specificity, and AUC values for all the prediction models using an MDDR1 dataset.

| Activity Index | CNNfp | | | NaïveB | | | RBFN | | | LSVM | | |
|---|---|---|---|---|---|---|---|---|---|---|---|---|
| | Sens | Spec | AUC | Sens | Spec | AUC | Sens | Spec | AUC | Sens | Spec | AUC |
| 7707 | 1.00 | 1.00 | 1.00 | 0.99 | 1.00 | 0.99 | 0.63 | 1.00 | 0.82 | 0.93 | 0.95 | 0.94 |
| 7708 | 1.00 | 1.00 | 1.00 | 0.97 | 1.00 | 0.99 | 0.51 | 1.00 | 0.75 | 0.96 | 0.96 | 0.96 |
| 31420 | 1.00 | 1.00 | 1.00 | 1.00 | 1.00 | 1.00 | 0.96 | 0.96 | 0.96 | 0.92 | 0.99 | 0.96 |
| 42710 | 0.99 | 0.99 | 0.99 | 0.94 | 1.00 | 0.97 | 0.43 | 1.00 | 0.72 | 0.95 | 0.99 | 0.97 |
| 64100 | 0.97 | 0.99 | 0.98 | 0.95 | 1.00 | 0.97 | 0.97 | 0.90 | 0.94 | 0.96 | 0.99 | 0.98 |
| 64200 | 0.96 | 0.99 | 0.98 | 0.87 | 0.95 | 0.91 | 0.43 | 1.00 | 0.71 | 0.94 | 1.00 | 0.97 |
| 64220 | 1.00 | 1.00 | 1.00 | 0.97 | 0.99 | 0.96 | 0.95 | 0.97 | 0.96 | 0.92 | 1.00 | 0.96 |
| 64500 | 1.00 | 1.00 | 1.00 | 0.91 | 0.93 | 0.92 | 0.44 | 1.00 | 0.72 | 0.84 | 0.95 | 0.90 |
| 64350 | 1.00 | 1.00 | 1.00 | 0.94 | 0.96 | 0.95 | 0.80 | 1.00 | 0.90 | 0.90 | 0.94 | 0.92 |
| 75755 | 1.00 | 1.00 | 1.00 | 0.94 | 0.98 | 0.96 | 0.76 | 1.00 | 0.88 | 0.94 | 0.97 | 0.96 |
| mean | 0.98 | 0.99 | 0.99 | 0.94 | 0.98 | 0.96 | 0.69 | 0.98 | 0.84 | 0.93 | 0.97 | 0.95 |

**Table 3.** Sensitivity, specificity, and AUC values for the prediction models using an MDDR2 dataset.

| Activity Index | CNNfp | | | NaïveB | | | RBFN | | | LSVM | | |
|---|---|---|---|---|---|---|---|---|---|---|---|---|
| | Sens | Spec | AUC | Sens | Spec | AUC | Sens | Spec | AUC | Sens | Spec | AUC |
| 9249 | 1.00 | 1.00 | 1.00 | 0.91 | 0.99 | 0.95 | 0.82 | 0.98 | 0.90 | 0.95 | 0.97 | 0.96 |
| 12455 | 1.00 | 1.00 | 1.00 | 0.88 | 0.97 | 0.92 | 0.66 | 0.98 | 0.82 | 0.93 | 0.96 | 0.94 |
| 12464 | 1.00 | 1.00 | 1.00 | 0.85 | 0.99 | 0.92 | 0.75 | 0.95 | 0.85 | 0.89 | 0.97 | 0.93 |
| 31281 | 1.00 | 1.00 | 1.00 | 0.94 | 1.00 | 0.97 | 0.53 | 1.00 | 0.76 | 0.95 | 0.97 | 0.96 |
| 43210 | 0.99 | 0.99 | 0.99 | 0.84 | 0.99 | 0.91 | 0.78 | 0.97 | 0.87 | 0.93 | 0.96 | 0.94 |
| 71522 | 1.00 | 1.00 | 1.00 | 0.82 | 0.99 | 0.91 | 0.75 | 0.97 | 0.86 | 0.91 | 0.97 | 0.94 |
| 75721 | 1.00 | 1.00 | 1.00 | 0.91 | 0.99 | 0.95 | 0.86 | 0.98 | 0.92 | 0.96 | 0.97 | 0.96 |
| 78331 | 0.98 | 0.99 | 0.99 | 0.81 | 0.96 | 0.89 | 0.79 | 0.93 | 0.86 | 0.81 | 0.96 | 0.88 |
| 78348 | 0.99 | 0.99 | 0.99 | 0.65 | 0.99 | 0.82 | 0.74 | 0.96 | 0.85 | 0.88 | 0.97 | 0.92 |
| 78351 | 0.99 | 0.99 | 0.99 | 0.82 | 0.94 | 0.88 | 0.59 | 0.96 | 0.78 | 0.91 | 0.95 | 0.93 |
| mean | 0.99 | 0.99 | 0.99 | 0.84 | 0.98 | 0.91 | 0.73 | 0.97 | 0.85 | 0.91 | 0.97 | 0.94 |

**Table 4.** Sensitivity, specificity, and AUC values for the prediction models using a Sutherland dataset.

| Activity Class | CNNfp | | | NaïveB | | | RBFN | | | LSVM | | |
|---|---|---|---|---|---|---|---|---|---|---|---|---|
| | Sens | Spec | AUC | Sens | Spec | AUC | Sens | Spec | AUC | Sens | Spec | AUC |
| Estrogen receptor | 1.00 | 1.00 | 1.00 | 1.00 | 1.00 | 1.00 | 0.62 | 0.70 | 0.64 | 0.98 | 1.00 | 0.99 |
| Dihydrofolate reductase | 0.99 | 0.99 | 0.99 | 0.99 | 1.00 | 0.99 | 0.86 | 0.80 | 0.84 | 0.90 | 0.98 | 0.94 |
| Cyclooxygenase-2 inhibitors | 1.00 | 1.00 | 1.00 | 1.00 | 0.99 | 1.00 | 0.93 | 0.76 | 0.84 | 1.00 | 0.99 | 0.99 |
| Benzodiazepine receptor | 1.00 | 1.00 | 1.00 | 0.94 | 0.61 | 0.78 | 0.99 | 0.65 | 0.82 | 0.95 | 0.92 | 0.93 |
| mean | 0.99 | 0.99 | 0.99 | 0.98 | 0.90 | 0.94 | 0.85 | 0.73 | 0.79 | 0.95 | 0.97 | 0.96 |

Boxplot charts were used to assess the performance of a set of fingerprints, ECFP4, EPFP4, and ECFC4, using three algorithms (RBFN, NaiveB, and LSVM).

Here, MDDR1, MDDR2, and the Sutherland datasets, with their activity classes described in Tables 5–7, were regarded as judges. In contrast, parameters such as sensitivity, specificity, and AUC, measured using different prediction algorithms, were regarded as objects. The outputs of this test included *p*-value, median, and variance. Figure 9 shows the results of the boxplot chart, where the sensitivity values of the six algorithms were compared. The results show a *p*-value of 0.008 based on the ANOVA significance test results, which revealed a significant difference between all algorithms. The CNNfp algorithm showed a high sensitivity of 0.985, while the NaiveB and LSVM ML algorithms showed a high variance of 0.15 and 0.23, respectively, compared to the CNNfp. Diversity in all sensitivity values was especially seen in the algorithms that displayed a variance of $10^{-4}$. Furthermore, these models showed a mean sensitivity of 0.90 and 0.74, respectively.

**Table 5.** MDDR activity classes for DS1 dataset.

| Activity Index | Activity Class | Active Molecules | Pairwise Similarity |
|---|---|---|---|
| 07707 | Adenosine agonists A1 | 207 | 0.229 |
| 07708 | Adenosine agonists A2 | 156 | 0.305 |
| 31420 | Rennin inhibitors | 1130 | 0.290 |
| 42710 | CCK agonists | 111 | 0.361 |
| 64100 | Monocyclic_-lactams | 1346 | 0.336 |
| 64200 | Cephalosporins | 113 | 0.322 |
| 64220 | Carbacephems | 1051 | 0.269 |
| 64500 | Carbapenems | 126 | 0.260 |
| 64350 | Tribactams | 388 | 0.305 |
| 75755 | Vitamin D analogues | 455 | 0.386 |

**Table 6.** MDDR activity classes for DS2 dataset.

| Activity Index | Activity Class | Active Molecules | Pairwise Similarity |
|---|---|---|---|
| 09249 | Muscarinic (M1) agonists | 900 | 0.111 |
| 12455 | NMDAreceptor antagonists | 1400 | 0.098 |
| 12464 | Nitric oxide synthase inhibitor | 505 | 0.102 |
| 31281 | Dopamine hydroxylase inhibitors | 106 | 0.125 |
| 43210 | Aldose reductase inhibitors | 957 | 0.119 |
| 71522 | Reverse transcriptase inhibitors | 700 | 0.103 |
| 75721 | Aromatase inhibitors | 636 | 0.110 |
| 78331 | Cyclooxygenase inhibitors | 636 | 0.108 |
| 78348 | Phospholipase A2 inhibitors | 617 | 0.123 |
| 78351 | Lipoxygenase inhibitors | 2111 | 0.113 |

**Table 7.** Sutherland activity classes.

| Activity Class | Active Molecules | Pairwise Similarity |
|---|---|---|
| Estrogen receptor | 141 | 0.468 |
| Ddihydrofolate reductase | 393 | 0.502 |
| Cyclooxygenase-2 inhibitors | 303 | 0.687 |
| Benzodiazepine receptor | 306 | 0.536 |

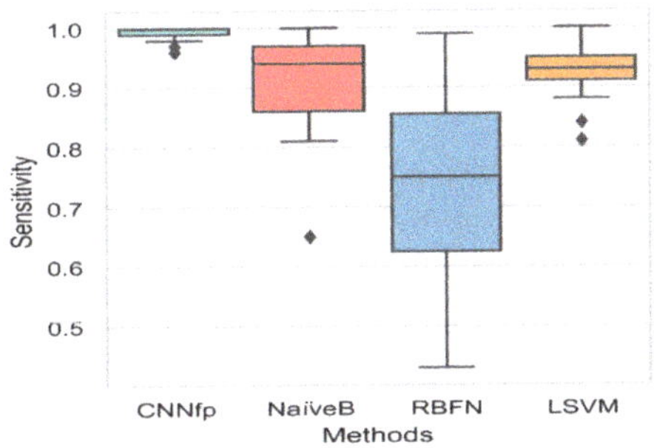


**Figure 9.** Boxplot chart results based on comparing the sensitivity values of different algorithms: CNNfp, NaiveB, RBFN, and LSVM.

Figure 10 shows the boxplot chart results after comparing the specificity values of the CNNfp, NaiveB, RBFN, and LSVM algorithms. The NaiveB and RBFN ML algorithms showed a higher variance of 0.01 and 0.04, respectively, compared to the CNNfp. This diversity in all specificity values was especially seen in the algorithms that displayed a variance of $2.5 \times 10^{-5}$. Furthermore, the CNNfp algorithm showed a high specificity value of 1.0, whereas the NaiveB and the RBFN algorithms displayed average specificity values of 0.99 and 0.98, respectively. The results showed a small $p$-value of $3.5 \times 10^{-5}$, highlighting a significant difference between all algorithms.

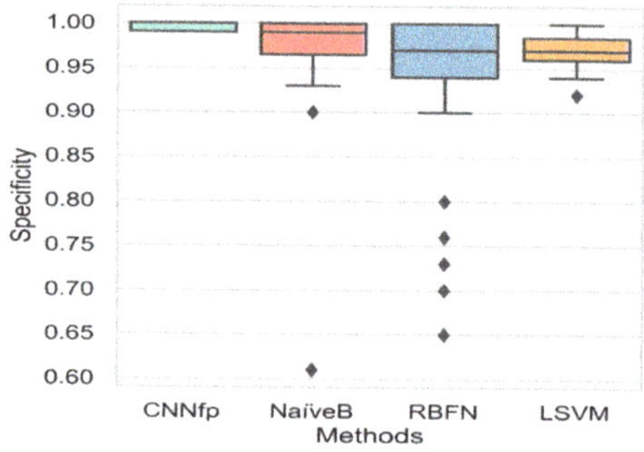


**Figure 10.** Boxplot chart results based on comparing the specificity values of different algorithms: CNNfp, NaiveB, RBFN, and LSVM.

Figure 11 describes the Boxplot chart results after comparing the AUC values of the CNNfp, NaiveB, RBFN, and LSVM algorithms. The LSVM, NaiveB, and RBFN ML algorithms showed a higher variance of 0.125, 0.083, and 0.033, respectively, compared to CNNfp. This diversity in all AUC values was especially seen in the algorithms that displayed a variance of $4.13 \times 10^{-5}$. A combination of the Mol2mat with the CNNfp algorithm showed an AUC value of 0.99, whereas the LSVM, NaiveB, and RBFN algorithms displayed higher average AUC values of 0.96, 0.99, and 0.85, respectively. The results showed a $p$-value of $4.2 \times 10^{-3}$, highlighting a significant difference between all algorithms.

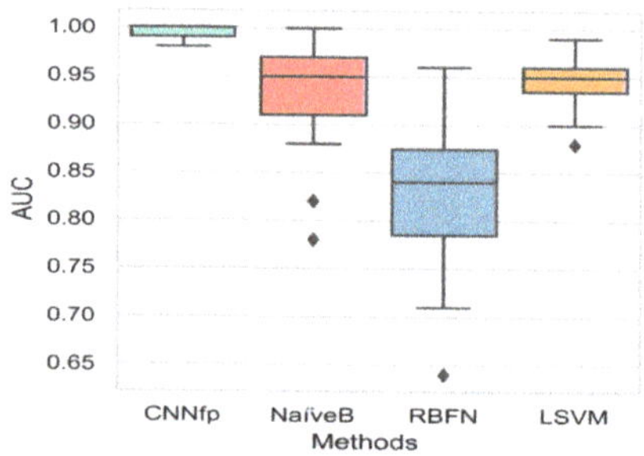


**Figure 11.** Boxplot chart results based on the comparison of the AUC values of different algorithms: CNNfp, NaiveB, RBFN, and LSVM.

The boxplot chart results (Figures 9–11) showed that the use of CNNfp was very efficient and convenient and presented less severe outliers in comparison to the NaiveB, RBFN, and LSVM algorithms, thereby indicating the effectiveness of this prediction approach. The results presented in Tables 2–4 for all three datasets show that the combination of ECFP4, EPFP4, and ECFC4 fingerprints with a CNN activity prediction method resulted in the lowest variance for the sensitivity, specificity, and AUC values for all activity classes compared to the traditional NaiveB, RBFN, and LSVM algorithms. These results suggest that a deep learning technique could be a promising, novel, and effective method of predicting the activities of a range of chemical compounds.

## 3. Discussion

### 3.1. Similarity Searching

Comparing unknown molecules to known chemical compounds allows us to predict the activities of targets that are unknown compounds. Thus, the target compounds will exhibit the activities of similar compounds. Several successful target prediction techniques have been proposed in the literature [11,23,24]. For example, the authors in [25] implemented a method for activity prediction using the Multi-level Neighbourhoods of Atoms (MNA) structural descriptor. This descriptor is generated based on the connection table and the table of atoms that represent each compound. A specific integer number is given to each descriptor according to its dictionary. The Tanimoto coefficient was effectively used to calculate the molecular similarity. The target compound activities were then predicted based on the activities of the most similar known compound.

A number of machine learning techniques have been used for activity prediction (target), including Binary Kernel Discrimination (BKD), Naive Bayesian Classifier (NBC), Artificial Neural Networks (ANN), and Support Vector Machines (SVM). The authors of [26] predicted five different ion channel targets using BKD and two different types of activity data. They found that the effectiveness of the model increased using highly similar activity classes. However, if this similarity was too low, the models would not work. As it is simple to build a network to include many sources of significant information about molecular structure, the authors of [27] used data fusion to aggregate the results of BIN searches using multiple reference structures. The authors in [28] presented a new classifier of Kinase Inhibitors using the NBC model. One advantage of this method that was noted is finding compounds that are structurally unrelated to known actives or novel targets for which there are inadequate data to develop a specific kinase model. In [29], the authors summarised how networks could conduct the equivalent of discriminant and regression analyses and underlined how initial overtraining and overfitting could lead to poor prediction performance. According to their predictions, the next revolution in QSAR will focus on developing better descriptors for connecting chemical structure to biological activity. The authors of [30] created a set of SVM classifiers that collectively account for 100 different forms of drug molecule action.

In their study, the multilabel-predicted chemical activity profiling was successfully accomplished by SVM classifiers, and they suggest that the proposed approach can forecast the biological activities of unidentified chemicals or signal negative consequences of drug candidates. In [11,31], the Bayesian belief network classifier was applied to predict the compound's target activities. The authors applied a novel technique to extend previous work, based on a convolutional neural network that uses the 2D fingerprint representation to predict the possibly bioactive molecules. The proposed CNN model for activity prediction also included the substructural information of the molecule.

### 3.2. Convolutional Neural Network for Biological Activity Prediction

In [32], the authors used Merck's drug discovery datasets and showed that Deep Neural networks (DNN) could obtain better prospective predictions than the existing machine learning methods. In addition, The Multi-Task Deep Neural Network (MT-DNN) model [33] demonstrated good performance by training the neural network with a number of output neurons, where the input molecule's activity is predicted by every neuron using different assays. In addition, [34–36] demonstrated how MT-DNN may be scaled to incorporate big databases such as PubChem Bioassays [37] and ChEMBL [38].

However, several issues and limitations still exist with the current methods. For instance, these methods work with targets that already have more available data and, thus, they cannot predict novel targets. Additionally, the current DL approaches rely on fingerprints, such as ECFP [39], which limit feature discovery to the composition of the particular chemical structures identified by the fingerprinting process [10,34,40]. This reduces their ability to discover arbitrary features. Moreover, the existing DL methods are blind to the target, as they are not able to elucidate the potential molecular interactions.

Another commonly used method is applying the similarity principle [41], which claims that substances with similar structures have similar biological characteristics. However, the authors in [42] discovered that it frequently fails because minor structural modifications can diminish the ligand's pharmacological activities that describe the molecular similarity within the substructures.

In order to address these issues and limitations, a novel Convolutional Neural Network (CNN)-based model using a 2D Fingerprint was proposed in this study for bioactivity prediction. This technique can be used for several applications such as bioactivity prediction, molecular searching, molecular classification, and virtual screening. The next section provides a description of how the suggested strategy was developed.

## 4. Materials and Methods

This section explains how this model is used for identifying and predicting the bioactivities of chemical compounds. First, we describe how various experimental benchmarks can be built and then utilised for system testing. Next, we discuss the systems for input representation and data encoding and deep convolutional network architecture.

### *4.1. Data Sets*

The proposed prediction model was experimentally evaluated using multiple datasets. This study used three datasets (Tables 5–7), which were described earlier in [43,44] and used in several studies for validating the ligand-based virtual screening methods [7,11,24,31,45,46].

The datasets used are disparate, including a structurally homogeneous dataset, as shown in Figure 12, and a structurally diverse dataset, as shown in Figure 13 [3].

The original version of the MDDR database includes 707 distinct activity classes. The mean pair-wise similarity (MPS) was then computed for each activity class. The mean pair-wise similarity (MPS) of each set of active molecules was used to estimate the diversity. The mean pairwise similarity (MPS) for 102,000 compounds selected randomly from MDDR was 0.200. Figure 14 presents how the MPS can divide the dataset into high- and low-diversity active classes, so that the cut-off point between the high- and low-diversity groups is equal to 0.200. This method is briefly explained and demonstrated in [3].

These datasets, MDDR1 and MDDR2, comprise 10 homogeneous and heterogeneous activity classes; the Sutherland dataset comprises four activity classes each. Tables 5–7 list the activity classes, molecules in each class, and diversity between classes. These tables were created using ECFP4 to estimate the mean pairwise Tanimoto similarity across all of the chemical pairs within each class (extended connectivity).

As noted above, the MPS values identify the diversity of activity classes that are used to evaluate the similarity search methods and biological activity prediction. Thus, the MPS values were used to compare the three used databases, as shown in Figure 15.

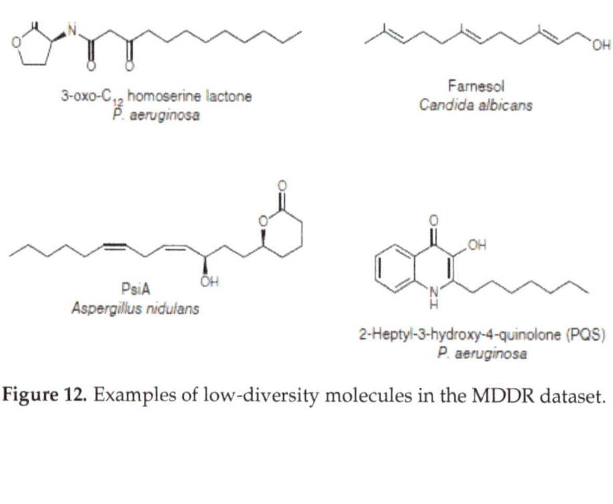


**Figure 12.** Examples of low-diversity molecules in the MDDR dataset.

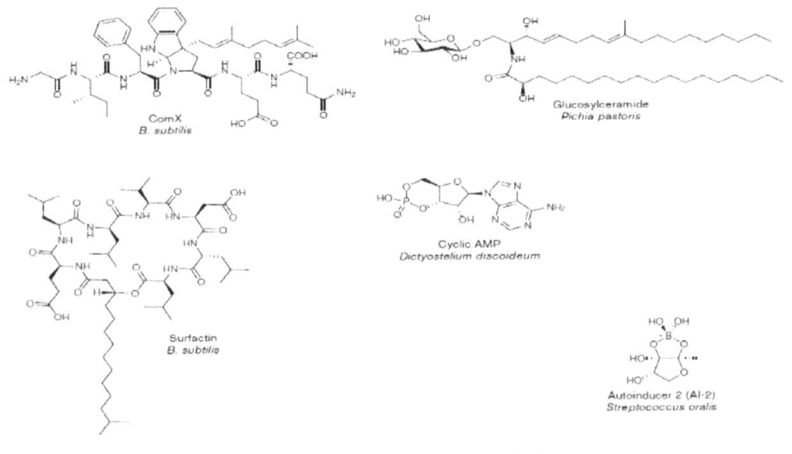


**Figure 13.** Examples of high-diversity molecules in the MDDR dataset.

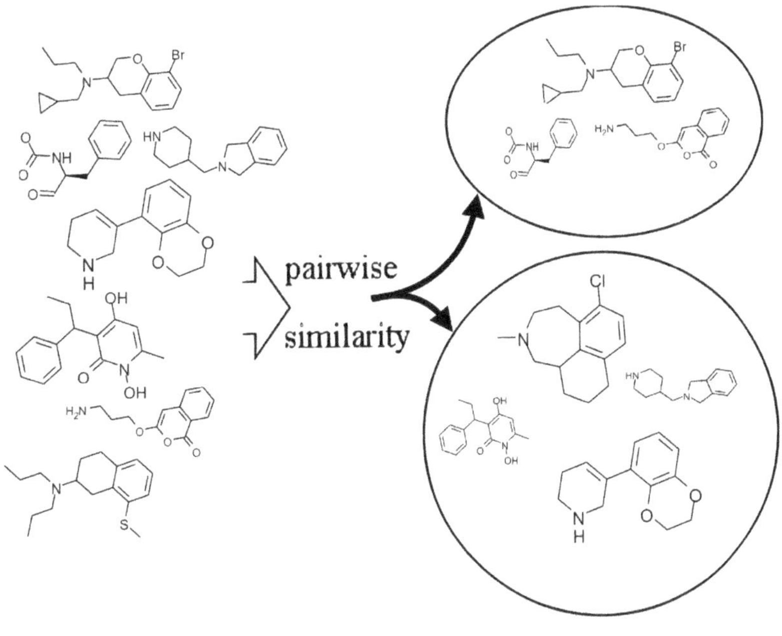


**Figure 14.** The average pairwise similarity (MPS) across each set of active molecules.

Box plots are the chart type that is used to visually present the distribution of all numerical data based on their average values and quartiles (or percentiles). Generally, box plots are applied in descriptive statistics since they help in overviewing the set of distributed data along with its range. The right-hand side of Figure 15 depicts the creation of a box, while the median MPS value is represented by the medium segment in the box. The first and third quartiles' MPS values are shown in the lower quartile and the upper quartile, while an empty circle represents the outlier.

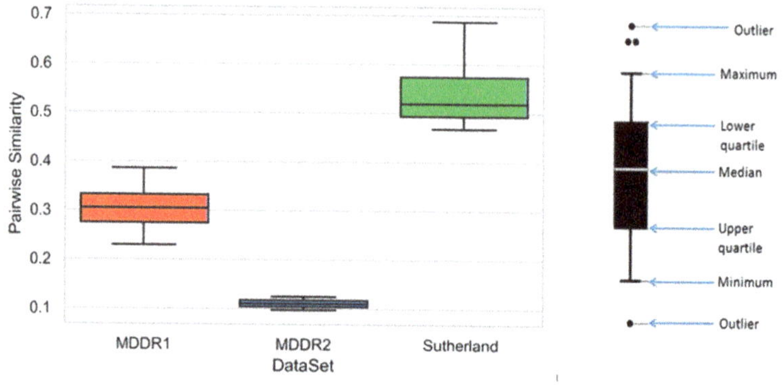


**Figure 15.** Comparison of MPS values of the three databases using boxplot.

## 4.2. Input Representation

One of the major issues affecting chemoinformatics and QSAR applications is the need for good input features. The general graph-based storage format for chemical compounds' numerical properties can be calculated using a variety of techniques. Fingerprints are a specific type of complex descriptor that detects the feature distribution from the bit string representations [3]. However, a feature extraction step was necessary to analyse the data in the machine learning technique. The performance of all learning algorithms is enhanced by this stage, which aids in expressing the interpretable data in the machines. Even the best algorithms may perform poorly if the wrong features are used, while simple techniques also perform well if suitable features are applied. Feature extraction techniques can be unsupervised or manually conducted. Here, the authors have presented a new molecular representation, Mol2mat (molecule to matrix), used to reshape each fingerprint molecule representation into a 2D array malleable for use in deep learning architecture.

In this study, the authors investigated eight different 2D fingerprints that were generated using Scitegics Pipeline Pilot software [47]. These included the 120-bit ALOGP, 1024-bit CDK (CDKFP), 1024-bit Path Fingerprints (EPFP4), 1024-bit ECFP4, 1024-bit ECFC4, 1024-bit Graph-Only Fingerprints (GOFP), 881-bit PubChem Fingerprints (PCFP), and the 166-bit Molecular Design Limited (MDL) fingerprints. Table 8 describes the storage of the fingerprint representatives for every molecule in a 2D array, with the help of the row-major order, and also describes every matrix representation Mol2mat size for each fingerprint.

**Table 8.** Details of every matrix size for every fingerprint.

| Fingerprint | Features Size | $\sqrt{\textit{Features Size}}$ | Mol2mat Size $n \times n$ |
|---|---|---|---|
| ALOGP | 120 | 10.95 | $11 \times 11$ |
| CDK | 1024 | 32 | $32 \times 32$ |
| ECFC4 | 1024 | 32 | $32 \times 32$ |
| ECFP4 | 1024 | 32 | $32 \times 32$ |
| EPFP4 | 1024 | 32 | $32 \times 32$ |
| GOFP | 1024 | 32 | $32 \times 32$ |
| PCFP | 881 | 29.68 | $30 \times 30$ |
| MDL | 166 | 12.88 | $13 \times 13$ |

To show the difference between different 2D fingerprint representations used in this paper, the authors plotted the scatter graphs in Figure 16 using 5083 molecules (from the MDDR dataset) that are grouped into ten activity classes. These scatter plots were used to establish the relationships between the various compounds belonging to the same class. The molecules were represented by different individual 2D fingerprints and descriptors. In addition, to represent their features, the representation was reduced to a 3D structure using the Principal Component Analysis (PCA) method.

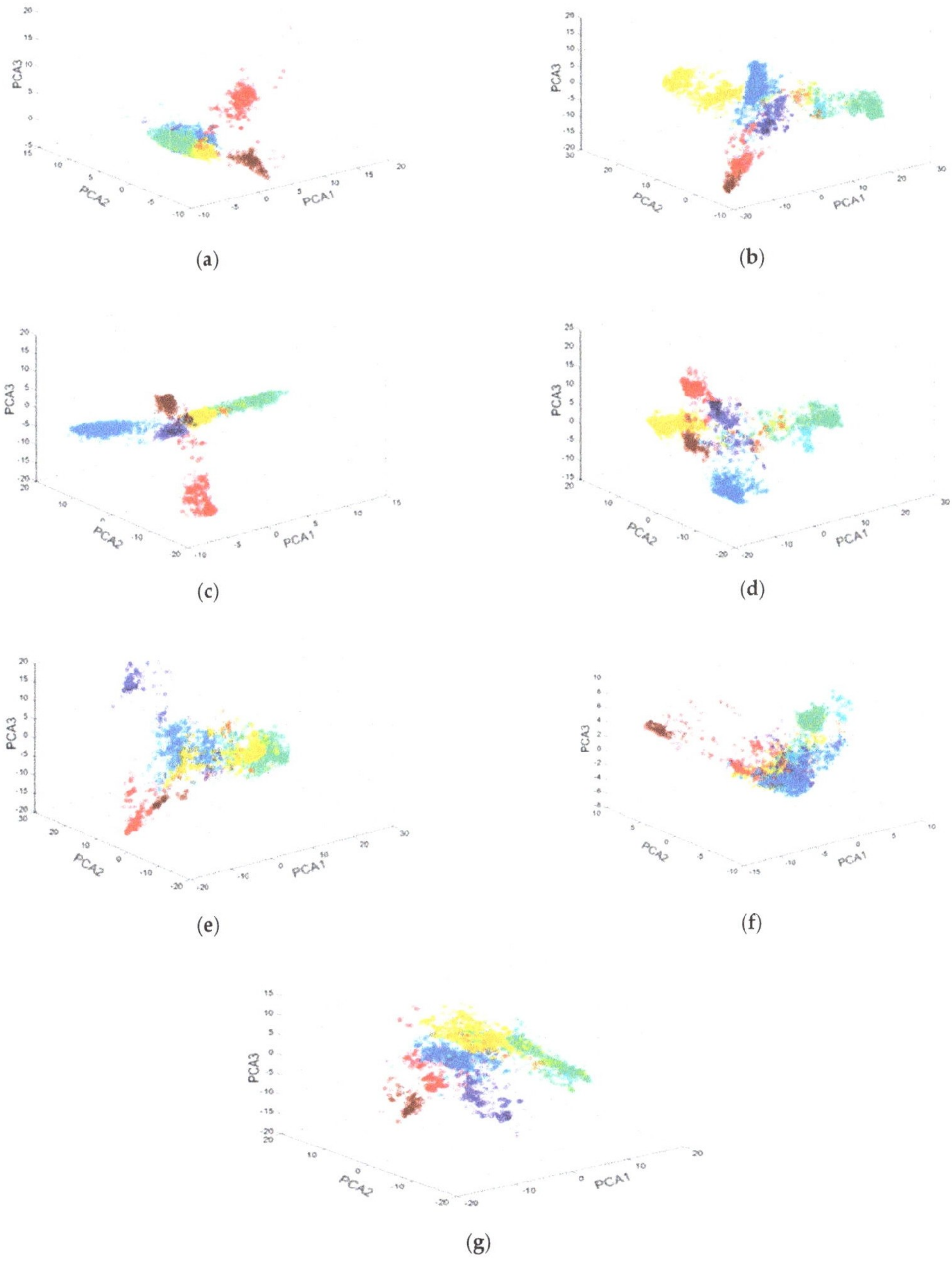


**Figure 16.** 3D−scatter plots based on seven fingerprints and representations of descriptors: (**a**) ALogP, (**b**) CDKFp, (**c**) ECFP4, (**d**) EPFP4, (**e**) GraphOnly, (**f**) MDL, and (**g**) PubchemFp of 5083 different molecules that were selected from the 10 biological activity classes of the MDDR dataset.

As shown in Figure 16, the ECFP4 2D fingerprint representation can be easily observed and was not overlapping. In addition, the molecules' biological activities can be segregated. This shows that the suggested 2D fingerprint representation may be successfully used for predicting the biological activity of various chemical substances.

After the generation of the eight fingerprints, the molecular fingerprints were stored in a 2D array using the row-major order, as shown in Algorithm 1.

Algorithm 1 is A summary of the storage of the fingerprints in a 2D array to yield the Mol2mat presentation.

**Algorithm 1**: Storing fingerprint in a 2D array

```
INPUT: FP:array of Fingerprint molecular representation.
          N:size of matrix
OUTPUT: Mol2Mat: matrix of FingerPrint representation;
Begin
P ← Get_fingerprint_features_size(FP) Using Table 4
Mol2Mat ← Create_matrix_of_size(N,N);
k ← 1;
for i = 1 to n do
    for j = 1 to n do
        Mol2Mat(i,j) ← FP(k);
        ← k +1 ;
    end
end
end.
```

Algorithm 1 summarises the storage of the fingerprint in a 2D array using the row-major order in pseudo-code form. The algorithm's output was a 2D array of Mol2mat representations of the input molecule. Figure 17 summarises the design of the Mol2mat presentation process.

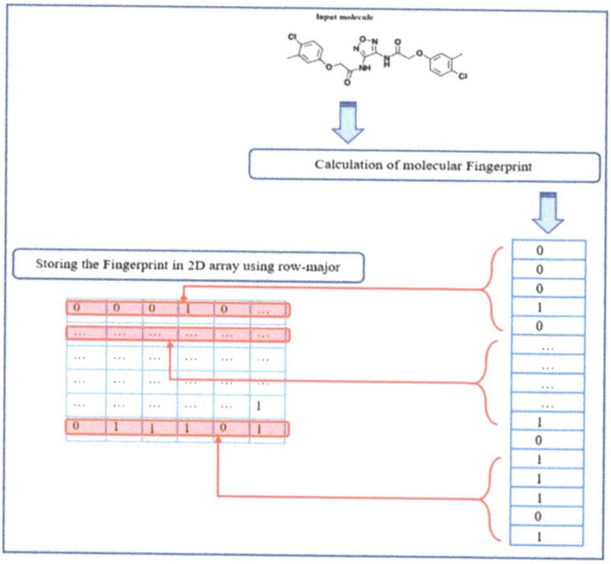


**Figure 17.** A summary of the newly proposed Mol2mat presentation process.

After evaluating each fingerprint, the authors assessed all the probable combinations based on the five best descriptors. The combinations were based on the fusion of the extracted feature levels. The combination of multi-CNN can be performed as illustrated in [48,49]. Initially, the combination cases for 2, 3, 4, and 5 were generated by selecting two fingerprints, then three, followed by four, and finally, all five. Thereafter, the best combination was chosen.

### 4.3. Convolutional Neural Network

The default architecture was seen to be a convolutional architecture with fully connected layers. The authors used the Krizhevsky principles [50] for designing the CNN model configuration that was used for viewing the source code [51]. This configuration followed the earlier generic design [50]. Figure 18 presents the general CNN configuration, where the image was passed through the stack of convolutional (conv.) layers. The convolution step employed a max-pooling layer. It was observed that this combination improved the accuracy model and enhanced the CNN configuration.

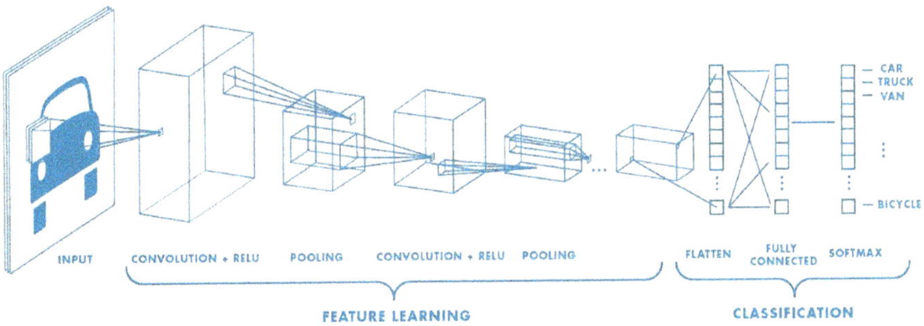


**Figure 18.** The general CNN configuration.

The flattened layer came after the max-pooling layer. This transformed the 2D matrix data into a single vector, assisting in processing the output that had dense layers, i.e., fully connected layers. The final layer was made of the classification Softmax layer [52,53].

Although CNN displayed good results for the feature learning and the prediction tasks, recent studies have shown a better performance by fusing different CNNs [20,21,54,55]. These combinations can be implemented using feature concatenation or by computing the average or output prediction scores derived from various CNNs.

Some studies [48,49] described the combination of 3 CNN models, as shown in Figure 19. These models were based on the fusion of the information level. Fusion could be performed early in the network after modifying the 1st-layer convolution filters for an extension of time, or it could be performed later, after placing 2 different single-frame networks and then fusing their outputs after the processing. The yellow, green, red, and blue boxes depict the fully connected, normalisation, convolution, and pooling layers, respectively. In a Slow Fusion model, the highlighted columns share the parameters.

In this stage, we used better techniques to combine the various sources of knowledge available in the area of deep learning [20–22]. Firstly, we proposed a feature extraction step for presenting every selected molecular fingerprint. This combination significantly improved the models, since they could benefit from every molecular fingerprint and then combine all the extracted features from various sources after a flattened layer, which followed the max-pooling layer. This helped them convert the 2D matrix data into the vector. As a result, they could process the output data using the fully connected layers, called the dense layers. In this section, we described the CNN architecture used in this research and how we can combine multi CNNs in one model. In the next section, we will describe the performance evaluation.

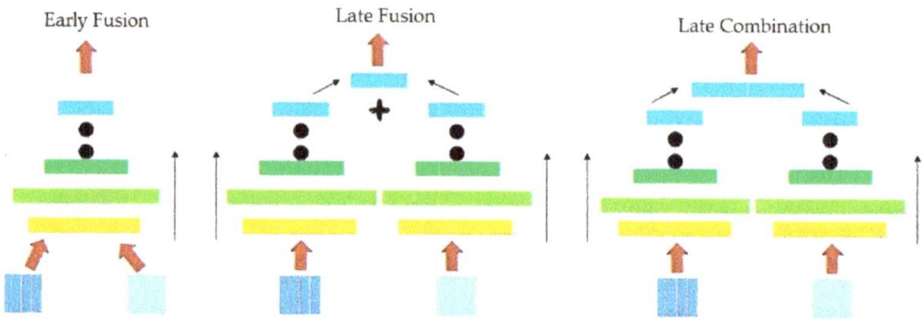


**Figure 19.** Different approaches used for fusing the information present in the CNN layers.

*4.4. Network Architecture*

As mentioned above, eight fingerprint representatives were generated using the Scitegics Pipeline Pilot software [47]. They were further stored in the 2D array with a row-major order for deriving a novel matrix representation Mol2mat, which used the above-mentioned algorithm.

As previously stated, a few fingerprints complemented one another, and their combination yielded good results. This indicated that different fingerprints generated differing results with regard to biological activity prediction or similarity searches. This further indicated that the different QSAR models could be developed based on different fingerprints with similar accuracy. Currently, researchers tend to combine and merge all fingerprints and descriptor sets, which comprise various types of fingerprints [3]. After evaluating each fingerprint, the authors assessed all the probable combinations based on the five best descriptors. The combinations were based on the fusion of the extracted feature levels.

In the present study, we used better techniques for combining the various sources of knowledge available in the area of deep learning [20–22]. Firstly, we proposed a feature extraction step for presenting every best molecular fingerprint in which all molecules were passed through 2 conv. layers, using a (3 × 3) feature map size for convolution and one max-pooling layer. This combination significantly improved the models since they could benefit from every molecular fingerprint and combine all the extracted features from various sources after a flattened layer. As a result, they could process the output data using the fully connected layers. The first two fully connected layers were built using a different number of nodes in every combination. Table 9 presents these node numbers in detail in every combination. The combination cases for 2, 3, 4, and 5 were generated by selecting two fingerprints, then three, followed by four, and finally, all five. The best combination was then chosen.

**Table 9.** Details of the first and second fully connected layers for every combination.

| Combined Case | Combined Layer Size | Number of Nodes in 1st Fully Connected Layer | Number of Nodes in 2nd Fully Connected Layer |
|---|---|---|---|
| 2 Fingerprints | 6272 | 128 | 64 |
| 3 Fingerprints | 9408 | 256 | 128 |
| 4 Fingerprints | 12,544 | 512 | 256 |
| 5 Fingerprints | 15,680 | 1024 | 512 |

The final layer included the Softmax layer [50,52,53]. Figure 20 describes the configuration of the combined CNN, which was used to assess 3 fingerprints.

The target was as follows: to predict if the specific chemical compound, $i$, showed activity for target, $t$. These data could be encoded in the binary form, $y_{it}$, where $y_{it} = 1$ for the active compound and $y_{it} = 0$ for the inactive compound. This also included the prediction of the compound's behaviour from targets, simultaneously. In the training stage, a general back-propagation algorithm was used to determine the CNN and decrease the cross-entropy of all targets and the activation of the output layer.

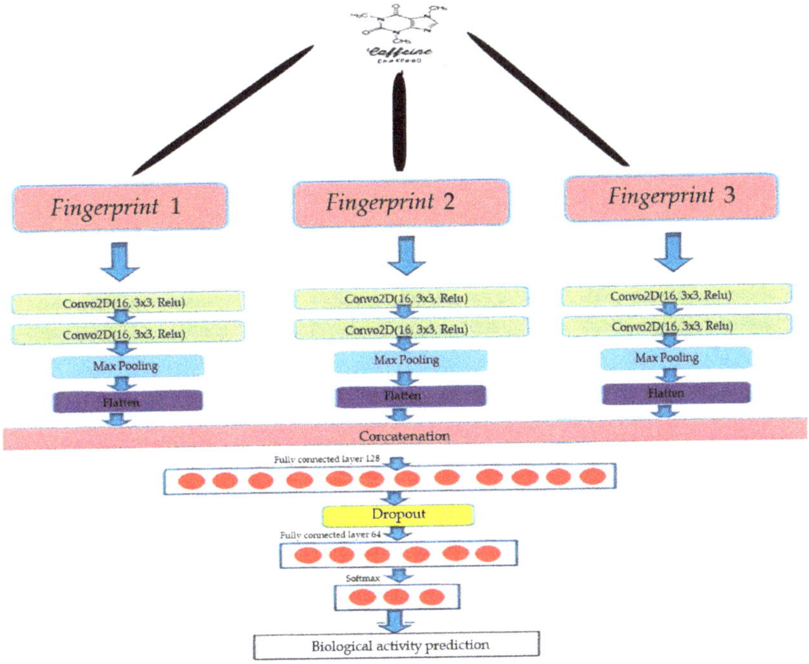


**Figure 20.** The configuration of the combined CNN that was used for 3 fingerprints.

## 5. Conclusions

This study has investigated the use of molecular fingerprinting in the Convolution Neural Network model to predict the activities of ligand-based targets. The results indicate that the combination of the ECFP4, EPFP4, and ECFC4 fingerprints with a CNN activity prediction method produced the lowest variance for the sensitivity, specificity, and AUC values for all the activity classes, when compared to the three traditional ML algorithms of NaiveB, LSVM, and RBFN, available in the WEKA-Workbench. The paper described a novel Mol2mat process, which showed low overlap and was able to segregate all the biological activities of the chemical compounds. A combination of three fingerprints with CNN was used on some popular datasets, and the performance of this combination was compared to that of three traditional ML algorithms. The proposed algorithm achieved good prediction rates (where the low- and high-diversity datasets displayed a 98% AUC value). The results also showed that combining the ECFP4, EPFP4, and ECFC4 fingerprints with CNN improved the performance of both the heterogeneous and homogeneous datasets. In this study, the authors have shown that this combination of fingerprints with the CNN technique is a convenient and stable prediction process, which could be used for determining the activities of unknown chemical compounds. However, this field needs to be investigated further, and better accuracy prediction processes must be developed for high-diversity activity compounds.

**Author Contributions:** Conceptualization, H.H. and N.S.; methodology, H.H., F.S. and N.S.; software, H.H.; validation, B.B., I.N., A.T. and M.N.; formal analysis, H.H., B.B., I.N., A.T. and M.N.; investigation, H.H., B.B., I.N., A.T. and M.N.; resources, H.H., F.S. and M.N.; data curation, H.H., B.B. and A.T.; writing—original draft preparation, H.H., N.S.; writing—review and editing, H.H., F.S, N.S, I.N. and M.N.; visualization, H.H.; supervision, N.S. and F.S.; project administration, N.S. and F.S.; funding acquisition, F.S., N.S. and I.N. All authors have read and agreed to the published version of the manuscript.

**Funding:** This research was funded by the Research Management Center at Universiti Teknologi Malaysia (Vot No: Q.J130000.21A6.00P48) and Ministry of Higher Education, Malaysia (JPT(BKPI) 1000/016/018/25(58)) through Malaysia Big Data Research Excellence Consortium (BiDaREC) (Vot No: R.J130000.7851.4L933), (Vot No: R.J130000.7851.4L942), (Vot No: R.J130000.7851.4L938), and (Vot No: R.J130000.7851.4L936). We are also grateful to (Project No: KHAS-KKP/2021/FTMK/C00003) and (Project No: KKP002-2021) for their financial support of this study.

**Institutional Review Board Statement:** Not applicable.

**Informed Consent Statement:** Not applicable.

**Data Availability Statement:** The MDL Drug Data Report (MDDR) dataset is owned by www.accelrys.com, accessed on 15 January 2020. A license is required to access the data.

**Acknowledgments:** The authors would like to thank the Research Management Center at Universiti Teknologi Malaysia for funding this research using (Vot No: Q.J130000.21A6.00P48) and Ministry of Higher Education, Malaysia (JPT(BKPI)1000/016/018/25(58)) through Malaysia Big Data Research Excellence Consortium (BiDaREC) (Vot No: R.J130000.7851.4L933), (Vot No: R.J130000.7851.4L942), (Vot No: R.J130000.7851.4L938), and (Vot No: R.J130000.7851.4L936). We are also grateful to (Project No: KHAS-KKP/2021/FTMK/C00003) and (Project No: KKP002-2021) for their financial support of this study.

**Conflicts of Interest:** The authors declare no conflict of interest.

## References

1. Willett, P.; Barnard, J.M.; Downs, G.M. Chemical Similarity Searching. *J. Chem. Inf. Comput. Sci.* **1998**, *38*, 983–996. [CrossRef]
2. Li, J.; Luo, D.; Wen, T.; Liu, Q.; Mo, Z. Representative Feature Selection of Molecular Descriptors in QSAR Modeling. *J. Mol. Struct.* **2021**, *1244*, 131249. [CrossRef]
3. Willett, P. Similarity-Based Virtual Screening Using 2D Fingerprints. *Drug Discov. Today* **2006**, *11*, 1046–1053. [CrossRef]
4. Du, Z.; Wang, D.; Li, Y. Comprehensive Evaluation and Comparison of Machine Learning Methods in QSAR Modeling of Antioxidant Tripeptides. *ACS Omega* **2022**, *7*, 25760–25771. [CrossRef]
5. Priyanka, D.; Supratik, K.; Pravin, A.; Kunal, R. Prediction Reliability of QSAR Models: An Overview of Various Validation Tools. *Arch. Toxicol.* **2022**, *96*, 1279–1295.
6. Huang, T.; Guohui, S.; Lijiao, Z.; Na, Z.; Rugang, Z.; Yongzhen, P. Quantitative Structure-Activity Relationship (QSAR) Studies on the Toxic Effects of Nitroaromatic Compounds. *Int. J. Mol. Sci.* **2021**, *22*, 8557. [CrossRef]
7. Hentabli, H.; Saeed, F.; Abdo, A.; Salim, N. A New Graph-Based Molecular Descriptor Using the Canonical Representation of the Molecule. *Sci. World J.* **2014**, *2014*, 286974. [CrossRef] [PubMed]
8. Wang, D.; Yu, J.; Chen, L.; Li, X.; Jiang, H.; Chen, K.; Zheng, M.; Luo, X. A Hybrid Framework for Improving Uncertainty Quantification in Deep Learning-Based QSAR Regression Modeling. *J. Cheminform.* **2021**, *13*, 69. [CrossRef]
9. Kumari, M.; Tiwari, N.; Chandra, S.; Subbarao, N. Comparative Analysis of Machine Learning Based QSAR Models and Molecular Docking Studies to Screen Potential Anti-Tubercular Inhibitors against InhA of Mycobacterium Tuberculosis. *Int. J. Comput. Biol. Drug Des.* **2018**, *11*, 209–235. [CrossRef]
10. Ding, Y.; Chen, M.; Guo, C.; Zhang, P.; Wang, J. Molecular Fingerprint-Based Machine Learning Assisted QSAR Model Development for Prediction of Ionic Liquid Properties. *J. Mol. Liq.* **2021**, *326*, 115212. [CrossRef]
11. Nasser, M.; Salim, N.; Saeed, F.; Basurra, S.; Rabiu, I.; Hamza, H.; Alsoufi, M.A. Feature Reduction for Molecular Similarity Searching Based on Autoencoder Deep Learning. *Biomolecules* **2022**, *12*, 508. [CrossRef] [PubMed]
12. Li, Y.; Xu, Y.; Yu, Y. CRNNTL: Convolutional Recurrent Neural Network and Transfer Learn-Ing for QSAR Modelling. *arXiv* **2021**, arXiv:2109.03309.
13. Chollet, F. Keras Documentation. Keras.Io. 2015. Available online: https://keras.io/ (accessed on 18 October 2020).
14. Bastien, F.; Lamblin, P.; Pascanu, R.; Bergstra, J.; Goodfellow, I.; Bergeron, A.; Bouchard, N.; Warde-Farley, D.; Bengio, Y. Theano: New Features and Speed Improvements. *arXiv* **2012**, arXiv:1211.5590.
15. Ammar, A.; Valérie, L.; Philippe, J.; Naomie, S.; Maude, P. Prediction of New Bioactive Molecules Using a Bayesian Belief Network. *J. Chem. Inf. Model.* **2014**, *54*, 30–36. [CrossRef]
16. Witten, I.H.; Frank, E.; Hall, M.A.; Pal, C.J. *Data Mining: Practical Machine Learning Tools and Techniques*; Morgan Kaufmann: Burlington, MA, USA, 2016.
17. John, G.H.; Langley, P. Estimating Continuous Distributions in Bayesian Classifiers. *arXiv* **2013**, arXiv:1302.4964.
18. Chih-Chung, C. LIBSVM: A Library for Support Vector Machines. *ACM Trans. Intell. Syst. Technol.* **2011**, *2*, 27.
19. Bugmann, G. Normalized Gaussian Radial Basis Function Networks. *Neurocomputing* **1998**, *20*, 97–110. [CrossRef]
20. Dolz, J.; Desrosiers, C.; Ayed, I. ben IVD-Net: Intervertebral Disc Localization and Segmentation in MRI with a Multi-Modal UNet. In Proceedings of the MICCAI 2018 IVD Challenge, Granada, Spain, 16 September 2018; pp. 1–7. [CrossRef]

21. Williams, J.; Comanescu, R.; Radu, O.; Tian, L. DNN Multimodal Fusion Techniques for Predicting Video Sentiment. In Proceedings of the Grand Challenge and Workshop on Human Multimodal Language (Challenge-HML), Melbourne, Australia, 20 July 2018; pp. 64–72.
22. Das, A.; Ghosh, S.; Sarkhel, R.; Choudhuri, S.; Das, N.; Nasipuri, M. Combining Multilevel Contexts of Superpixel Using Convolutional Neural Networks to Perform Natural Scene Labeling. *Adv. Intell. Syst. Comput.* **2019**, *740*, 297–306. [CrossRef]
23. Lagos, C.F.; Segovia, G.F.; Nu ez-Navarro, N.; Faúndez, M.A.; Zacconi, F.C. Novel FXa Inhibitor Identification through Integration of Ligand- and Structure-Based Approaches. *Molecules* **2017**, *22*, 1588. [CrossRef]
24. Hentabli, H.; Naomie, S.; Saeed, F. An activity prediction model using shape-based descriptor method. *J. Teknol.* **2016**, *78*, 1–8.
25. Filimonov, D.; Poroikov, V.; Borodina, Y.; Gloriozova, T. Chemical Similarity Assessment through Multilevel Neighborhoods of Atoms: Definition and Comparison with the Other Descriptors. *J. Chem. Inf. Comput. Sci.* **1999**, *39*, 666–670. [CrossRef]
26. Willett, P.; Wilton, D.; Hartzoulakis, B.; Tang, R.; Ford, J.; Madge, D. Prediction of Ion Channel Activity Using Binary Kernel Discrimination. *J. Chem. Inf. Model.* **2007**, *47*, 1961–1966. [CrossRef] [PubMed]
27. Chen, B.; Mueller, C.; Willett, P. Evaluation of a Bayesian Inference Network for Ligand-Based Virtual Screening. *J. Cheminform.* **2009**, *1*, 5. [CrossRef]
28. Xia, X.; Maliski, E.G.; Gallant, P.; Rogers, D. Classification of Kinase Inhibitors Using a Bayesian Model. *J. Med. Chem.* **2004**, *47*, 4463–4470. [CrossRef]
29. Winkler, D.A.; Burden, F.R. Application of Neural Networks to Large Dataset QSAR, Virtual Screening, and Library Design. *Methods Mol. Biol.* **2002**, *201*, 325–367. [CrossRef] [PubMed]
30. Kawai, K.; Fujishima, S.; Takahashi, Y. Predictive Activity Profiling of Drugs by Topological-Fragment-Spectra-Based Support Vector Machines. *J. Chem. Inf. Model.* **2008**, *48*, 1152–1160. [CrossRef]
31. Berrhail, F.; Belhadef, H. Genetic Algorithm-Based Feature Selection Approach for Enhancing the Effectiveness of Similarity Searching in Ligand-Based Virtual Screening. *Curr. Bioinform.* **2019**, *15*, 431–444. [CrossRef]
32. Ma, J.; Sheridan, R.P.; Liaw, A.; Dahl, G.E.; Svetnik, V. Deep Neural Nets as a Method for Quantitative Structure-Activity Relationships. *J. Chem. Inf. Model.* **2015**, *55*, 263–274. [CrossRef]
33. Dahl, G.E.; Jaitly, N.; Salakhutdinov, R. Multi-Task Neural Networks for QSAR Predictions. *arXiv* **2014**, arXiv:1406.1231.
34. Unterthiner, T.; Mayr, A.; Klambauer, G.; Hochreiter, S. Toxicity Prediction Using Deep Learning. *arXiv* **2015**, arXiv:1503.01445. [CrossRef]
35. Unterthiner, T.; Mayr, A.; Klambauer, G.; Steijaert, M.; Wegner, J.K.; Ceulemans, H. Deep Learning as an Opportunity in Virtual Screening. In Proceedings of the Deep Learning and Representation Learning Workshop: NIPS, Montreal, QC, Cananda, 12 December 2014; pp. 1–9.
36. Ramsundar, B.; Kearnes, S.; Riley, P.; Webster, D.; Konerding, D.; Pande, V. Massively Multitask Networks for Drug Discovery. *arXiv* **2015**, arXiv:1502.02072.
37. Wang, Y.; Xiao, J.; Suzek, T.O.; Zhang, J.; Wang, J.; Zhou, Z.; Han, L.; Karapetyan, K.; Dracheva, S.; Shoemaker, B.A.; et al. PubChem's BioAssay Database. *Nucleic Acids Res.* **2011**, *40*, D400–D412. [CrossRef] [PubMed]
38. Bento, A.P.; Gaulton, A.; Hersey, A.; Bellis, L.J.; Chambers, J.; Davies, M.; Krüger, F.A.; Light, Y.; Mak, L.; McGlinchey, S.; et al. The ChEMBL Bioactivity Database: An Update. *Nucleic Acids Res.* **2014**, *42*, D1083–D1090. [CrossRef] [PubMed]
39. Rogers, D.; Hahn, M. Extended-Connectivity Fingerprints. *J. Chem. Inf. Model.* **2010**, *50*, 742–754. [CrossRef] [PubMed]
40. Dana, D.; Gadhiya, S.V.; St Surin, L.G.; Li, D.; Naaz, F.; Ali, Q.; Paka, L.; Yamin, M.A.; Narayan, M.; Goldberg, I.D.; et al. Deep Learning in Drug Discovery and Medicine; Scratching the Surface. *Molecules* **2018**, *23*, 2384. [CrossRef]
41. MLA, J.; Maggiora, G.M. Concepts and Application of Molecular Similarity. *Wiley Interdiscip. Rev. Comput. Mol. Sci.* **1990**, *50*, 376–377.
42. Martin, Y.C.; Kofron, J.L.; Traphagen, L.M. Do Structurally Similar Molecules Have Similar Biological Activity? *J. Med. Chem.* **2002**, *45*, 4350–4358. [CrossRef]
43. Sci Tegic Accelrys Inc. Available online: http://accelrys.com/products/collaborative-science/databases/bioactivity-databases/mddr.html (accessed on 15 January 2020).
44. Sutherland, J.J.; O'Brien, L.A.; Weaver, D.F. Spline-Fitting with a Genetic Algorithm: A Method for Developing Classification Structure-Activity Relationships. *J. Chem. Inf. Comput. Sci.* **2003**, *43*, 1906–1915. [CrossRef]
45. Hentabli, H.; Salim, N.; Abdo, A.; Saeed, F. LINGO-DOSM: LINGO for Descriptors of Outline. In *Intelligent Information and Database Systems*; Springer: Berlin/Heidelberg, Germany, 2013; pp. 315–324.
46. Hentabli, H.; Salim, N.; Abdo, A.; Saeed, F. LWDOSM: Language for Writing Descriptors. In *Advanced Machine Learning Technologies and Applications*; Springer: Berlin/Heidelberg, Germany, 2012; pp. 247–256.
47. Stevenson, J.M.; Mulready, P.D. Pipeline Pilot 2.1 By Scitegic, 9665 Chesapeake Drive, Suite 401, San Diego, CA 92123-1365. *J. Am. Chem. Soc.* **2003**, *125*, 1437–1438. [CrossRef]
48. Simonyan, K.; Zisserman, A. Two-Stream Convolutional Networks for Action Recognition in Videos. *Adv. Neural Inf. Process. Syst.* **2014**, *27*, 568–576. [CrossRef]
49. Karpathy, A.; Toderici, G.; Shetty, S.; Leung, T.; Sukthankar, R.; Fei-Fei, L. Large-Scale Video Classification with Convolutional Neural Networks. In Proceedings of the International Computer Vision and Pattern Recognition (CVPR 2014), Columbus, ON, USA, 23–28 June 2014. [CrossRef]

50. Krizhevsky, A.; Sutskever, I.; Hinton, G.E. ImageNet Classification with Deep Convolutional Neural Networks. *Commun. ACM* **2017**, *60*, 84–90. [CrossRef]
51. Gupta, V. Image Classification Using Convolutional Neural Networks in Keras. Available online: https://www.learnopencv.com/image-classification-using-convolutional-neural-networks-in-keras/ (accessed on 18 October 2020).
52. Angermueller, C.; Pärnamaa, T.; Parts, L.; Stegle, O. Deep Learning for Computational Biology. *Mol. Syst. Biol* **2016**, *12*, 878. [CrossRef] [PubMed]
53. Wang, H.; Raj, B. On the Origin of Deep Learning. *ArXiv* **2017**, arXiv:1702.07800. [CrossRef]
54. Su, H.; Maji, S.; Kalogerakis, E.; Learned-Miller, E. Multi-View Convolutional Neural Networks for 3D Shape Recognition. In Proceedings of the IEEE International Conference on Computer Vision, Santiago, Chile, 7–13 December 2015; Volume 1, pp. 945–953. [CrossRef]
55. Cheng, Y.; Wang, F.; Zhang, P.; Hu, J. Risk Prediction with Electronic Health Records: A Deep Learning Approach. In Proceedings of the 2016 SIAM International Conference on Data Mining, Miami, FL, USA, 5–7 May 2016; pp. 432–440. [CrossRef]

International Journal of *Molecular Sciences*

MDPI

*Article*

# In Vitro Conditioning of Adipose-Derived Mesenchymal Stem Cells by the Endothelial Microenvironment: Modeling Cell Responsiveness towards Non-Genetic Correction of Haemophilia A

**Silvia Barbon [1,2], Elena Stocco [1,2], Senthilkumar Rajendran [3], Lorena Zardo [4], Veronica Macchi [1], Claudio Grandi [2], Giuseppe Tagariello [4], Andrea Porzionato [1,2], Paolo Radossi [4,*], Raffaele De Caro [1,2] and Pier Paolo Parnigotto [2]**

1 Section of Human Anatomy, Department of Neuroscience, University of Padova, 35121 Padova, Italy; silvia.barbon@unipd.it (S.B.); elena.stocco@unipd.it (E.S.); veronica.macchi@unipd.it (V.M.); andrea.porzionato@unipd.it (A.P.); raffaele.decaro@unipd.it (R.D.C.)
2 Foundation for Biology and Regenerative Medicine, Tissue Engineering and Signaling—TES, Onlus, 35030 Padova, Italy; claudio.grandi@unipd.it (C.G.); pierpaolo.parnigotto@unipd.it (P.P.P.)
3 Department of Surgery Oncology and Gastroenterology, University of Padova, 35124 Padova, Italy; senthilstem@gmail.com
4 Haematology and Haemophilia Centre, Castelfranco Veneto Hospital, 31033 Castelfranco Veneto, Italy; lorena.zardo@aulss2.veneto.it (L.Z.); tagariello.giuseppe@gmail.com (G.T.)
* Correspondence: paolo.radossi@aulss2.veneto.it

**Citation:** Barbon, S.; Stocco, E.; Rajendran, S.; Zardo, L.; Macchi, V.; Grandi, C.; Tagariello, G.; Porzionato, A.; Radossi, P.; De Caro, R.; et al. In Vitro Conditioning of Adipose-Derived Mesenchymal Stem Cells by the Endothelial Microenvironment: Modeling Cell Responsiveness towards Non-Genetic Correction of Haemophilia A. *Int. J. Mol. Sci.* **2022**, 23, 7282. https://doi.org/10.3390/ijms23137282

Academic Editors: Ralf Schubert and Patrick C. Baer

Received: 18 May 2022
Accepted: 27 June 2022
Published: 30 June 2022



**Abstract:** In recent decades, the use of adult multipotent stem cells has paved the way for the identification of new therapeutic approaches for the treatment of monogenic diseases such as Haemophilia A. Being already studied for regenerative purposes, adipose-derived mesenchymal stem cells (Ad-MSCs) are still poorly considered for Haemophilia A cell therapy and their capacity to produce coagulation factor VIII (FVIII) after proper stimulation and without resorting to gene transfection. In this work, Ad-MSCs were in vitro conditioned towards the endothelial lineage, considered to be responsible for coagulation factor production. The cells were cultured in an inductive medium enriched with endothelial growth factors for up to 21 days. In addition to significantly responding to the chemotactic endothelial stimuli, the cell populations started to form capillary-like structures and up-regulated the expression of specific endothelial markers (CD34, PDGFRα, VEGFR2, VE-cadherin, CD31, and vWF). A dot blot protein study detected the presence of FVIII in culture media collected from both unstimulated and stimulated Ad-MSCs. Remarkably, the activated partial thromboplastin time test demonstrated that the clot formation was accelerated, and FVIII activity was enhanced when FVIII deficient plasma was mixed with culture media from the untreated/stimulated Ad-MSCs. Overall, the collected evidence supported a possible Ad-MSC contribution to HA correction via specific stimulation by the endothelial microenvironment and without any need for gene transfection.

**Keywords:** adipose-derived stem cells; Haemophilia A; coagulation factor VIII; endothelial differentiation; stem cell therapy; regenerative medicine

## 1. Introduction

Stem cell therapy research is advancing rapidly as a novel regenerative approach to treat both acquired and genetic diseases [1]. The rationale behind this is that patients suffering from organ and tissue dysfunctions can be treated with transplanted stem cells which have the potential to restore specific tissue compartments [2].

Over the last few decades, Haemophilia A (HA) has been investigated among the genetic diseases that could be treated with this regenerative strategy. There is not yet a definitive cure for this pathology, and current therapy consists of the so-called replacement treatment, which aims to sufficiently increase the concentration of the missing coagulation factor VIII (FVIII) to prevent or stop spontaneous and traumatic hemorrhages [3,4].

With the progress in biomedical engineering, preclinical and clinical research has started to conceptualize and develop long-term treatments for HA by using gene and cell therapy. Although recently receiving promising results in phase I/II trials [5,6], adeno-associated virus (AAV) gene therapy is not exempt from important concerns, including capsid-specific T-cell responses, which reduce transgene expression and cause hepatotoxicity. Another consideration to be made is in regards to the fact that viral vector production on a clinical scale is costly and variable [7]. Furthermore, the clinical development of AAV-FVIII still appears to be more complicated in comparison with AAV-factor IX (FIX) due to (1) the difficulty of efficient AAV vector packaging for large FVIII cDNA, and (2) the uncertain choice of liver target cell [8]. Notably, phase I/II studies reporting long-term follow-up highlighted an unexpected decline in FVIII levels after 1 year and continuously during the following 5 years post AAV-mediated expression [9,10].

According to experimental and clinical experience, the cell compartments that produce FVIII are mainly represented by endothelial sinusoidal cells. In fact, it is well known that liver-transplanted patients resolve the hemorrhagic disease [11,12]. Considering that, HA cell therapy is based on the hypothesis that the transplantation of exogenous cells capable of releasing FVIII could become a functional cure for coagulation deficiency. A variety of genetically modified cells has been tested so far as delivery vehicles for FVIII, including hematopoietic stem/progenitor cells [13–15], bone marrow-derived mesenchymal stem cells [7,16,17], endothelial cells [18], human placental cells [19], and adipose stromal cells [20]. Despite promising pre-clinical results, the development of effective cell therapy approaches for HA treatment may benefit from providing insights into non-genetically modified cell populations able to produce FVIII and endowed with high engraftment potential. Indeed, pre-clinical studies have already demonstrated the therapeutic potential of non-transduced cell populations in correcting the Haemophilia A phenotype, resorting to the transplantation into animal HA models of tissue-specific cells such as Liver Endothelial Sinusoidal Cells (LESCs) [21] and human pluripotent stem cell (hPSC)-derived LSEC progenitors [22]. Based on the previous research, either hepatocytes or LSECs could be candidate cells for the therapeutic intervention of HA; nevertheless, the transplantation of stem cells rather than already specialized cells seems to be the most promising approach due to the greater proliferative and differentiation potential of cell populations with stem/progenitor characteristics [23]. For Haemophilia A stem cell therapy, the efficacy of bone marrow-derived cell transplantation has already been demonstrated by Follenzi and collaborators [24]. Bone marrow has been considered for ages the main multipotent stem cell source for regenerative therapies, but its clinical use faces some practical difficulties, such as extraction procedures, which are highly invasive for the donor, and cell yield, which is highly variable and dependent on the age of the donor [25]. Thus, more accessible stem cell sources have recently been studied, such as peripheral blood [26–28] and adipose tissue [29,30]. In recent years, Adipose-derived Stromal Cells (ADSCs) have proven to have ideal characteristics for regenerative purposes [31]. In addition to decreased sampling risk for individual donors, easier method for isolation, and a higher number of isolated stromal cells, ADSCs are superior to BMSCs in some biological features, including the immunomodulatory properties and the secretion of a variety of growth factors and cytokines, as well as anti-apoptosis and anti-inflammation potential [32]. What is more, in vitro ADSC cultures are identified by an active proliferation rate, a multidifferentiative capacity for mesodermal, endodermal, and ectodermal lineages [33], and a lack of major histocompatibility class II (MHC II) [34]. This last feature appears to be fundamental for the treatment of genetic pathologies such as Haemophilia A since the patient cells are affected by the mutation. Using non-immunogenic stem cells, the isolation source can safely be allogenic without having to fall into the genetic modification of autologous cells.

This work investigated the in vitro conditioning of a commercial adipose-derived stromal cell line of human origin, called Adipose-derived Mesenchymal Stem Cells (Ad-

MSCs), to in vitro model cell responsiveness towards the endothelial microenvironment, which may resemble the sinusoidal endothelial niche. Evidence has already been reported in the literature about ADSC differentiative potential towards the endothelial lineage [35–37], with scant studies having also preliminarily demonstrated that undifferentiated and non-transfected human adipose stem cells express FVIII [38] at low but significant levels (0.03 μg/mL per $10^5$ cells in 48 h; ~30% of normal human plasma FVIII level) [20]. However, a systematic in vitro characterization of ADSC responsiveness towards the endothelial microenvironment in terms of FVIII expression is currently missing. Thus, for the first time in ADSC literature, we herein focused on their ability to express and release FVIII after specific endothelial stimulation and without any genetic manipulation in order to make them acquire LESC-like functionality.

## 2. Results

### 2.1. Characterization of Ad-MSC Cultures

First of all, the commercial Ad-MSC primary cell line was analyzed for its stemness features. A morphological characterization of Ad-MSC cells by optical microscopy highlighted their spindle-shaped morphology, which is typical of mesenchymal/stromal cells and was maintained over passages in sub-confluent culture conditions (Figure 1a,b). In parallel, the specific immunophenotype of Ad-MSCs was characterized by flow cytometry and defined to be $CD90^+/CD105^+/CD44^+/CD29^+/CD14^-/CD45^-$ (Figure 1c).

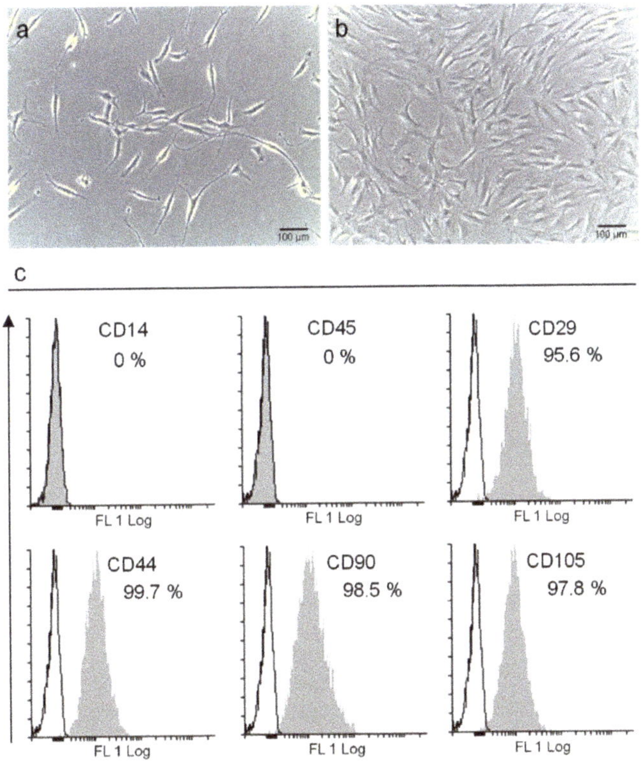


**Figure 1. Characterization of Ad-MSC cultures.** Morphological analysis by optical microscopy of Ad-MSCs at passage 3, in (**a**) sub-confluence and (**b**) confluence conditions (Scale bar: 100 μm). (**c**) Cytometric evaluation of Ad-MSC immunophenotype. Data are presented as percentage of positive cells (gray profile) in comparison with the isotype control (white profile).

*2.2. Cell Migration Capacity*

The capacity of Ad-MSC to respond to the endothelial microenvironment was evaluated by testing their migration across a porous filter under the stimulation of endothelial factors [namely, vascular endothelial growth factor (VEGF), human basic fibroblast growth factor (hFGF-b), epidermal growth factor (EGF), insulin-like growth factor-1 (IGF-1), heparin, and ascorbic acid], which were present in the upper or in the bottom side of the membrane. Figure 2 shows the images of the untreated cultures (Figure 2a) and the cells treated with the endothelial medium (Figure 2b,c). The nuclei count demonstrated that Ad-MSCs are responsive to the endothelial microenvironment, as evidenced by the significant ($p \leq 0.05$) increase in cells that crossed the filter with respect to the control cultures (Figure 2d). In particular, specific endothelial growth factors demonstrated the ability to both stimulate cell migration and exert a strong chemotactic effect on the treated cultures.

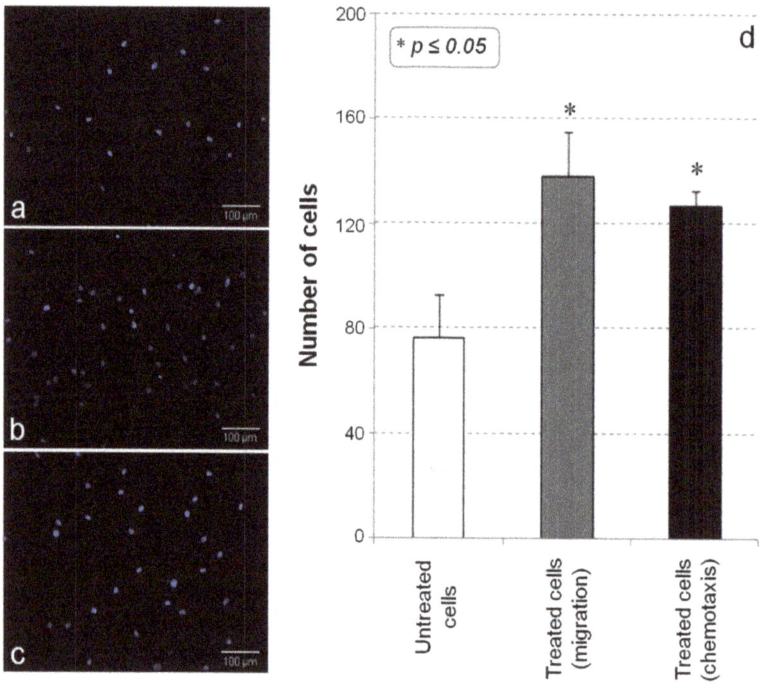


**Figure 2. Cell migration assay.** Optical microscopy of Ad-MSCs migrated through the porous filter after 24 h of incubation: (**a**) untreated cells vs. cultures treated with differentiation medium for (**b**) migration or (**c**) chemotaxis evaluation (Scale bar: 100 µm). (**d**) Effects on cell migration and chemotaxis of endothelial growth factors. The results are reported as mean of three different experiments ± standard deviation. (* $p < 0.05$).

*2.3. Endothelial Differentiation*

To confirm Ad-MSC responsiveness to an endothelial-like microenvironment in vitro, the cells underwent a specific differentiation treatment through a 7-, 14- and 21-day stimulation with endothelial growth medium containing specific inductive factors, (i.e., VEGF, hFGF-b, EGF and IGF-1). In parallel, Ad-MSCs grown in a proliferation medium were considered as the undifferentiated control.

This differentiation treatment was demonstrated to induce changes in Ad-MSC morphology during the first seven days of stimulation. When cultured in an endothelial medium, the cells stopped proliferating and acquired a more elongated morphology, and started to organize into capillary-like structures (Figure 3a).

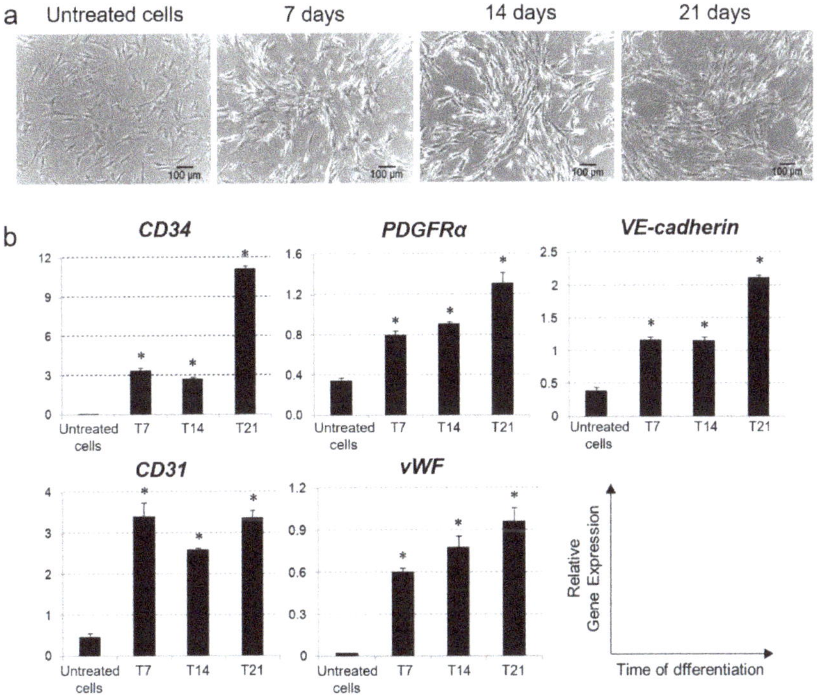


**Figure 3. Morphology and gene expression profile of differentiated Ad-MSCs.** (**a**) Phase-contrast micrographs of Ad-MSC population cultured in proliferative medium (untreated cells) or in endothelial differentiative medium for 7, 14 and 21 days (Scale bar: 100 µm). (**b**) Real Time PCR analysis of endothelial mRNAs expression on untreated cells or in Ad-MSCs cultured in induction medium for 7 (T7), 14 (T14) and 21 (T21) days. Relative gene expression was defined as an n-fold difference referred to the calibrator sample (HUVEC cultures); the housekeeping gene (GAPDH) expression was considered to validate the amplification efficiency of target genes. In the right bottom of the figure, the measured units on the x-axis and y-axis of other graphs are indicated. (* $p < 0.01$). **Abbreviations:** *PDGFRα, Platelet Derived Growth Factor Receptor Alpha; VE-cadherin, vascular endothelial cadherin; vWF, von Willebrand factor.*

Gene expression analysis revealed that the specific induction treatment significantly stimulated the transcription of genes associated with the endothelial lineage, i.e., *CD34*, *Platelet-Derived Growth Factor Receptor Alpha* (*PDGFRα*), *vascular endothelial (VE)-cadherin*, *CD31*, and *von Willebrand factor (vWF)*. This confirmed the high responsiveness of Ad-MSCs to the endothelial microenvironment in vitro (Figure 3b).

The flow cytometry study of the differentiated Ad-MSCs and untreated controls corroborated the gene expression results, demonstrating that the stimulation with specific growth factors enhanced the expression of vascular-endothelial proteins (Figure 4). In particular, Ad-MSC cultures were found to be negative for CD31, CD34, VEGFR2, and VE-cadherin when maintained in basal conditions (Figure 4a) but were able to respond to the endothelial stimulation by activating the expression of these specific lineage proteins (Figure 4b).

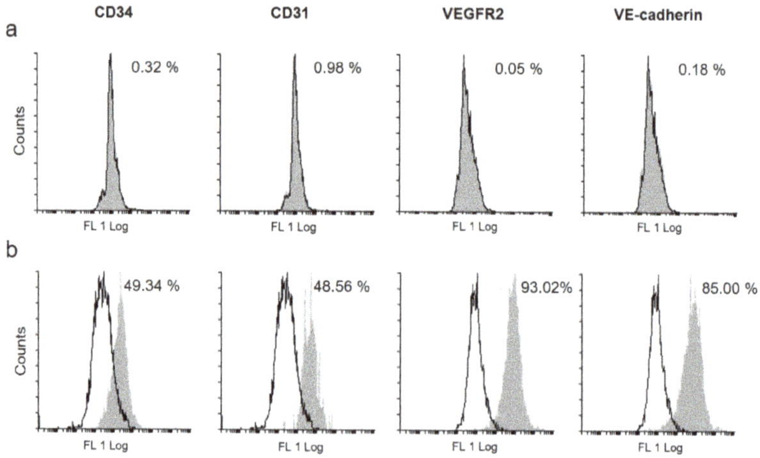


**Figure 4. Protein expression study.** Flow cytometry study of endothelial protein expression in (**a**) untreated cells or (**b**) Ad-MSCs cultured in induction medium for 21 days. Data are presented as percentage of positive cells (gray profile) in comparison with the isotype control (white profile). **Abbreviations:** *VEGFR2, Vascular Endothelial Growth Factor Receptor 2; VE-cadherin, vascular endothelial cadherin.*

### 2.4. FVIII Release

The final aim of Ad-MSC endothelial differentiation was to verify their capacity to increase the expression and release of FVIII in response to specific stimulation treatment. Both gene (Figure 5a) and protein (Figure 5b) expression studies highlighted that Ad-MSCs expressed the FVIII gene and protein in basal conditions. When cultured in an endothelial inductive medium, they significantly up-regulated the transcription of the *FVIII* gene ($p < 0.01$) in comparison with the untreated control, whereas no significant differences were detected among the treated samples. (Figure 5a). In addition, a sustained synthesis and release of the corresponding protein over time were demonstrated following dot blot analysis on cell supernatants (Figure 5b).

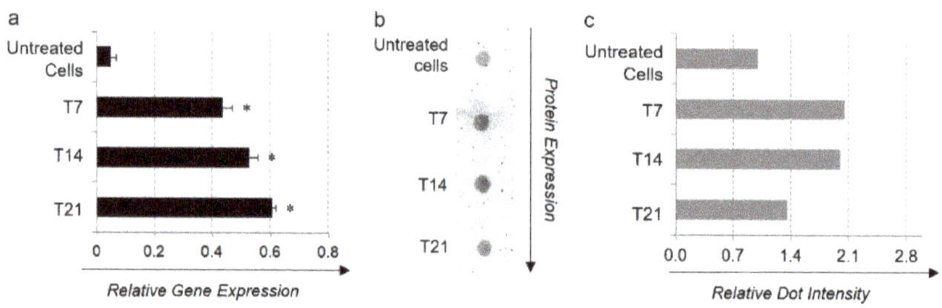


**Figure 5. FVIII expression.** (**a**) Real-Time PCR analysis of FVIII mRNA expression in untreated cells or in Ad-MSCs cultured in induction medium for 7 (T7), 14 (T14), and 21 (T21) days (*: $p < 0.01$ in comparison with the untreated cells). (**b**) Dot blot evaluation of FVIII release into the culture media of untreated cells and 7 (T7), 14 (T14), 21 (T21) day-differentiated Ad-MSCs. (**c**) Dot blot densitometry and relative intensity of protein dots calculated using Image J software. Results are normalized towards the undifferentiated sample.

### 2.5. Coagulation Assay

The activated partial thromboplastin time (aPTT) test was performed to investigate any effect of the FVIII released by untreated/differentiated Ad-MSCs on blood coagulation. Culture medium collected from undifferentiated cells (Test sample 1) or Ad-MSCs stimulated with endothelial medium for 7 (Test sample 2) and 14 days (Test sample 3) was mixed with FVIII deficient plasma, and the aPTT time was measured in seconds. The aPTT correction by this mixing suggested that the added test sample likely contained the missing coagulation factor (FVIII). In addition, the correction of the clotting time of the deficient plasma is proportional to the concentration (% activity) of FVIII in the sample (Figure 6b), interpolated from a calibration curve constructed using pooled human plasma [39]. The results of the coagulation test were also reported as the aPTT ratio (aPTT(r)) of patient-to-normal clotting time (Figure 6b), which is often used to improve the comparability of the results between different clinical laboratories [40].

a

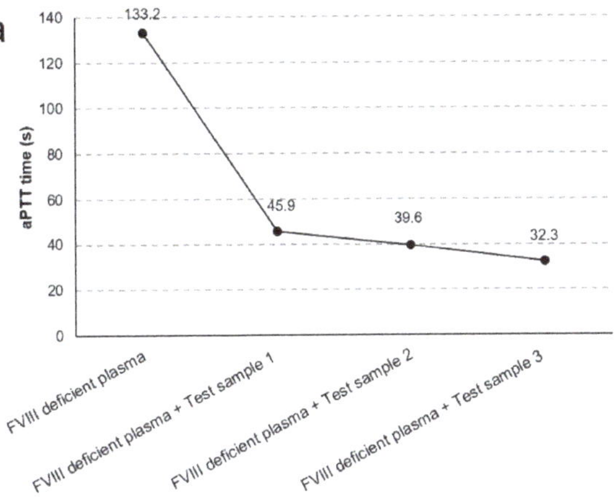


b

| Sample | aPTT ratio | FVIII activity (%) |
|---|---|---|
| FVIII deficient plasma | 4.47 | 0.2 |
| FVIII deficient plasma + Test sample 1 | 1.54 | 4.1 |
| FVIII deficient plasma + Test sample 2 | 1.33 | 6.9 |
| FVIII deficient plasma + Test sample 3 | 1.08 | 26.2 |

**Figure 6. Coagulation assay.** (**a**) aPTT reduction after mixing FVIII deficient plasma with culture medium collected from undifferentiated cells (Test sample 1) or Ad-MSCs stimulated with endothelial medium for 7 (Test sample 2), 14 (Test sample 3), and 21 days (Test sample 4). (**b**) Results of the aPTT assay reported as aPTT ratio and FVIII activity.

As shown in Figure 6, a reduction in aPTT time and aPTT(r), together with a progressive increase in FVIII activity, were registered in supplemented vs. normal deficient plasma.

## 3. Discussion

As a monogenic disease, Haemophilia A seems to be an ideal candidate to be treated with stem cell therapy. Since the genetic mutation affects the activity of coagulation factor VIII, advanced therapies may be based on the replacement of the deficient gene with the healthy gene that generates the functional protein (gene therapy), as well as on the incorporation of a full array of wild type genes and proteins through the transplantation of healthy cells, which may be able to restore the altered ones (cell therapy) [41]. Regarding cell therapy, both pre-clinical [42–44] and clinical [45,46] trials have proven to successfully correct Haemophilia A phenotype by means of lentiviral and adeno-associated vector-mediated expression of FVIII in adult stem cells, autologous fibroblasts, platelets, or hematopoietic stem cells. Significant breakthroughs regarding advanced therapies for Haemophilia A are related to pre-clinical trials in the field of stem cell transplantation [21,24,47,48]. Unlike gene transfection-based strategies, the administration of non-genetically modified cells ensures to overcome the risks of DNA mutation and immune response in the patient. On the other hand, the possibility of using adult stem cell populations assures more abundant, accessible, and rich cell sources than embryonic/induced pluripotent stem cells or tissue-specific cells. In particular, adipose tissue has been recognized as the ideal source for stem population extraction in terms of cell abundance and tissue accessibility. Despite having already been investigated for cell therapy applications [49], adipose-derived stem cells are still little considered for Haemophilia A treatment [20,38].

This work investigated the endothelial induction of an Ad-MSC primary cell line, with the final aim to demonstrate their capability to secrete FVIII after specific stimulation. To the best of our knowledge, this is the first report about adipose stem cell capacity to produce FVIII in vitro after endothelial stimulation and without resorting to gene transfection. Based on this preliminary in vitro study, Ad-MSCs will be tested for the transplantation into the liver circulation of HA rat models in order to correct the pathologic phenotype.

As already demonstrated in the literature, stem cells can respond to specific stimuli, such as tissue damage, by migrating from their niche and entering the bloodstream to reach the site where they are recruited for tissue regeneration (homing) [50]. In the case of cell therapy strategies, cell migration from the site of the implant to the damaged tissue is a fundamental step of the regenerative process. Herein, the response of Ad-MSC to endothelial growth factors was assessed, demonstrating increased migration/chemotactic activity when the cells were conditioned by an endothelial milieu. In particular, soluble factors which were used to stimulate Ad-MSCs are proangiogenic mediators which regulate the proliferation and migration of endothelial cells and promote angiogenesis [51,52]. Since these microenvironmental cues may be present in the vascular niche associated with the endothelial sinusoid compartment of the liver, it is tempting to hypothesize an instructive effect on Ad-MSC recruitment during local and systemic administration of the cell therapy.

The differentiative potential of adipose-derived MSCs toward the endothelial lineage has already been widely demonstrated by a variety of studies focused on promoting the angiogenesis of injured tissues by stem cell-based therapy to develop new endothelial cells [35–37]. On the contrary, very few investigations have thought to take advantage of this property to contribute to FVIII production and restore the endothelial cell component that is not functional in Haemophilia A patients [20,38]. In this study, the acquisition of endothelial-like characteristics by induced Ad-MSCs was demonstrated at the gene and protein levels. Both early and late endothelial markers were taken into consideration. In particular, *CD34* and *PDGFRα* are generally regarded as vascular endothelial progenitor markers [53,54], while *VE-cadherin*, *CD31*, and *vWF* are expressed by mature endothelial cells [55]. As evidenced by the qPCR results, in most cases, the expression of the target genes and differentiation period were directly proportional. Only *CD31* expression was up-regulated by endothelial induction and seemed to be maintained at the same level for all the stimulation period. In parallel, flow cytometry confirmed the acquired expression of CD34, CD31, and VE-cadherin in endothelial-induced Ad-MSCs, also highlighting the positivity of stimulated cultures to VEGFR2, which was indicated as a marker of functional endothelial precursors [56].

A further element pointing to stem cell therapy for Haemophilia A treatment is the fact that a modest increase in coagulation factor expression levels is enough to obtain a healthy phenotype [57]. In this in vitro study, Ad-MSCs were confirmed to express the FVIII gene at lower levels in the undifferentiated state, demonstrating an ability to increase gene expression along time in response to the specific stimuli provided by the endothelial microenvironment. In addition to FVIII mRNA expression, it was fundamental to demonstrate protein production and secretion within the cell culture medium. The paracrine activity of undifferentiated Ad-MSCs was largely documented in the literature, their secretome being enriched with a variety of growth factors [i.e., hepatocyte growth factor (HGF), granulocyte and macrophage colony-stimulating factors, tumor necrosis factor-α (TNF-α), VEGF, brain-derived neurotrophic factor (BDNF), nerve growth factor (NGF), interleukins (ILs) 6, 7, 8 and 11], as well as extracellular vesicles [58,59]. This secretome is responsible for the trophic effects exerted by Ad-MSCs on the protection, survival, and differentiation of several endogenous cells/tissues, in addition to their immunomodulatory properties [59]. On the other hand, no studies investigated the secretome of these cells after endothelial differentiation, but it is likely to speculate that this kind of stimulation may induce an increase in the secretion of molecules/factors related to angiogenesis processes. Regarding the release of coagulation factors for Hemophilia cell therapy, Ad-MSCs were induced to secrete factors VIII and IX by gene transfer approaches [20,33,60], whereas, to our knowledge, no evidence has been reported so far about the enhancement of coagulation factor release after endothelial differentiation of these cells. In this study, only FVIII release was assessed into differentiated cell culture media since the main focus was the possible contribution of Ad-MSCs to the correction of the coagulation factor deficiency in Hemophilia A patients. The protein expression study allowed for the detection of secreted FVIII in the culture media collected by basal and conditioned Ad-MSC cultures. Remarkably, the functionality of secreted FVIII was assessed by the aPTT test.

Together, the collected data pointed out the high responsiveness of Ad-MSCs to the endothelial microenvironment, offering preliminary directions to approach the restoration of the vascular compartment with transplanted Ad-MSCs capable of secreting FVIII. Further validation of cell homing and engraftment into the liver sinusoidal compartment will be performed by in vivo studies of Ad-MSC administration to animal models affected by Hemophilia A.

Lastly, substantial limitations of this in vitro modeling study need to be mentioned, being mainly represented by the failure to replicate the innate complexity of in vivo organ systems. For example, a multitude of soluble factors and signaling molecules mediate the effects of hepatic vascular niche on stem/progenitor cell recruitment [61]. This means that the heterogeneity of the in vivo microenvironment can hardly be reproduced ex vivo. Additionally, interactions between different cell types, cell-matrix interplay, or biochemical processes of cell turnover and metabolism can hardly be captured by in vitro models, which are still far from simulating three-dimensional microarchitecture and the compartmentation of organs [62]. Other drawbacks may regard difficulties in converting in vivo doses to in vitro concentrations, as well as problems derived from long-term in vitro culture. However, this study may be considered a valid starting point for the pre-clinical translation of Ad-MSC therapy into animal HA models.

## 4. Materials and Methods

### *4.1. In Vitro Culture of Human Adipose-Derived Mesenchymal Stem Cells (Ad-MSCs)*

The primary cell cultures of human Ad-MSCs were purchased from Tebu bio (Le Perray-en-Yvelines, France) and grown in complete Human Mesenchimal Stem Cell (hMSC) Expansion Media, according to the protocol recommended by the company. Briefly, the vial of cells was defrosted in a 37 °C water bath with constant, moderate agitation until the ice in the ampoule was no longer visible. Under a laminar flow hood, the vial content was transferred to a sterile 15 mL tube and centrifuged at 1500 rpm for 10 min (min) after the addition of the complete culture medium (10 mL). After discarding the supernatant, the pellet was re-suspended in 5 mL of hMSC Expansion Media, seeded into a T-25 (25 $cm^2$) culture flask, and incubated at 37 °C, 5% $CO_2$, and 90% humidity. The cells were ready to pass between 3 and 7 days after seeding and were sub-cultured at a density of 7000 cells/$cm^2$. Ad-MSCs were purchased at passage two and grown until passage four/five before preparing the following experiments for specific differentiation.

### *4.2. Morphological Evaluation by Optical Microscopy*

Ad-MSC cultures were observed daily by optical microscope DM/IL (Leica, Wetzlar, Germany) to estimate cell viability and investigate their specific morphology at different sub-confluence stages. Pictures were taken with a Nikon Digital Sight Ds-SMCc camera (Nikon Corporation, Tokyo, Japan).

### 4.3. Immunophenotype Characterization by Flow Cytometry

After expansive culture, the cells were harvested by a treatment with EDTA/tripsyn, centrifuged at 1500 rpm for 5 min and re-suspended in Phosphate-Buffered Saline (PBS) (Sigma Aldrich, Darmstadt, Germany) and 0.2% Bovine Serum Albumin (BSA) was added (Sigma-Aldrich). The cells were then labeled for 15 min at room temperature (RT) in the dark with 5 µL of the following primary antibodies: Fluorescein isothiocyanate (FITC)-conjugated anti-CD14, anti-CD45, and anti-CD105 (Santa Cruz Biotechnologies, Dallas, TX, USA); Phycoerythrin (PE)-conjugated anti-CD29, anti-CD44 and anti-CD90 (BioLegend, San Diego, CA, USA). The negative controls were prepared by labeling cells with 5 µL of FITC- and PE-conjugated isotypic antibodies (Santa Cruz and BioLegend) for 15 min at RT in the dark. The data for 10,000 total events were acquired by a FACS Canto cytometer (Becton Dickinson, Franklin Lakes, NJ, USA) and analyzed by Flowing Software 2 Turku Bioscience Centre, Turku, Finland); the results were expressed as a percentage of positive cells compared to the negative isotype control.

### 4.4. Endothelial Differentiation

For induction of endothelial differentiation, Ad-MSCs at passage four were seeded on 6-well plates at a density of 10,000 cells/cm$^2$. Upon reaching about 70–80% confluence, the cells were treated for specific endothelial induction. To this end, Ad-MSCs were grown for 7, 14, or 21 days (d) in Endothelial Growth Medium (EGM)-2 (Lonza, Basel, Switzerland) with SingleQuots containing vascular endothelial growth factor (VEGF), human basic fibroblast growth factor (hFGF-b), epidermal growth factor (EGF), insulin-like growth factor-1 (IGF-1), heparin, ascorbic acid, and 2% fetal bovine serum. In parallel, the undifferentiated control was prepared by treating cells with an expansion medium only.

### 4.5. Cell Migration Assay

The stimulating effect of the endothelial microenvironment on cell migration was evaluated by the Boyden chamber experiment. Ad-MSCs ($5 \times 10^4$ cells/cm$^2$) were seeded on the upper side of a 5.0 µm pore Transwell insert (Corning Inc., Corning, NY, USA) in differentiation medium EGM-2. Inserts were placed in a 24-well plate containing an expansion medium (lower chamber). In parallel, the chemotactic effect induced by the endothelial treatment was investigated by seeding cells in the expansion medium into the upper side of the insert and by adding EGM-2 into the lower chamber. As a control, the cells were maintained in an expansion medium in both the insert and the lower chamber. After 24 h (h), Ad-MSC cultures were fixed in 4% paraformaldehyde (PFA) (Sigma-Aldrich), and the upper membrane of the insert was swabbed to remove non-migrated cells. The membrane was cut from the insert and mounted with DAPI. Ad-MSC migration was quantified by fluorescence microscopy: the number of nuclei in the lower side of the membrane was counted in five random fields per insert (magnification $\times$ 100) by using Image J Software (U. S. National Institutes of Health, Bethesda, MD, USA).

### 4.6. Acquisition of Endothelial-like Characteristics

#### 4.6.1. Gene Expression Study by qPCR

The acquisition of endothelial-like characteristics was first investigated at the mRNA level, verifying the up-regulated expression of specific markers after cell differentiation. To this end, total RNA was extracted from both differentiated cells and undifferentiated control by using Trizol® Reagent (Sigma-Aldrich), according to the manufacturer's protocol. After quantification by the Eppendorf BioPhotometer Spectrophotometer (Eppendorf, Hamburg, Germany), the isolated RNA was analyzed for the expression of CD34, Platelet-derived Growth Factor Receptor alpha (PDGFR$\alpha$), Vascular Endothelial cadherin (VE-cadherin), CD31, von Willebrand Factor (vWF), and FVIII.

For relative quantification, the target gene expression for all experimental samples is expressed as an n-fold difference referred to as a base sample called the calibrator. In this study, Human Umbilical Vein Endothelial Cells (HUVECs) were considered as the calibrator sample. After stabilization in culture, total RNA from HUVECs was isolated and quantitated as previously described.

For qPCR, the total RNA (500 ng per 100 μL final reaction volume) from the experimental and calibrator samples was reverse-transcribed using random primers and Multi-Scribe Reverse Transcriptase (High-Capacity cDNA reverse transcription kit, Applied Biosystems, Foster City, CA, USA). The reaction mixture was incubated for 10 min at 25 °C and then at 37 °C for 120 min. cDNA was stored at −80 °C until use. The transcriptional levels of the above-mentioned genes were measured by means of quantitative real-time PCR (qPCR) using the relative quantification method ($2^{-\Delta\Delta Ct}$ method) [63]. The method was validated for our experimental system by verifying that the efficiencies of amplification of the targets and the glyceraldehyde 3-phosphate dehydrogenase (GAPDH) genes were similar. TaqMan® Gene Expression Assays specific for the above-mentioned endothelial markers were purchased from Applied Biosystems (Table 1).

**Table 1. Oligonucleotides for qPCR analysis.** TaqMan® Gene Expression Assays for relative gene expression analysis. **Abbreviations:** *PDGFRα, Platelet Derived Growth Factor Receptor Alpha; VE-cadherin, vascular endothelial cadherin; vWF, von Willebrand factor, FVIII, factor VIII.*

| Target Gene | TaqMan® Probe ID | Species | Amplicon Length (bp) |
|---|---|---|---|
| *CD34* | Hs02576480_m1 | *Human* | 63 |
| *PDGFRα* | Hs00998026_m1 | *Human* | 60 |
| *VE-cadherin* | Hs00170986_m1 | *Human* | 70 |
| *CD31* | Hs01065282_m1 | *Human* | 67 |
| *vWF* | Hs01109446_m1 | *Human* | 56 |
| *FVIII* | Hs00252034_m1 | *Human* | 127 |

To avoid amplifying contaminated genomic DNA, the primer pair was placed at the junction between two exons. The qPCR assay was performed using the ABI PRISM 7300 Sequence Detection system. The PCR reaction was run in a mixture (30 μL) containing 15 μL of 2× TaqMan Universal PCR Master Mix, 1.5 μL of 20× TaqMan Gene Expression assay (all reagents from Applied Biosystems), 12.5 μL of water, and 1 μL of cDNA template. Fifty cycles of amplification were performed at 95 °C (15 s) and 60 °C (1 min), and mRNA expression levels were normalized against quantified GAPDH expression for each sample.

#### 4.6.2. Protein Expression Study by Flow Cytometry

After qPCR analysis, the activation of specific endothelial markers by differentiated Ad-MSCs was also assessed by a protein expression study. The flow cytometry analysis was performed as already described in paragraph 2.3, labeling both 21-day-differentiated cells and undifferentiated cultures with the following primary antibodies: FITC-conjugated anti-CD34 and anti-CD31; PerCP-Cy™5.5 (PC 5.5)-conjugated anti-Vascular Endothelial Growth Factor Receptor (VEGFR) and anti-VE-cadherin (Cell Signaling Technology, Danvers, MA, USA).

### *4.7. FVIII Protein Study*

The capacity of Ad-MSC to release FVIII after endothelial stimulation was assessed by a dot blot assay. Culture media of 7, 14, and 21 days differentiated cells and undifferentiated control were collected and preserved at −80 °C until the time of analysis.

For Dot Blot evaluation, the samples (8 μL) were spotted onto an Immobilon® membrane (Merck-Millipore, Darmstadt, Germany) and allowed to stand for 90 min at RT. The membrane was then saturated for 2 h with PBS, containing 5.0% (*w*/*w*) skim milk, and challenged for 2 h with a sheep anti-human FVIII (Affinity Biologicals, Hamilton, ON, Canada) in PBS-skim milk. After washing (3 × 5 mL, for 10 min) with PBS, the membrane was incubated for 2 h with a peroxidase-conjugated donkey anti-sheep IgG (Affinity Biologicals) as the secondary antibody. The membrane was finally stained with the ECL® staining kit (GE-Healthcare, Buckinghamshire, United Kingdom) using a Molecular Imager VersaDoc MP4000 instrument (BioRad, Hercules, CA, USA). For semiquantitative analysis of the protein expression, dot blot densitometry and relative dot intensity were calculated using ImageJ software and normalizing data towards the untreated sample.

### *4.8. Coagulation Assay*

The presence of biologically active FVIII in Ad-MSC supernatants was determined by performing the aPTT test, which defines the partial thromboplastin time as the time it takes for a clot to form, measured in seconds.

The aPTT was determined routinely using a HemosIL aPTT SynthASil with synthetic phospholipids and silica as an activator on an Automated Coagulation Laboratory (ACL) analyzer (ACL TOP 700 and ACL TOP 500, Instrumentation Laboratory SpA, Milan, Italy). The range of normal aPTT ratios was 0.82–1.20.

The test is based on a comparison of the ability of the different test samples to correct the aPTT of a plasma totally deficient in the coagulation factor to be measured but containing normal amounts of the other factors (i.e., FVIII deficient plasma). Thus, aPTT was measured on human FVIII-deficient plasma or mixtures (1:1) of deficient plasma and cell culture medium collected from untreated Ad-MSCs (Test sample 1) and cells treated with endothelial induction factors for 7 (Test sample 2) and 14 days (Test sample 3). An anticoagulant (sodium citrate) was also added to each test sample (9:1). Results were presented as aPTT time (seconds), aPTT ratio, and FVIII activity (%).

### *4.9. Statistics*

All of the data are expressed as means of at least three different experiments $\pm$ Standard Deviation (SD). Statistical significance was determined by an unpaired *t*-test. The differences were considered to be significant with $p$ values of $\leq$0.05.

## 5. Conclusions

This preliminary in vitro study demonstrated that Ad-MSCs are conditioned by the endothelial microenvironment by enhancing their migration/chemotaxis response to specific soluble factors and upregulating specific endothelial markers at both the gene and protein level. For the first time, we demonstrated that these cells, when properly stimulated, are able to upregulate the gene and protein expression of the coagulation factor VIII without the need to resort to gene transfection strategies. This evidence makes Ad-MSCs promising candidates to be used in Haemophilia A cell therapy and lays the groundwork for the future pre-clinical investigation of Ad-MSCs for cell therapy approaches on haemophilic animal models.

**Author Contributions:** Conceptualization, S.B., A.P., P.R. and P.P.P.; Data curation, S.B., E.S., S.R. and L.Z.; Formal analysis, S.B. and S.R.; Funding acquisition, C.G., G.T. and P.P.P.; Investigation, S.B., E.S. and S.R.; Methodology, S.B., E.S., S.R. and L.Z.; Supervision, P.R., R.D.C. and P.P.P.; Writing—original draft, S.B. and P.R.; Writing—review and editing, V.M., C.G., G.T. and A.P. All authors have read and agreed to the published version of the manuscript.

**Funding:** This research was financially supported by liberal donations from T.E.S. Foundation, Padova, Italy and A.P.E. (Avis per il Progresso Ematologico) Association, Castelfranco Veneto (TV), Italy.

**Institutional Review Board Statement:** Not applicable.

**Informed Consent Statement:** Not applicable.

**Data Availability Statement:** Not applicable.

**Acknowledgments:** The authors would like to thank T.E.S. Foundation and A.P.E. (Avis per il Progresso Ematologico) Association for supporting the present research by donations in kind.

**Conflicts of Interest:** The authors declare no conflict of interest.

## References

1. Steward, C.G.; Jarisch, A. Haemopoietic Stem Cell Transplantation for Genetic Disorders. *Arch. Dis. Child.* **2005**, *90*, 1259–1263. [CrossRef]
2. Uzarski, J.S.; Xia, Y.; Belmonte, J.C.I.; Wertheim, J.A. New Strategies in Kidney Regeneration and Tissue Engineering. *Curr. Opin. Nephrol. Hypertens.* **2014**, *23*, 399–405. [CrossRef]
3. Cafuir, L.A.; Kempton, C.L. Current and Emerging Factor VIII Replacement Products for Hemophilia, A. *Ther. Adv. Hematol.* **2017**, *8*, 303–313. [CrossRef]
4. Stocco, E.; Barbon, S.; Radossi, P.; Rajendran, S.; Dalzoppo, D.; Bortolami, M.; Bagno, A.; Grandi, F.; Gamba, P.G.; Parnigotto, P.P.; et al. Autologous chondrocytes as a novel source for neo-chondrogenesis in haemophiliacs. *Cell Tissue Res.* **2016**, *366*, 51–61. [CrossRef]
5. Viiala, N.; Larsen, S.; Rasko, J. Gene Therapy for Hemophilia: Clinical Trials and Technical Tribulations. *Semin. Thromb. Hemost.* **2009**, *35*, 81–92. [CrossRef]
6. Butterfield, J.S.S.; Hege, K.M.; Herzog, R.W.; Kaczmarek, R. A Molecular Revolution in the Treatment of Hemophilia. *Mol. Ther.* **2020**, *28*, 997–1015. [CrossRef]
7. Lee, S.S.; Sivalingam, J.; Nirmal, A.J.; Ng, W.H.; Kee, I.; Song, I.C.; Kiong, C.Y.; Gales, K.A.; Chua, F.; Pena, E.M.; et al. Durable engraftment of genetically modified FVIII-secreting autologous bone marrow stromal cells in the intramedullary microenvironment. *J. Cell. Mol. Med.* **2018**, *22*, 3698–3702. [CrossRef]
8. Arruda, V.R. Why is AAV FVIII gene therapy not approved by the US Food and Drug Administration yet? *Blood Adv.* **2021**, *5*, 4313. [CrossRef]
9. Pasi, K.J.; Rangarajan, S.; Mitchell, N.; Lester, W.; Symington, E.; Madan, B.; Laffan, M.; Russell, C.B.; Li, M.; Pierce, G.F.; et al. Multiyear Follow-up of AAV5-hFVIII-SQ Gene Therapy for Hemophilia, A. *N. Engl. J. Med.* **2020**, *382*, 29–40. [CrossRef]
10. Rangarajan, S.; Walsh, L.; Lester, W.; Perry, D.; Madan, B.; Laffan, M.; Yu, H.; Vettermann, C.; Pierce, G.F.; Wong, W.Y.; et al. AAV5-Factor VIII Gene Transfer in Severe Hemophilia, A. *N. Engl. J. Med.* **2017**, *377*, 2519–2530. [CrossRef]
11. Ragni, M.V.; Humar, A.; Stock, P.G.; Blumberg, E.A.; Eghtesad, B.; Fung, J.J.; Stosor, V.; Nissen, N.; Wong, M.T.; Sherman, K.E.; et al. Hemophilia Liver Transplantation Observational Study. *Liver Transplant.* **2017**, *23*, 762–768. [CrossRef]
12. Shahani, T.; Covens, K.; Lavend'homme, R.; Jazouli, N.; Sokal, E.; Peerlinck, K.; Jacquemin, M. Human Liver Sinusoidal Endothelial Cells but Not Hepatocytes Contain Factor VIII. *J. Thromb. Haemost.* **2014**, *12*, 36–42. [CrossRef]
13. Doering, C.B.; Denning, G.; Shields, J.E.; Fine, E.J.; Parker, E.T.; Srivastava, A.; Lollar, P.; Spencer, H.T. Preclinical Development of a Hematopoietic Stem and Progenitor Cell Bioengineered Factor VIII Lentiviral Vector Gene Therapy for Hemophilia, A. *Hum. Gene Ther.* **2018**, *29*, 1183–1201. [CrossRef]
14. Merlin, S.; Famà, R.; Borroni, E.; Zanolini, D.; Bruscaggin, V.; Zucchelli, S.; Follenzi, A. FVIII expression by its native promoter sustains long-term correction avoiding immune response in hemophilic mice. *Blood Adv.* **2019**, *3*, 825–838. [CrossRef]
15. Russell, A.L.; Prince, C.; Lundgren, T.S.; Knight, K.A.; Denning, G.; Alexander, J.S.; Zoine, J.T.; Spencer, H.T.; Chandrakasan, S.; Doering, C.B. Non-genotoxic conditioning facilitates hematopoietic stem cell gene therapy for hemophilia A using bioengineered factor VIII. *Mol. Ther. Methods Clin. Dev.* **2021**, *21*, 710–727. [CrossRef]
16. Porada, C.D.; Sanada, C.; Kuo, C.J.; Colletti, E.; Mandeville, W.; Hasenau, J.; Zanjani, E.D.; Moot, R.; Doering, C.; Spencer, H.T.; et al. Phenotypic correction of hemophilia A in sheep by postnatal intraperitoneal transplantation of FVIII-expressing MSC. *Exp. Hematol.* **2011**, *39*, 1124–1135.e4. [CrossRef]
17. Ohmori, T.; Mizukami, H.; Katakai, Y.; Kawai, S.; Nakamura, H.; Inoue, M.; Shu, T.; Sugimoto, H.; Sakata, Y. Safety of intra-articular transplantation of lentivirally transduced mesenchymal stromal cells for haemophilic arthropathy in a non-human primate. *Int. J. Hematol.* **2018**, *108*, 239–245. [CrossRef]
18. Olgasi, C.; Borsotti, C.; Merlin, S.; Bergmann, T.; Bittorf, P.; Adewoye, A.B.; Wragg, N.; Patterson, K.; Calabria, A.; Benedicenti, F.; et al. Efficient and safe correction of hemophilia A by lentiviral vector-transduced BOECs in an implantable device. *Mol. Ther. Methods Clin. Dev.* **2021**, *23*, 551–566. [CrossRef]
19. El-Akabawy, N.; Rodriguez, M.; Ramamurthy, R.; Rabah, A.; Trevisan, B.; Morsi, A.; George, S.; Shields, J.; Meares, D.; Farland, A.; et al. Defining the Optimal FVIII Transgene for Placental Cell-Based Gene Therapy to Treat Hemophilia A. *Mol. Ther. Methods Clin. Dev.* **2020**, *17*, 465–477. [CrossRef]
20. Rosen, E.D.; Liang, Z.; Song, H.; Bischoff, D.; Kamocka, M.; Zollman, A. Expression of Coagulation FVIII and FIX in Human Adipose Stromal Cells (hASC) and hASC-Derived Cells—Potential Autologous Cell Therapy for Hemophilia. *Blood* **2007**, *110*, 767. [CrossRef]
21. Follenzi, A.; Benten, D.; Novikoff, P.; Faulkner, L.; Raut, S.; Gupta, S. Transplanted Endothelial Cells Repopulate the Liver Endothelium and Correct the Phenotype of Hemophilia A Mice. *J. Clin. Investig.* **2008**, *118*, 935–945. [CrossRef]
22. Gage, B.K.; Merlin, S.; Olgasi, C.; Follenzi, A.; Keller, G.M. Therapeutic correction of hemophilia A by transplantation of hPSC-derived liver sinusoidal endothelial cell progenitors. *Cell Rep.* **2022**, *39*, 110621. [CrossRef]

23. Kuci, S.; Kuci, Z.; Latifi-Pupovci, H.; Niethammer, D.; Handgretinger, R.; Schumm, M.; Bruchelt, G.; Bader, P.; Klingebiel, T. Adult Stem Cells as an Alternative Source of Multipotential (Pluripotential) Cells in Regenerative Medicine. *Curr. Stem Cell Res. Ther.* **2009**, *4*, 107–117. [CrossRef]
24. Follenzi, A.; Raut, S.; Merlin, S.; Sarkar, R.; Gupta, S. Role of Bone Marrow Transplantation for Correcting Hemophilia A in Mice. *Blood* **2012**, *119*, 5532–5542. [CrossRef]
25. Stenderup, K.; Justesen, J.; Clausen, C.; Kassem, M. Aging Is Associated with Decreased Maximal Life Span and Accelerated Senescence of Bone Marrow Stromal Cells. *Bone* **2003**, *33*, 919–926. [CrossRef]
26. Di Liddo, R.; Aguiari, P.; Barbon, S.; Bertalot, T.; Mandoli, A.; Tasso, A.; Schrenk, S.; Iop, L.; Gandaglia, A.; Parnigotto, P.P.; et al. Nanopatterned acellular valve conduits drive the commitment of blood-derived multipotent cells. *Int. J. Nanomed.* **2016**, *11*, 5041–5055. [CrossRef]
27. Barbon, S.; Stocco, E.; Grandi, F.; Rajendran, S.; Borean, A.; Pirola, I.; Capelli, S.; Bagno, A.; Tavano, R.; Contran, M.; et al. Biofabrication of a novel leukocyte-fibrin-platelet membrane as a cells and growth factors delivery platform for tissue engineering applications. *J. Tissue Eng. Regen. Med.* **2018**, *12*, 1891–1906. [CrossRef]
28. Barbon, S.; Rajendran, S.; Bertalot, T.; Piccione, M.; Gasparella, M.; Parnigotto, P.P.; Di Liddo, R.; Conconi, M.T. Growth and Differentiation of Circulating Stem Cells After Extensive Ex Vivo Expansion. *Tissue Eng. Regen. Med.* **2021**, *18*, 411–427. [CrossRef]
29. Zuk, P.A.; Zhu, M.; Mizuno, H.; Huang, J.; Futrell, J.W.; Katz, A.J.; Benhaim, P.; Lorenz, H.P.; Hedrick, M.H. Multilineage Cells from Human Adipose Tissue: Implications for Cell-Based Therapies. *Tissue Eng.* **2001**, *7*, 211–228. [CrossRef]
30. Zuk, P.A.; Zhu, M.; Ashjian, P.; De Ugarte, D.A.; Huang, J.I.; Mizuno, H.; Alfonso, Z.C.; Fraser, J.K.; Benhaim, P.; Hedrick, M.H. Human Adipose Tissue Is a Source of Multipotent Stem Cells. *Mol. Biol. Cell* **2002**, *13*, 4279–4295. [CrossRef]
31. Sheykhhasan, M.; Qomi, R.T.; Ghiasi, M. Fibrin Scaffolds Designing in Order to Human Adipose-Derived Mesenchymal Stem Cells Differentiation to Chondrocytes in the Presence of TGF-β3. *Int. J. Stem Cells* **2015**, *8*, 219–227. [CrossRef]
32. Zhao, Q.; Ren, H.; Han, Z. Mesenchymal Stem Cells: Immunomodulatory Capability and Clinical Potential in Immune Diseases. *J. Cell. Immunother.* **2016**, *2*, 3–20. [CrossRef]
33. Watanabe, N.; Ohashi, K.; Tatsumi, K.; Utoh, R.; Shim, I.K.; Kanegae, K.; Kashiwakura, Y.; Ohmori, T.; Sakata, Y.; Inoue, M.; et al. Genetically Modified Adipose Tissue-Derived Stem/Stromal Cells, Using Simian Immunodeficiency Virus-Based Lentiviral Vectors, in the Treatment of Hemophilia B. *Hum. Gene Ther.* **2013**, *24*, 283–294. [CrossRef]
34. Álvarez-Viejo, M.; Menéndez-Menéndez, Y.; Otero-Hernández, J. CD271 as a Marker to Identify Mesenchymal Stem Cells from Diverse Sources before Culture. *World J. Stem Cells* **2015**, *7*, 470. [CrossRef]
35. Cao, Y.; Sun, Z.; Liao, L.; Meng, Y.; Han, Q.; Zhao, R.C. Human Adipose Tissue-Derived Stem Cells Differentiate into Endothelial Cells in Vitro and Improve Postnatal Neovascularization in Vivo. *Biochem. Biophys. Res. Commun.* **2005**, *332*, 370–379. [CrossRef]
36. Fischer, L.J.; McIlhenny, S.; Tulenko, T.; Golesorkhi, N.; Zhang, P.; Larson, R.; Lombardi, J.; Shapiro, I.; DiMuzio, P.J. Endothelial Differentiation of Adipose-Derived Stem Cells: Effects of Endothelial Cell Growth Supplement and Shear Force. *J. Surg. Res.* **2009**, *152*, 157–166. [CrossRef]
37. Konno, M.; Hamazaki, T.S.; Fukuda, S.; Tokuhara, M.; Uchiyama, H.; Okazawa, H.; Okochi, H.; Asashima, M. Efficiently Differentiating Vascular Endothelial Cells from Adipose Tissue-Derived Mesenchymal Stem Cells in Serum-Free Culture. *Biochem. Biophys. Res. Commun.* **2010**, *400*, 461–465. [CrossRef]
38. Miguelino, M.; Sahar, D.; Powell, J. Abstract 134. Adipose Derived Stem Cells Express von Willebrand Factor and Factor VIII. *Plast. Reconstr. Surg.* **2014**, *133*, 150. [CrossRef]
39. Sanada, C.; Kuo, C.J.; Colletti, E.J.; Soland, M.; Mokhtari, S.; Knovich, M.A.; Owen, J.; Zanjani, E.D.; Poradax, C.D.; Almeida-Porada, G. Mesenchymal stem cells contribute to endogenous FVIII:c production. *J. Cell. Physiol.* **2013**, *228*, 1010–1016. [CrossRef]
40. Bronić, A.; Margetić, S.; Coen Herak, D.; Milić, M.; Krešić, B.; Radišić Biljak, V.; Leniček Krleža, J. Reporting of activated partial thromboplastin time (aPTT): Could we achieve better comparability of the results? *Biochem. Med.* **2021**, *31*, 020708. [CrossRef]
41. Sokal, E.M.; Lombard, C.; Mazza, G. Mesenchymal Stem Cell Treatment for Hemophilia: A Review of Current Knowledge. *J. Thromb. Haemost.* **2015**, *13*, S161–S166. [CrossRef] [PubMed]
42. Brown, B.D.; Cantore, A.; Annoni, A.; Sergi, L.S.; Lombardo, A.; della Valle, P.; D'Angelo, A.; Naldini, L. A MicroRNA-Regulated Lentiviral Vector Mediates Stable Correction of Hemophilia B Mice. *Blood* **2007**, *110*, 4144–4152. [CrossRef]
43. Jeon, H.J.; Oh, T.K.; Kim, O.H.; Kim, S.T. Delivery of Factor VIII Gene into Skeletal Muscle Cells Using Lentiviral Vector. *Yonsei Med. J.* **2010**, *51*, 52. [CrossRef]
44. Ramezani, A.; Zweier-Renn, L.; Hawley, R. Factor VIII Delivered by Haematopoietic Stem Cell-Derived B Cells Corrects the Phenotype of Haemophilia A Mice. *Thromb. Haemost.* **2011**, *105*, 676–687. [CrossRef]
45. Matsui, H. Endothelial Progenitor Cell-Based Therapy for Hemophilia A. *Int. J. Hematol.* **2012**, *95*, 119–124. [CrossRef]
46. Ward, N.J.; Buckley, S.M.K.; Waddington, S.N.; VandenDriessche, T.; Chuah, M.K.L.; Nathwani, A.C.; McIntosh, J.; Tuddenham, E.G.D.; Kinnon, C.; Thrasher, A.J.; et al. Codon Optimization of Human Factor VIII CDNAs Leads to High-Level Expression. *Blood* **2011**, *117*, 798–807. [CrossRef]
47. Xu, D.; Alipio, Z.; Fink, L.M.; Adcock, D.M.; Yang, J.; Ward, D.C.; Ma, Y. Phenotypic Correction of Murine Hemophilia A Using an IPS Cell-Based Therapy. *Proc. Natl. Acad. Sci. USA* **2009**, *106*, 808–813. [CrossRef]
48. Yadav, N.; Kanjirakkuzhiyil, S.; Kumar, S.; Jain, M.; Halder, A.; Saxena, R.; Mukhopadhyay, A. The Therapeutic Effect of Bone Marrow–Derived Liver Cells in the Phenotypic Correction of Murine Hemophilia A. *Blood* **2009**, *114*, 4552–4561. [CrossRef]

49. Gir, P.; Oni, G.; Brown, S.A.; Mojallal, A.; Rohrich, R.J. Human Adipose Stem Cells. *Plast. Reconstr. Surg.* **2012**, *129*, 1277–1290. [CrossRef]
50. Kolf, C.M.; Cho, E.; Tuan, R.S. Mesenchymal Stromal Cells: Biology of Adult Mesenchymal Stem Cells: Regulation of Niche, Self-Renewal and Differentiation. *Arthritis Res. Ther.* **2007**, *9*, 204. [CrossRef]
51. Lamalice, L.; Le Boeuf, F.; Huot, J. Endothelial Cell Migration During Angiogenesis. *Circ. Res.* **2007**, *100*, 782–794. [CrossRef] [PubMed]
52. Rema, R.B.; Rajendran, K.; Ragunathan, M. Angiogenic efficacy of Heparin on chick chorioallantoic membrane. *Vasc. Cell.* **2012**, *4*, 8. [CrossRef] [PubMed]
53. Sidney, L.E.; Branch, M.J.; Dunphy, S.E.; Dua, H.S.; Hopkinson, A. Concise Review: Evidence for CD34 as a Common Marker for Diverse Progenitors. *Stem Cells* **2014**, *32*, 1380–1389. [CrossRef]
54. Chong, J.J.H.; Reinecke, H.; Iwata, M.; Torok-Storb, B.; Stempien-Otero, A.; Murry, C.E. Progenitor Cells Identified by PDGFR-Alpha Expression in the Developing and Diseased Human Heart. *Stem Cells Dev.* **2013**, *22*, 1932–1943. [CrossRef]
55. Sen, S.; McDonald, S.P.; Coates, P.T.H.; Bonder, C.S. Endothelial Progenitor Cells: Novel Biomarker and Promising Cell Therapy for Cardiovascular Disease. *Clin. Sci.* **2011**, *120*, 263–283. [CrossRef] [PubMed]
56. Peichev, M.; Naiyer, A.J.; Pereira, D.; Zhu, Z.; Lane, W.J.; Williams, M.; Oz, M.C.; Hicklin, D.J.; Witte, L.; Moore, M.A.; et al. Expression of VEGFR-2 and AC133 by circulating human CD34(+) cells identifies a population of functional endothelial precursors. *Blood* **2000**, *95*, 952–958. [CrossRef]
57. Liras, A.; Garcia-Arranz, M.; Vega, L.; Garcia-Olmo, D.; Olmedillas López, S. Factor IX secretion in human adipose-derived stem cells by non-viral gene transfer. *Haemophilia* **2012**, *18*, A65.
58. Trzyna, A.; Banaś-Ząbczyk, A. Adipose-Derived Stem Cells Secretome and Its Potential Application in "Stem Cell-Free Therapy". *Biomolecules* **2021**, *11*, 878. [CrossRef]
59. Salgado, A.J.; Reis, R.L.; Sousa, N.J.; Gimble, J.M. Adipose tissue derived stem cells secretome: Soluble factors and their roles in regenerative medicine. *Curr. Stem Cell Res. Ther.* **2010**, *5*, 103–110. [CrossRef]
60. Lei, Z.; Hu, X.; Wu, Y.; Fu, L.; Lai, S.; Lin, J.; Li, X.; Lv, Y. The Role and Mechanism of the Vascular Endothelial Niche in Diseases: A Review. *Front. Physiol.* **2022**, *13*, 863265. [CrossRef]
61. Olmedillas López, S.; Garcia-Arranz, M.; Garcia-Olmo, D.; Liras, A. Preliminary study on non-viral transfection of F9 (factor IX) gene by nucleofection in human adipose-derived mesenchymal stem cells. *PeerJ.* **2016**, *4*, e1907. [CrossRef] [PubMed]
62. Ghallabx, A. In vitro test systems and their limitations. *EXCLI J.* **2013**, *12*, 1024–1026.
63. Livak, K.J.; Schmittgen, T.D. Analysis of Relative Gene Expression Data Using Real-Time Quantitative PCR and the $2^{-\Delta\Delta CT}$ Method. *Methods* **2001**, *25*, 402–408. [CrossRef] [PubMed]

International Journal of *Molecular Sciences*

MDPI

*Article*

# A New Perspective for Bone Tissue Engineering: Human Mesenchymal Stromal Cells Well-Survive Cryopreservation on β-TCP Scaffold and Show Increased Ability for Osteogenic Differentiation

**Liudmila Leppik [1,*], Anna Gempp [1], Zyrafete Kuçi [2], Selim Kuçi [2], Peter Bader [2], Halvard Bönig [3], Ingo Marzi [1] and Dirk Henrich [1]**

[1] Department of Trauma-, Hand- and Reconstructive Surgery, University Hospital Frankfurt, Goethe-University, 60590 Frankfurt am Main, Germany; gempp.anna@gmail.com (A.G.); Ingo.Marzi@kgu.de (I.M.); d.henrich@trauma.uni-frankfurt.de (D.H.)
[2] Department for Children and Adolescents, Division for Stem Cell Transplantation and Immunology, University Hospital Frankfurt, 60590 Frankfurt am Main, Germany; Zyrafete.Kuci@kgu.de (Z.K.); Selim.Kuci@kgu.de (S.K.); Peter.Bader@kgu.de (P.B.)
[3] Institute for Transfusion Medicine and Immunohematology, Goethe University, German Red Cross Blood Service BaWüHe, 60528 Frankfurt am Main, Germany; Halvard.Boenig@kgu.de
* Correspondence: Liudmila.Leppik@kgu.de

**Citation:** Leppik, L.; Gempp, A.; Kuçi, Z.; Kuçi, S.; Bader, P.; Bönig, H.; Marzi, I.; Henrich, D. A New Perspective for Bone Tissue Engineering: Human Mesenchymal Stromal Cells Well-Survive Cryopreservation on β-TCP Scaffold and Show Increased Ability for Osteogenic Differentiation. *Int. J. Mol. Sci.* **2022**, 23, 1425. https://doi.org/10.3390/ijms23031425

Academic Editor: Carlo Ventura

Received: 16 December 2021
Accepted: 25 January 2022
Published: 26 January 2022



**Abstract:** The clinical breakthrough of bone tissue engineering (BTE) depends on the ability to provide patients routinely with BTE products of consistent pharmacological quality. The bottleneck of this approach is the availability of stem cells. To avoid this, we suggest immobilization of random-donor-derived heterologous osteoinductive MSCs onto osteoconductive matrices. Such BTE products could then be frozen and, after thawing, could be released as ready-to-use products for permanent implantation during surgery. For this purpose, we developed a simple protocol for cryopreservation of BTE constructs and evaluated the effects of this procedure on human MSC (hMSCs) metabolic and osteogenic activity in vitro. Our findings show that hMSCs can be freeze-thawed on a β-TCP scaffold through a technically simple procedure. Treated cells sustained their metabolic activity and showed favorable osteogenic potential. Mechanistically, HIF1α and YBX1 genes were activated after freeze-thawing, and supposed to be linked to enhanced osteogenesis. However, the detailed mechanisms as to how the cryopreservation procedure beneficially affects the osteogenic potential of hMSCs remains to be evaluated. Additionally, we demonstrated that our BTE products could be stored for 3 days on dry ice; this could facilitate the supply chain management of cryopreserved BTE constructs from the site of manufacture to the operating room.

**Keywords:** human MSCs; cryopreservation; osteogenic differentiation; scaffold; 3D culture

## 1. Introduction

Large non-healing bone defects, largely irrespective their etiology, constitute a major challenge for patients and physicians [1–3]. Treatment of choice is with autologous bone grafts [4]; however, several other treatments are also used [5–7]. These methods are not always effective, but even where they are, the morbidity associated with autologous tissue harvesting is considerable, fueling the search for alternatives.

Bone tissue engineering (BTE) is one such alternative approach, which holds great promise for promoting bone healing and regeneration while overcoming some of the drawbacks of current techniques. BTE approaches typically combine bone forming stem or progenitor cells with scaffolds, which restore missing bone volume and signaling molecules, which control cell–cell and cell–scaffold interactions in the bone defect. Among many different materials used to construct scaffolds, the biodegradable ceramic scaffolds with desirable pore size, porosity, and mechanical properties are probably the most preferable in BTE, since these materials generally show better tissue responses compared to polymers and metals [8]. Tricalcium phosphate (TCP) ceramics in contrast to other ceramics (for example hydroxyapatite) found wide application in the clinic due to their osteoconductivity and bioresorbability [9]. Clinical BTE approaches have demonstrated encouraging early outcomes [10]. However, in these types of approaches availability of stem cells is a bottleneck of the whole procedure. Isolation, characterization, and expanding of autologous mesenchymal stem cells (MSCs) or use of cryopreserved dissociated stem cell populations from cryobanks [11] need significant investment of time and resources that limits translation of BTE approaches into clinic. To be able to provide BTE products of consistent medicinal quality, conducive to routine clinical use, we proposed to immobilize random-donor-derived osteoinductive MSCs onto osteoconductive matrices. These BTE products were then frozen and released ready-to-use for permanent implantation during surgery after thawing.

The work builds on earlier work which has definitively established that cryopreserved MSCs maintain potential for proliferation and osteogenic differentiation in vitro [12–15] and in vivo [16,17]. The methods used, however, were not conducive to clinical application. A recent pilot study of Hernandez–Tapia showed that osteoblasts have good survival rate after cryopreservation on β-TCP scaffold [18]. The equally recent study of Mutsenko et al. [19] showed that MSCs could be frozen in 3D collagen-hydroxyapatite scaffold without significant loss of cell viability. Although these studies use technically challenging methodology, they represent invaluable first steps towards the creation of a tissue-engineered constructs biobanks.

In the present study, we developed a simple protocol for cryopreservation of BTE constructs (human MSCs seeded on 3D β-TCP scaffold) and evaluated the effects this procedure has on human MSCs metabolic and osteogenic activity in vitro.

## 2. Results

### 2.1. *Effect of Cryopreservation on hMSCs Metabolic Activity*

First, we analyzed how cryopreservation affects the metabolic activity of cells frozen on scaffold granules following common freezing protocol and whether the cells could recover their activity with time. Our results show that MSCs retained approximately 40% of their metabolic activity at first two days after cell thawing as compared to control (not frozen) cells (Figure 1A), and at later time points cells increased their activity to the level of control cells. We also compared recovery of cells frozen in this way with the cells frozen by mean of "air-dry" freezing protocol. No significant difference in metabolic activity of these two groups of cells post-thaw was detected for all time points (Figure 1B). Based on this the "air-dry" protocol was chosen for the following experiments.

In addition, we analyzed if "air-dry" 3D cryopreserved hMSCs could be stored at −20 °C, in dry ice or in a combination of both. We compared the metabolic activity of cryopreserved cells stored in (1) liquid nitrogen, (2) liquid nitrogen and −20 °C, (3) liquid nitrogen and dry ice (3 days), and (4) liquid nitrogen, dry ice (3 days) and −20 °C (1 day) after thawing. Our results show that cells did not survive storage at −20 °C (Figure 1D); however, storage in dry ice for 3 days did not negatively affect cell metabolic activity after thawing. The combination of both storage conditions, dry ice, and −20 °C had a significant negative effect on cell metabolic activity at 2 and 7 days after thawing (Figure 1D).

To compare the cell distribution on a scaffold before and after freezing, live cells were visualized with Calcein AM and DAPI staining. As can be seen from Figure 2, there were less cells present on a granule surface and more cells present inside pores in the samples 1-h post-thaw (Figure 2C,D) as compared to control samples without freezing.

### 2.2. *Cryopreserved hMSCs Show Stronger Osteogenic Activity Than Control Cells*

In order to analyze if cryopreservation has effect on osteogenic potential of hMCS post-thawing and therefore on future use of these cells in BTE approaches, cells after cryopreservation were thawed and cultured for two weeks in osteogenic-supplemented medium

(Figure 1C). There was no difference in the level of alkaline phosphatase (ALP) expression among control and cryopreserved cells at day 7 of culture in osteogenic-supplemented medium. In both type of cells expression of ALP was significantly higher than in cells cultured in growth medium (Figure 1E).

The results of osteogenic gene expression analysis are presented in Figure 3. Surprisingly, the expression of all tested genes was enhanced in the experimental group as compared to control (not frozen) cells. Expression of early markers of osteogenesis RunX2, Col1, and ALPL in the experimental group peaked on day 7 of osteogenic culture, whereas in the control group it did not peak until day 14. Col1 expression was significantly higher ($p < 0.05$) in experimental group cells as compared to control cells at day 7 of osteogenic culture. On the contrary, its expression was significantly ($p < 0.05$) higher in control group cells at day 14 of culture. Expression of the SPP1 gene was low in both groups of cells on day 7, but was significantly increased on day 14 in the experimental group as compared to control group cells ($p < 0.01$). None of osteogenic marker genes was expressed in cells cultured in growth medium except for SPP1, which showed strong up regulation on day 14 of culture (Supplementary Figure S1). In the control-DMSO group, in which cells were treated with freezing medium but were not frozen, expression of osteogenic marker genes was similar as in control cells except for Col1. This gene was highly upregulated in these cells at day 7 of osteogenic culture (Supplementary Figure S1).

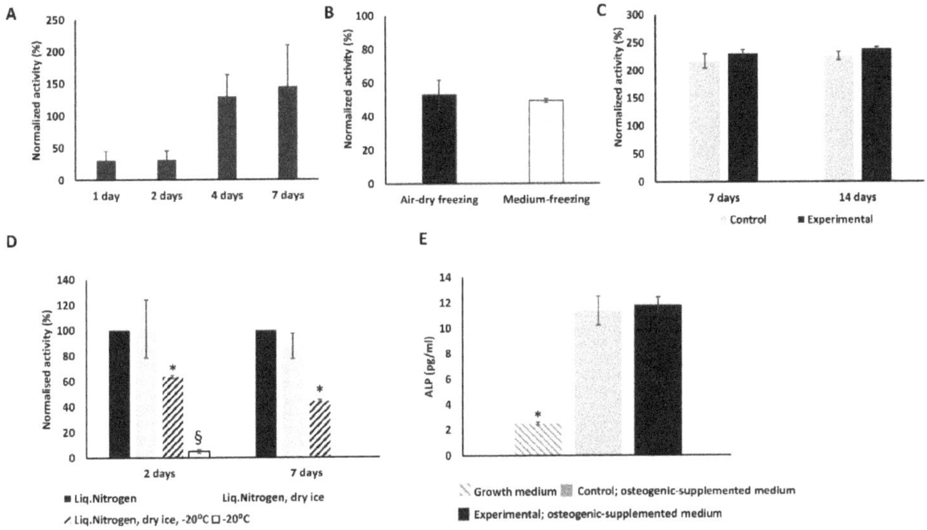


**Figure 1.** hMSCs preseeded on scaffolds retain their metabolic activity after cryopreservation. (**A**) Metabolic activity of MSCs was measured 1, 2, 4, and 7 days after cryopreservation, shown in % of control cells metabolic activity. MSCs were preseeded on scaffolds 48 h prior to cryopreservation. (**B**) There was no significant difference in metabolic activity of hMSCs processed with two different cryopreservation protocols and measured 2 days after cryopreservation. (**C**) Metabolic activity of control and experimental cells at 7 and 14 days of culture in osteogenic supplemented medium showed no significant differences. (**D**) hMSCs were first cryopreserved in liquid nitrogen and either stored for 3 days on dry ice or stored for 3 days on dry ice and 1 day at −20 °C. The metabolic activity of cells was measured at days 2 and 7 post-thaw and is shown in % of metabolic activity of cells stored in liquid nitrogen only. (**E**) ALP expression is significantly enhanced in both, control and experimental group cells at day 7 of culture in osteogenic-supplemented medium, as compared to cells cultured in growth medium. (* $p < 0.05$, § $p < 0.01$).

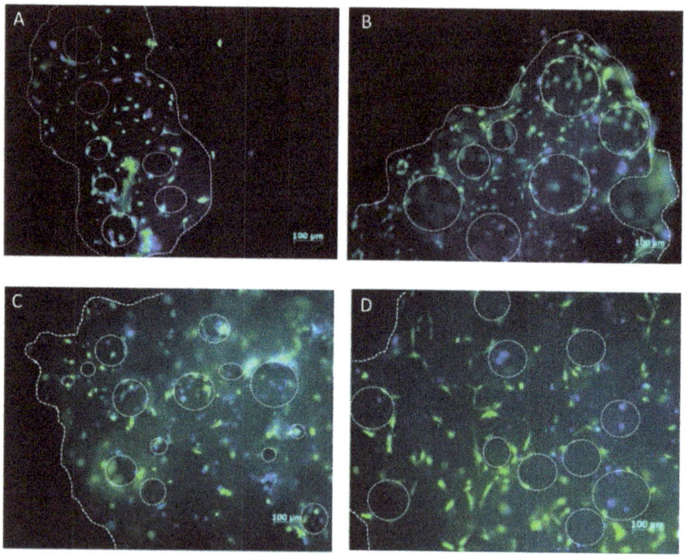


**Figure 2.** hMSCs distribution on a β-TCP scaffold granule before and post-thawing. Control hMSCs were seeded on a β-TCP scaffold and cultured either for 24 (**A**) or 72 h (**B**). Experimental hMSCs were seeded on a β-TCP scaffold, cultured for 24 h, frozen in liquid nitrogen for seven days, thawed, and cultured for 1 (**C**) and 48 (**D**) hours. All cells are stained with Calcein AM and DAPI. Dotted line shows the edges and pores of β-TCP granule. (5× magnification; scale bar = 100 μM).

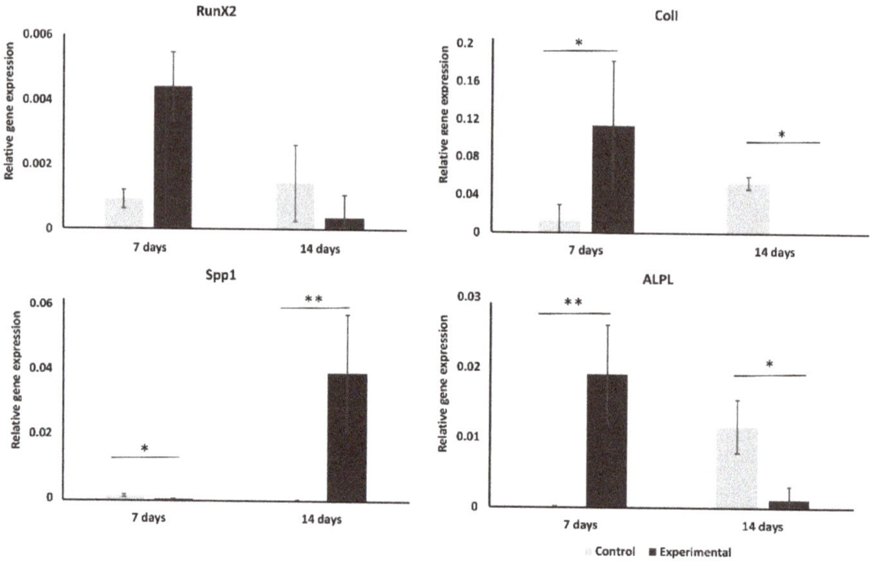


**Figure 3.** Osteogenic marker genes mRNAs expression in control and experimental cells. Expression of osteogenic marker genes was evaluated in experimental and control group hMSCs at days 7 and 14 of culture in osteogenic-supplemented medium. (* $p < 0.05$; ** $p < 0.01$).

### 2.3. Cryopreservation Has an Effect on Expression of Hypoxia-Related Genes

Hypothesizing that the positive effect of cryopreservation on hMSCs osteogenic differentiation could be promoted by the difference of oxygen concentration developed due to temperature difference, we compared expression of hypoxia-related genes in experimental and control cells (Figure 4). The results showed that expression of the major hypoxia marker gene HIF1α and several of its target genes was affected by cryopreservation (Figure 4 and Supplementary Figure S2). Expression of HIF1α in the experimental group cells was significantly ($p < 0.05$) higher than in the control cells one day post-thaw. On day seven, expression of HIF1α and its target genes PDK1, SLC2A1, EGLN1, and BNIP3 was higher in cells which had undergone the cryopreservation procedure compared to control cells, albeit not significantly. Expression of VEGF mRNA was significantly lower on days 1 and 7, and significantly higher on day 14 of culture in the experimental group cells as compared to controls (Figure 4A).

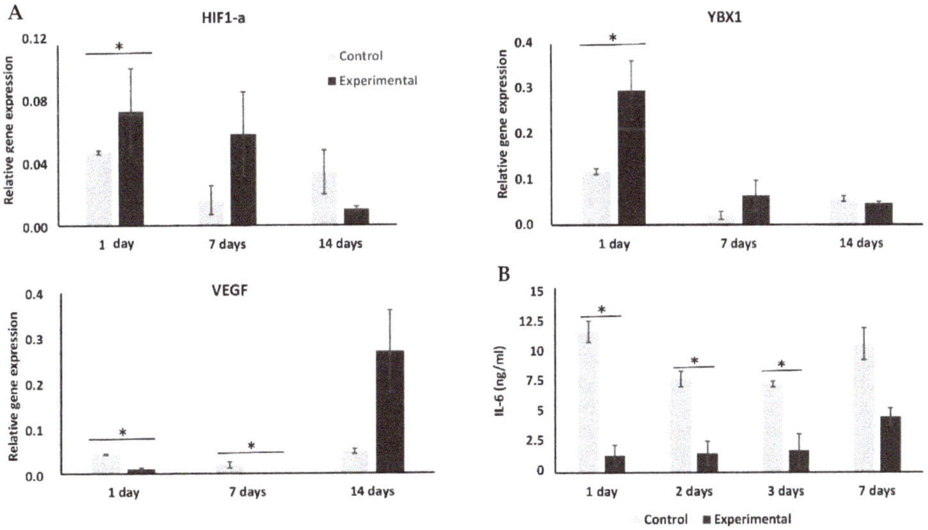


**Figure 4.** mRNA levels of gene expression of HIF1-α, YBX1, and VEGF and IL-6 protein secretion at days 1, 7, and 14 of culture. (**A**) Expression of HIF1α, YBX1, and VEGF genes was evaluated in experimental and control group hMSCs at days 1, 7, and 14 of culture in osteogenic-supplemented medium. (**B**) Concentration of IL-6 was measured in medium supernatants of both groups at days 1, 2, 3, and 7 of culture. (* $p < 0.05$).

### 2.4. Cryopreservation Enhances YBX1 Expression but Has No Effect on Wnt/Smad Pathways

We analyzed if the freezing procedure has an effect on expression of the cold shock protein YBX1, known to react to low temperature and to be an important regulator of transcription [20]. Expression of this gene was significantly ($p < 0.05$) enhanced as early as 1 day after thawing in the experimental group and remained high until day 7 of culture. On day 14 of culture, expression of this gene was the same in both experimental and control group cells (Figure 4).

In order to investigate if the freezing procedure could have effect on key signaling (Wnt and SMAD) pathways in hMSCs and thereby influence hMSCs osteogenic potential, we analyzed expression of Wnt3a, SMAD5, MAPK8, and MAPK14 genes early after the freezing and thawing event (1 day) as well as at later time points (7 and 14 days). No significant difference in expression of these genes among experimental and control group cells was found at any time point (Supplementary Figure S3).

### 2.5. Effect of Cryopreservation on Interleukin Expression

We evaluated if cryopreservation has an effect on hMSC's IL-10 and IL-6 expression 1–7 days into osteogenic culture, as the mediators could be important for future use of these cells in BTE approaches (Figure 4B). In both control and experimental cells, expression of IL-10 was below detection level of our method (data not shown). Concentration of IL-6 was lower in the experimental group on days 1, 2, and 3 as compared to control-group medium supernatant. On day 7 of osteogenic culture, the difference in expression of IL-6 between two groups of cells lost statistical significance.

## 3. Discussion

The clinical breakthrough of bone tissue engineering depends on the ability routinely to provide patients with BTE products of consistent pharmacological quality. For this reason, the ability to batch-produce and cryogenically preserve tissue engineered construct represents major progress. Cryopreservation is the use of very low temperature to maintain living cells and tissues in a quiescent status for a long period, without losing their structure and function [21]. Several works have been devoted to the cryopreservation of engineered biological constructs [18,19,22–24], which could guide our experimentation on the cryopreservation of biofabricated BTE constructs.

In the present study we evaluated whether hMSCs retain their activity after cryopreservation on β-TCP scaffolds. Our results showed that despite the blunted metabolic activity at early time points after thaw and presumably representing a modicum of cell attrition, hMSCs quantitatively regain their metabolic activity later. This temporal loss of metabolic activity could be explained by the loss of cells from the scaffold granule surface, as was shown by Calcein and DAPI staining of cells one-hour post-thaw (Figure 2C, Supplementary Figure S4). These findings are in accordance with other studies, showing that *Callithrix jacchus* MSCs and cells of human osteoblastic cell lines sustain their activity after being frozen on hydroxyapatite or β-TCP scaffolds, respectively [18,19]. In contrast to both these studies, we employed the simplest cell culture and cryopreservation procedures in order to develop protocols which can be later easily translated to clinical application. Therefore, we used a commonly used β-TCP scaffold, MSCs generated from pooled bone marrow mononuclear cells of eight third-party donors and an "air-dry" cryopreservation protocol, which involved the use of standard cell culture equipment and reagents. We also evaluated whether 3D-cryopreserved hMSCs could survive storage at −20 °C, dry ice, or a combination of both. Our results clearly demonstrate that cells seeded on β-TCP scaffold and stepwise cryopreserved could not be stored at −20 °C as that dramatically effects metabolic activity of cells post-thaw. However, cryopreserved BTE constructs could be stored at least for 3 days on dry ice without an apparent impact on post-thaw cell activity. Cryopreserved hMSCs-coated matrix also survived the storage on dry ice followed by 1 day at −20 °C, albeit with significant loss of metabolic activity post-thaw. This information will guide the supply chain management of cryopreserved BTE constructs from the site of manufacture to the operating room.

Osteogenic potential is arguably one of the most important characteristics of biofabricated BTE constructs; therefore, we evaluated how cryopreservation on 3D scaffold affects hMSCs osteogenic potential. Our results showed that osteogenic potential of frozen cells was rather improved. ALP activity of control and experimental cells was the same at day 7 of osteogenic culture (Figure 3). According to the osteogenic marker gene expression analysis, hMSCs osteogenic differentiation was accelerated after freezing/thawing procedure, as most of the analyzed markers were expressed earlier in experimental cells than in control cells. It has been previously shown that MSC retain osteogenic differentiation ability after cryopreservation [25,26], but for hMCSs cultured and cryopreserved on β-TCP scaffold this was not shown before. The effect of increased osteogenic potential was not observed in our "DMSO-control" cells, demonstrating that increased osteogenic gene expression is rather induced by physical effects and not by components of the cryopreservation medium (e.g., DMSO).

We hypothesized that increased osteogenic gene expression could either be a result of unspecific up-regulation of major pathways due to the cryopreservation procedure and/or reaction to the oxygen gradient induced by temperature changes [27,28]. To validate this hypothesis, we analyzed expression of key signaling stem cell fate marker genes and expression of hypoxia-related genes. Results of this analysis showed that expression of Wnt3A, MAPK8, MAPK14, and SMAD5 was not significantly altered in cryopreserved cells (Supplementary Figure S1), whereas the expression pattern of hypoxia related genes differed between control and experimental cells. Expression of HIF1α gene and its target genes PDK1, SLC2A1, EGLN1, and BNIP3 was upregulated in experimental group cells on day 7, whereas in control cells those genes were only upregulated on day 14 of culture. It was previously shown that hypoxia promotes osteogenesis of human MSCs in a HIF-1-dependent manner [29]. We speculate that the cryopreservation procedure affects HIF1α expression, and this mechanism stimulates osteogenesis. In addition, VEGF, a known target of HIF1α, was also significantly upregulated in experimental group cells (Figure 4). A previous study had shown that RunX2 acts together with HIF1α to stimulate angiogenic gene expression in bone cells [30]. Our results are in accordance with it.

We also verified if transcription of cold shock protein YBX1, known to regulate expression of HIF1α [31], was activated after cryopreservation (Figure 4). YBX1 protein possess a cold shock domain and stimulates translation of proteins under low temperature [20,32]. YBX1 protein was shown to be a critical regulator of HIF1α expression in sarcoma cells [33]; however, less is known about the role of this protein in MSCs. Our results show that cryopreservation affects YBX1 gene expression in hMSCs, which would be compatible with a role in the response of the MSCs to cryopreservation and enhancement of their osteogenic potential.

## 4. Materials and Methods

### 4.1. hMSCs Culture and Seeding on β-TCP Granules

Human bone marrow MSC cells (hMSCs) were provided by Department for Children and Adolescents, Division for Stem Cell Transplantation and Immunology, University Hospital Frankfurt. The MSCs were generated from pooled, previously isolated, and cryopreserved mononuclear cells from eight random bone marrow donors by plastic adherence, which have been expanded to near-confluence and cryopreserved in small aliquots as described in detail elsewhere [34]. It has already been demonstrated that these MSCs were effective in the treatment of patients with acute steroid-resistant acute GvHD [35,36]. For subcultures, 3.5 × 105 hMSCs were seeded in T175 cell culture flasks (Sarstedt, Nümbrecht, Germany) and cultured in animal-free growth medium (10% CruxRufa Media Supplement; (TrinovaBiochem, Giessen, Germany); 1 IU/mL of heparin (Ratiopharm, Ulm, Germany); GlutaMax Gibco™ DMEM, (Thermofischer, Dreieich, Germany)) at 37 °C, 5% $CO_2$ in a humidified incubator until 80–90% confluence. The culture medium was changed every three to four days and the cells were expanded until passage 5. These MSCs were used either for experiments or stored in liquid nitrogen for future use. Before cryopreservation, hMSC was assessed for expression of typical MSC cell surface markers by using flow cytometry analysis. Phenotypical analysis showed that these cells express high levels of surface proteins CD90 and CD105, but very low or no expression of hematopoietic cell markers CD45 and CD34 (data not shown [34,37]).

For 3D hMSC culture, 500 µL of β-TCP scaffold granules (ChronOS Granules, 1.4–2.8 mm, 60% porosity; Synthes, Oberdorf, Switzerland) were placed in 6.5-mm Transwell ® membrane inserts (3.0 µm pore, polycarbonate, Corning, Wiesbaden, Germany) allocated in individual wells of a 24-well plate. Then 2 × $10^5$ hMSCs in 150 µL of PBS (Gibco) were dripped slowly on the scaffolds and incubated for 10 min at 37 °C. After incubation, the cell suspension not absorbed by the scaffold was removed, dripped again over the material, and incubated for another 10 min. This step was repeated twice. Afterwards, 1 mL of pre-warmed growth medium was added to each well and cells were further incubated at 37 °C, 5% $CO_2$ in a humidified incubator for 24 h.

### 4.2. Effect of Cryopreservation on Cell Activity

In order to evaluate the effect of cryopreservation on cell metabolic activity, osteogenic differentiation, and interleukin expression, the scaffold granules with seeded cells were frozen, thawed, and cultured for two weeks in either growth or osteogenic-supplemented medium (experimental group). In the control group, cells were treated the same way as in experimental group except for freeze-thawing.

### 4.3. Cryopreservation and Thawing of hMSCs on β-TCP-Scaffold

For "standard protocol" cryopreservation, scaffold granules seeded with cells and cultured for 24 h were transferred into a cryovial, and entire culture medium was exchanged with 1 mL of cold fresh-prepared freezing medium (10% $v/v$ DMSO, 25% human AB serum (Sigma–Aldrich, Heidelberg, Germany) in DMEM Glutamax). The cryovial with cells was stored at −80 °C for 24 h in a controlled rate freezing container (Nalgene ® Mr. Frosty, Merck, Darmstadt, Germany) and then transferred into a liquid nitrogen for at least 7 days' storage. For thawing, a cryovial with hMSCs-seeded β-TCP scaffold granules was removed from the nitrogen tank, briefly warmed in a water bath (37 °C) for 1 min. The cryovial content was transferred into one well of a 24-well-plate and the entire freezing medium was aspirated. Cells were washed with 1 mL of fresh, pre-warmed growth medium, and cultured in 1 mL of growth medium (at 37 °C; 5% $CO_2$) for 48 h at which time the cell metabolic activity was assayed.

For the "air-dry protocol" cryopreservation [19], scaffold granules seeded with cells were transferred into a cryovial and the entire culture medium was exchanged with 1 mL of cold fresh-prepared freezing medium. After 15 min incubation on ice, medium was soaked out and cryovial with cell-coated scaffold (without medium) was first frozen at 80 °C in a controlled rate freezing container for 24 h, and subsequently stored in liquid nitrogen for at least 7 days. For thawing, a cryovial with hMSCs-seeded β-TCP scaffold granules was removed from the nitrogen tank, briefly warmed in a water bath (37 °C) for 1 min. One mL of pre-warmed growth medium was quickly added and the complete cryovial content was transferred into one well of a 24-well-plate. Cells were incubated at 37 °C, 5% $CO_2$ for 48 h before cell metabolic activity was assayed.

In order to control for possible effects of DMSO in freezing medium on cell activity after thaw, cells in "DMSO control" group were treated the same way as described in "air-dry protocol", but the freezing step (−80 °C, 24 h, and liquid nitrogen storage) was omitted.

### 4.4. Effect of Storage Temperature

To analyze if cryostorage temperature has an effect on the functionalized β-TCP scaffold, cryovials with 3D cryopreserved cells were first stored in liquid nitrogen (7 days) and then either directly thawed and cultured or stored for 1 day at: −20 °C (freezer); for 3 days in dry ice (Styrofoam container filled with dry ice = −78.5 °C); or for 3 days in dry ice and 1 day at 20 °C. Thereafter, cells were thawed as described and cultured for 48 h or 7 days before metabolic activity of cells was measured.

### 4.5. Cell Metabolic Activity Measurements

Metabolic activity of cells cultured on scaffold granules was measured by means of alamarBlue assay (Bio-Rad, Germany) according to the manufacturer's protocol 48 h after thaw in experimental or 72 h after seeding in control groups, respectively. Briefly, growth medium was soaked out and 400 μL of fresh growth medium and 40 μL of alamarBlue reagent were added to the cell-seeded scaffold granules. After 4 h of incubation at 37 °C, absorbance of conditioned medium was measured at 570 and 600 nm by means of Tecan Plate reader (Tecan, Crailsheim, Germany) and the percentage reduction of alamarBlue was calculated according to the manufacturer's instructions. Three samples were analyzed for each group and the mean value and standard deviation were calculated.

### 4.6. hMSCs Distribution on Scaffold Granules

To visualize hMSCs, seeded on β-TCP scaffold, before and after freezing-thawing procedure, Calcein AM (BD Pharmingen, Heidelberg, Germany) and DAPI (Life Technologies, Darmstadt, Germany) staining were applied. Scaffold granules with cells were transferred into a well of 24-well plate and 1 mL of pre-warmed growth medium with 20 μM Calcein AM was added. Cells were incubated for 40 min at 37 °C; washed several times with PBS; and DAPI (1 μg/mL in PBS) was added. After 10 min of incubation, cells were washed several times with PBS. hMSCs adherence and distribution on a scaffold was then assessed by directly viewing the samples with fluorescence microscopy using a Zeiss Axioobserver Z1 (Zeiss, Göttingen, Germany).

### 4.7. Osteogenic Differentiation

In order to evaluate the effect of cryopreservation on hMSCs osteogenic differentiation, cells after thawing were cultured for 24 h in growth medium and then growth medium was supplemented with 10-7 M of dexamethasone, 10 mM of β-glycerophosphate, and 0.05 mM of ascorbic acid-2-phosphate (Osteogenic medium, OM), all obtained from Sigma-Aldrich (Heidelberg, Germany). Cells from the control group were cultured in growth medium for 48 h after seeding, whereupon medium was changed to osteogenic-supplemented medium. Cells from both groups were cultured for another two weeks in osteogenic-supplemented medium and medium was changed every 3–4 days.

### 4.8. Gene Expression Analysis

For gene expression analysis, total RNA from cells was isolated using RNeasy Mini Kit (Qiagen, Hilden, Germany) following the manufacturer's instructions. Purity and quantity of RNA were measured using an Infinite 200PRO NanoQuant device (Tecan, München, Germany). DNase-treated RNA samples were reverse-transcribed using iScript Select cDNA Synthesis Kit (Bio-Rad, Feldkirchen, Germany) according to the manufacturer's instructions. The RT-qPCR reaction was performed using the cDNA equivalent of 10 ng RNA and the RT2 SYBR Green qPCR Mastermix (Qiagen, Hilden, Germany). All samples were amplified in duplicates using a CFX96 Touch Real-Time PCR Detection System (Bio-Rad, Feldkirchen, Germany) with human gene specific primers (RT2 qPCR Primer Assays, Qiagen, Hilden, Germany), and thermal profile of 1 cycle with 10 min of 95 °C, 40 cycles with 15 s of 95 °C, and 1 min of 60 °C followed by dissociation curve. GAPDH was used as reference gene in each experiment. A melting curve analysis was applied to ensure the specificity of the PCR procedure. Relative quantification of messenger RNA (mRNA) levels of the target genes was analyzed using the comparative CT (threshold cycle values) method ($2^{-\Delta Ct}$) [38]. The results are presented as relative quantification (RQ), which is expression fold change compared to the reference gene. Three samples were analyzed for each group and mean value and standard deviation were calculated for further analysis.

### 4.9. Alkaline Phosphatase Expression Assay

As a marker of osteogenesis, ALP expression was assessed on day 7 of osteogenic culture according to the manufacturer's protocol (Sensolyte pNPP Alkaline Phosphatase Detection Kit, Anaspec Inc., Köln, Germany). Granules with seeded cells were washed twice with 1x assay buffer and cells were lysed with lysis buffer (0.02% Triton X-100 in 1× assay buffer). Cell lysates were collected into fresh micro tubes and processed for ALP activity measurements. Absorbance was measured at 405 nm by means of Infinite 200PRO NanoQuant plate reader (Tecan, Crailsheim, Germany). The absolute ALP value for each sample was calculated against an alkaline phosphatase standard curve.

### 4.10. ELISA

The effect of cryopreservation on interleukin (IL)-6 and IL-10 expression in hMSCs was evaluated using ELISA. The concentration of interleukins was measured in cell conditioned medium collected at days 1, 2, 3, and 7 post-thawing by means of human IL-6 DuoSet ®

ELISA and human IL-10 DuoSet ® ELISA (R&D Systems, Wiesbaden-Nordenstadt, Germany) in accordance with the manufacturer's instructions. The absorbance was measured at 450 nm and cytokine concentrations were calculated based on calibration curve plotted with different standards concentrations by means of Magellan Software.

### 4.11. Statistical Analysis

All experiments were done in triplicates. The data are presented as mean ± SD and significance level was set at $p < 0.05$. Nonparametric Kruskal–Wallis test followed by Bonferroni-corrected multiple Conover–Iman post hoc analysis was consequently applied. A $p < 0.05$ indicated statistical significance. Statistics were calculated using the software Bias 11.12 (Epsilon-Verlag, Darmstadt, Germany).

## 5. Conclusions

Our findings show that hMSCs can be freeze-thawed on β-TCP scaffold through a technically simple procedure and thus provide a critical step towards the development of clinical-grade BTEs for functional bone augmentation. Cells treated in this fashion sustained their metabolic activity and showed favorable osteogenic potential after cryopreservation. The detailed mechanisms as to how the cryopreservation procedure beneficially affects the osteogenic potential of hMSCs, specifically the roles of YBX1 and HIF1α genes, should be investigated in future studies.

**Supplementary Materials:** The following are available online at https://www.mdpi.com/article/10.3390/ijms23031425/s1.

**Author Contributions:** Conceptualization, P.B., H.B. and D.H.; Data curation, L.L. and A.G.; Formal analysis, L.L. and A.G.; Funding acquisition, D.H.; Investigation, L.L. and A.G.; Methodology, L.L., Z.K., S.K., H.B., I.M. and D.H.; Resources, Z.K. and S.K.; Supervision, P.B., H.B., I.M. and D.H.; Visualization, L.L. and A.G.; Writing—original draft, L.L.; Writing—review and editing, L.L., A.G., Z.K., S.K., P.B., H.B., I.M. and D.H. All authors have read and agreed to the published version of the manuscript.

**Funding:** This research was funded by the Federal Republic of Germany/Federal Ministry for Economic Affairs and Energy on the basis of a resolution of the German Bundestag (03THW10L27).

**Institutional Review Board Statement:** Not applicable.

**Informed Consent Statement:** Not applicable.

**Data Availability Statement:** Not applicable.

**Acknowledgments:** We would like to acknowledge Kathrin Rottluff and Alexander Schaible for technical support.

**Conflicts of Interest:** The authors declare no conflict of interest.

## References

1. Giannoudis, P.V.; Atkins, R. Management of long-bone non-unions. *Injury* **2007**, *38* (Suppl. 2), S1–S2. [CrossRef]
2. Ashman, O.; Phillips, A.M. Treatment of non-unions with bone defects: Which option and why? *Injury* **2013**, *44*, S43–S45. [CrossRef]
3. Kinaci, A.; Neuhaus, V.; Ring, D.C. Trends in bone graft use in the United States. *Orthopedics* **2014**, *37*, e783–e788. [CrossRef] [PubMed]
4. Pape, H.C.; Evans, A.; Kobbe, P. Autologous bone graft: Properties and techniques. *J. Orthop. Trauma* **2010**, *24* (Suppl. 1), S36–S40. [CrossRef] [PubMed]
5. Rigal, S.; Merloz, P.; Le Nen, D.; Mathevon, H.; Masquelet, A.-C. Bone transport techniques in posttraumatic bone defects. *Orthop. Traumatol. Surg. Res.* **2012**, *98*, 103–108. [CrossRef]
6. Malizos, K.N.; Zalavras, C.G.; Soucacos, P.N.; Beris, A.E.; Urbaniak, J.R. Free vascularized fibular grafts for reconstruction of skeletal defects. *J. Am. Acad. Orthop. Surg.* **2004**, *12*, 360–369. [CrossRef]
7. Roberts, T.T.; Rosenbaum, A.J. Bone grafts, bone substitutes and orthobiologics: The bridge between basic science and clinical advancements in fracture healing. *Organogenesis* **2012**, *8*, 114–124. [CrossRef]
8. Hench, L.L. Bioceramics. *J. Am. Ceram. Soc.* **1998**, *81*, 1705–1728. [CrossRef]

9. Seebach, C.; Schultheiss, J.; Wilhelm, K.; Frank, J.; Henrich, D. Comparison of six bone-graft substitutes regarding to cell seeding efficiency, metabolism and growth behaviour of human mesenchymal stem cells (MSC) in vitro. *Injury* **2010**, *41*, 731–738. [CrossRef]
10. Seebach, C.; Henrich, D.; Meier, S.; Nau, C.; Bonig, H.; Marzi, I. Safety and feasibility of cell-based therapy of autologous bone marrow-derived mononuclear cells in plate-stabilized proximal humeral fractures in humans. *J. Transl. Med.* **2016**, *14*, 314. [CrossRef]
11. Yong, K.W.; Choi, J.R.; Wan Safwani, W.K.Z. Biobanking of Human Mesenchymal Stem Cells: Future Strategy to Facilitate Clinical Applications. *Adv. Exp. Med. Biol.* **2016**, *951*, 99–110. [CrossRef] [PubMed]
12. Ginis, I.; Grinblat, B.; Shirvan, M.H. Evaluation of bone marrow-derived mesenchymal stem cells after cryopreservation and hypothermic storage in clinically safe medium. *Tissue Eng. Part C Methods* **2012**, *18*, 453–463. [CrossRef] [PubMed]
13. Yalvaç, M.E.; Ramazanoglu, M.; Tekguc, M.; Bayrak, O.F.; Shafigullina, A.K.; Salafutdinov, I.I.; Blatt, N.L.; Kiyasov, A.P.; Sahin, F.; Palotás, A.; et al. Human tooth germ stem cells preserve neuro-protective effects after long-term cryo-preservation. *Curr. Neurovasc. Res.* **2010**, *7*, 49–58. [CrossRef] [PubMed]
14. Woods, E.J.; Perry, B.C.; Hockema, J.J.; Larson, L.; Zhou, D.; Goebel, W.S. Optimized cryopreservation method for human dental pulp-derived stem cells and their tissues of origin for banking and clinical use. *Cryobiology* **2009**, *59*, 150–157. [CrossRef]
15. Gonda, K.; Shigeura, T.; Sato, T.; Matsumoto, D.; Suga, H.; Inoue, K.; Aoi, N.; Kato, H.; Sato, K.; Murase, S.; et al. Preserved proliferative capacity and multipotency of human adipose-derived stem cells after long-term cryopreservation. *Plast. Reconstr. Surg.* **2008**, *121*, 401–410. [CrossRef] [PubMed]
16. Li, H.; Yan, F.; Lei, L.; Li, Y.; Xiao, Y. Application of autologous cryopreserved bone marrow mesenchymal stem cells for periodontal regeneration in dogs. *Cells Tissues Organs* **2009**, *190*, 94–101. [CrossRef] [PubMed]
17. Liu, G.; Shu, C.; Cui, L.; Liu, W.; Cao, Y. Tissue-engineered bone formation with cryopreserved human bone marrow mesenchymal stem cells. *Cryobiology* **2008**, *56*, 209–215. [CrossRef]
18. Hernández-Tapia, L.G.; Fohlerová, Z.; Žídek, J.; Alvarez-Perez, M.A.; Čelko, L.; Kaiser, J.; Montufar, E.B. Effects of Cryopreservation on Cell Metabolic Activity and Function of Biofabricated Structures Laden with Osteoblasts. *Materials* **2020**, *13*, 1966. [CrossRef]
19. Mutsenko, V.; Knaack, S.; Lauterboeck, L.; Tarusin, D.; Sydykov, B.; Cabiscol, R.; Ivnev, D.; Belikan, J.; Beck, A.; Dipresa, D.; et al. Effect of 'in air' freezing on post-thaw recovery of Callithrix jacchus mesenchymal stromal cells and properties of 3D collagen-hydroxyapatite scaffolds. *Cryobiology* **2020**, *92*, 215–230. [CrossRef]
20. Lindquist, J.A.; Mertens, P.R. Cold shock proteins: From cellular mechanisms to pathophysiology and disease. *Cell Commun. Signal.* **2018**, *16*, 63. [CrossRef]
21. Hunt, C.J. Cryopreservation of Human Stem Cells for Clinical Application: A Review. *Transfus. Med. Hemother.* **2011**, *38*, 107–123. [CrossRef] [PubMed]
22. Pogozhykh, O.; Prokopyuk, V.; Prokopyuk, O.; Kuleshova, L.; Goltsev, A.; Figueiredo, C.; Pogozhykh, D. Towards biobanking technologies for natural and bioengineered multicellular placental constructs. *Biomaterials* **2018**, *185*, 39–50. [CrossRef] [PubMed]
23. Malpique, R.; Osório, L.M.; Ferreira, D.S.; Ehrhart, F.; Brito, C.; Zimmermann, H.; Alves, P.M. Alginate encapsulation as a novel strategy for the cryopreservation of neurospheres. *Tissue Eng. Part C Methods* **2010**, *16*, 965–977. [CrossRef] [PubMed]
24. Kuleshova, L.L.; Gouk, S.S.; Hutmacher, D.W. Vitrification as a prospect for cryopreservation of tissue-engineered constructs. *Biomaterials* **2007**, *28*, 1585–1596. [CrossRef]
25. Antebi, B.; Asher, A.M.; Rodriguez, L.A.; Moore, R.K.; Mohammadipoor, A.; Cancio, L.C. Cryopreserved mesenchymal stem cells regain functional potency following a 24-h acclimation period. *J. Transl. Med.* **2019**, *17*, 297. [CrossRef] [PubMed]
26. Bahsoun, S.; Coopman, K.; Akam, E.C. The impact of cryopreservation on bone marrow-derived mesenchymal stem cells: A systematic review. *J. Transl. Med.* **2019**, *17*, 397. [CrossRef]
27. Han, P.; Bartels, D.M. Temperature Dependence of Oxygen Diffusion in $H_2O$ and $D_2O$. *J. Phys. Chem.* **1996**, *100*, 5597–5602. [CrossRef]
28. Place, T.L.; Domann, F.E.; Case, A.J. Limitations of oxygen delivery to cells in culture: An underappreciated problem in basic and translational research. *Free Radic. Biol. Med.* **2017**, *113*, 311–322. [CrossRef]
29. Wagegg, M.; Gaber, T.; Lohanatha, F.L.; Hahne, M.; Strehl, C.; Fangradt, M.; Tran, C.L.; Schönbeck, K.; Hoff, P.; Ode, A.; et al. Hypoxia promotes osteogenesis but suppresses adipogenesis of human mesenchymal stromal cells in a hypoxia-inducible factor-1 dependent manner. *PLoS ONE* **2012**, *7*, e46483. [CrossRef]
30. Kwon, T.-G.; Zhao, X.; Yang, Q.; Li, Y.; Ge, C.; Zhao, G.; Franceschi, R.T. Physical and functional interactions between Runx2 and HIF-1$\alpha$ induce vascular endothelial growth factor gene expression. *J. Cell. Biochem.* **2011**, *112*, 3582–3593. [CrossRef]
31. Rauen, T.; Frye, B.C.; Wang, J.; Raffetseder, U.; Alidousty, C.; En-Nia, A.; Floege, J.; Mertens, P.R. Cold shock protein YB-1 is involved in hypoxia-dependent gene transcription. *Biochem. Biophys. Res. Commun.* **2016**, *478*, 982–987. [CrossRef] [PubMed]
32. Kretov, D.A.; Clément, M.-J.; Lambert, G.; Durand, D.; Lyabin, D.N.; Bollot, G.; Bauvais, C.; Samsonova, A.; Budkina, K.; Maroun, R.C.; et al. YB-1, an abundant core mRNA-binding protein, has the capacity to form an RNA nucleoprotein filament: A structural analysis. *Nucleic Acids Res.* **2019**, *47*, 3127–3141. [CrossRef] [PubMed]
33. El-Naggar, A.M.; Veinotte, C.J.; Cheng, H.; Grunewald, T.G.P.; Negri, G.L.; Somasekharan, S.P.; Corkery, D.P.; Tirode, F.; Mathers, J.; Khan, D.; et al. Translational Activation of HIF1$\alpha$ by YB-1 Promotes Sarcoma Metastasis. *Cancer Cell* **2015**, *27*, 682–697. [CrossRef]

34. Kuçi, Z.; Bönig, H.; Kreyenberg, H.; Bunos, M.; Jauch, A.; Janssen, J.W.G.; Škifić, M.; Michel, K.; Eising, B.; Lucchini, G.; et al. Mesenchymal stromal cells from pooled mononuclear cells of multiple bone marrow donors as rescue therapy in pediatric severe steroid-refractory graft-versus-host disease: A multicenter survey. *Haematologica* **2016**, *101*, 985–994. [CrossRef]
35. Bader, P.; Kuçi, Z.; Bakhtiar, S.; Basu, O.; Bug, G.; Dennis, M.; Greil, J.; Barta, A.; Kállay, K.M.; Lang, P.; et al. Effective treatment of steroid and therapy-refractory acute graft-versus-host disease with a novel mesenchymal stromal cell product (MSC-FFM). *Bone Marrow Transplant.* **2018**, *53*, 852–862. [CrossRef] [PubMed]
36. Bonig, H.; Kuçi, Z.; Kuçi, S.; Bakhtiar, S.; Basu, O.; Bug, G.; Dennis, M.; Greil, J.; Barta, A.; Kállay, K.M.; et al. Children and Adults with Refractory Acute Graft-versus-Host Disease Respond to Treatment with the Mesenchymal Stromal Cell Preparation "MSC-FFM"-Outcome Report of 92 Patients. *Cells* **2019**, *8*, 1577. [CrossRef] [PubMed]
37. Kuçi, Z.; Seiberth, J.; Latifi-Pupovci, H.; Wehner, S.; Stein, S.; Grez, M.; Bönig, H.; Köhl, U.; Klingebiel, T.; Bader, P.; et al. Clonal analysis of multipotent stromal cells derived from $CD271^+$ bone marrow mononuclear cells: Functional heterogeneity and different mechanisms of allosuppression. *Haematologica* **2013**, *98*, 1609–1616. [CrossRef] [PubMed]
38. Livak, K.J.; Schmittgen, T.D. Analysis of relative gene expression data using real-time quantitative PCR and the 2(-Delta Delta C(T)) Method. *Methods* **2001**, *25*, 402–408. [CrossRef]

International Journal of
*Molecular Sciences*

MDPI

*Communication*

# Lung Extracellular Matrix Hydrogels Enhance Preservation of Type II Phenotype in Primary Alveolar Epithelial Cells

**Esther Marhuenda [1,2,†], Alvaro Villarino [1,†], Maria Leonor Narciso [1,3], Marta Camprubí-Rimblas [2,4], Ramon Farré [1,2,5], Núria Gavara [1,2,3], Antonio Artigas [2,4], Isaac Almendros [1,2,5] and Jorge Otero [1,2,3,*]**

1 Unitat de Biofísica i Bioenginyeria, Facultat de Medicina i Ciències de la Salut, Universitat de Barcelona, 08036 Barcelona, Spain; marhuenda.esther@gmail.com (E.M.); alvaro.villarino.romero@gmail.com (A.V.); mnarciso@ibecbarcelona.eu (M.L.N.); rfarre@ub.edu (R.F.); ngavara@ub.edu (N.G.); ialmendros@ub.edu (I.A.)
2 CIBER de Enfermedades Respiratorias, 28029 Madrid, Spain; mcamprubi@tauli.cat (M.C.-R.); aartigas@tauli.cat (A.A.)
3 The Institute for Bioengineering of Catalonia (IBEC), The Barcelona Institute of Science and Technology (BIST), 08028 Barcelona, Spain
4 Corporació Sanitària Universitària Parc Tauli, I3PT, Universitat Autònoma de Barcelona, 08193 Sabadell, Spain
5 Institut d'Investigacions Biomèdiques August Pi i Sunyer, 08036 Barcelona, Spain
* Correspondence: jorge.otero@ub.edu
† These authors contributed equally to this work.

**Abstract:** One of the main limitations of in vitro studies on lung diseases is the difficulty of maintaining the type II phenotype of alveolar epithelial cells in culture. This fact has previously been related to the translocation of the mechanosensing Yes-associated protein (YAP) to the nuclei and Rho signaling pathway. In this work, we aimed to culture and subculture primary alveolar type II cells on extracellular matrix lung-derived hydrogels to assess their suitability for phenotype maintenance. Cells cultured on lung hydrogels formed monolayers and maintained type II phenotype for a longer time as compared with those conventionally cultured. Interestingly, cells successfully grew when they were subsequently cultured on a dish. Moreover, cells cultured on a plate showed the active form of the YAP protein and the formation of stress fibers and focal adhesions. The results of chemically inhibiting the Rho pathway strongly suggest that this is one of the mechanisms by which the hydrogel promotes type II phenotype maintenance. These results regarding protein expression strongly suggest that the chemical and biophysical properties of the hydrogel have a considerable impact on the transition from ATII to ATI phenotypes. In conclusion, culturing primary alveolar epithelial cells on lung ECM-derived hydrogels may facilitate the prolonged culturing of these cells, and thus help in the research on lung diseases.

**Keywords:** extracellular matrix; hydrogels; alveolar cells; type II phenotype; YAP

**Citation:** Marhuenda, E.; Villarino, A.; Narciso, M.L.; Camprubí-Rimblas, M.; Farré, R.; Gavara, N.; Artigas, A.; Almendros, I.; Otero, J. Lung Extracellular Matrix Hydrogels Enhance Preservation of Type II Phenotype in Primary Alveolar Epithelial Cells. *Int. J. Mol. Sci.* **2022**, 23, 4888. https://doi.org/10.3390/ijms23094888

Academic Editors: Athanasios G. Papavassiliou, Patrick C. Baer and Ralf Schubert

Received: 27 March 2022
Accepted: 26 April 2022
Published: 28 April 2022



## 1. Introduction

The fundamental questions on the precise mechanisms underlying alveolar epithelial cells (AECs) damage and epithelium repair in relevant diseases, such as acute respiratory distress syndrome or chronic obstructive pulmonary disease, are still unsolved. Although it is well-known that the alveolar epithelium is repaired by the proliferation of type II AECs (ATIIs), which differentiate into type I phenotype cells (ATIs), the involved mechanisms are still poorly understood [1]. In fact, a limitation hampering translational studies in lung diseases is the difficulty of maintaining the type II phenotype of primary AECs in vitro. Indeed, it is widely known that ATII-to-ATI transdifferentiation occurs very quickly in vitro; thus, primary type II AECs neither adequately proliferate, nor can be subcultured under conventional culture conditions [2].

To extend the maintenance of the type II AECs proliferative phenotype in vitro, it was proposed to coat the culture plates with hydrogels, such as Matrigel, resembling the extracellular matrix (ECM) [3]. Moreover, pioneering studies on subculturing strategies for primary ATII cells [4] employed inhibitors of Rho kinases, as the connection between the activation of the Rho pathway and cell mechanosensing of the ECM is well-established [5]. However, although the relationship between ATII phenotype maintenance in vitro and culturing cells on ECM-like substrates was proposed three decades ago [6], the problem remains open, probably because no hydrogels derived from lung ECM were available.

Interestingly, based on an initial report for obtaining hydrogels from the ECM of decellularized lungs [7], we have recently described a procedure for preparing such hydrogels by exclusively using lung ECM, with no need to add potentially toxic external cross-linkers [8]. If used as a culture substrate, this hydrogel, which realistically mimics the native lung ECM, could be particularly well-suited for providing a physiomimetic microenvironment to primary AECs. Therefore, we hypothesized that a lung ECM-based hydrogel would slow down the ATII-to-ATI transdifferentiation mediated by the inhibition of the Rho pathway, resulting in a decrease in F-actin polymerization and the formation of focal adhesions, as well as nuclear YAP activation [9–11].

## 2. Results

### 2.1. Primary Alveolar Epithelial Cells Form Monolayers on Lung-Derived Hydrogels

Isolated AECs were grown either in lung-derived hydrogels or plates for four days. Cells were able to form a monolayer on a lung-derived hydrogel, as shown in bright-field images in Figure 1. Differences in the morphology of the cells as a function of the substrate were also noticeable as shown by confocal images in Figure 2. Throughout the culture, cells on plates started to show a more flattened shape as well as larger cytoplasms (cells cultured on plates presented, in general, sizes about 5 times larger than cells cultured on hydrogels), and the presence of vacuoles was noticeable. On the contrary, AECs cultured on hydrogels showed cuboidal shapes, and even their microvilli, a phenotypical characteristic of ATII cells [12], could be distinguished. Cells cultured on plates formed monolayers faster (day 3 vs. day 4).

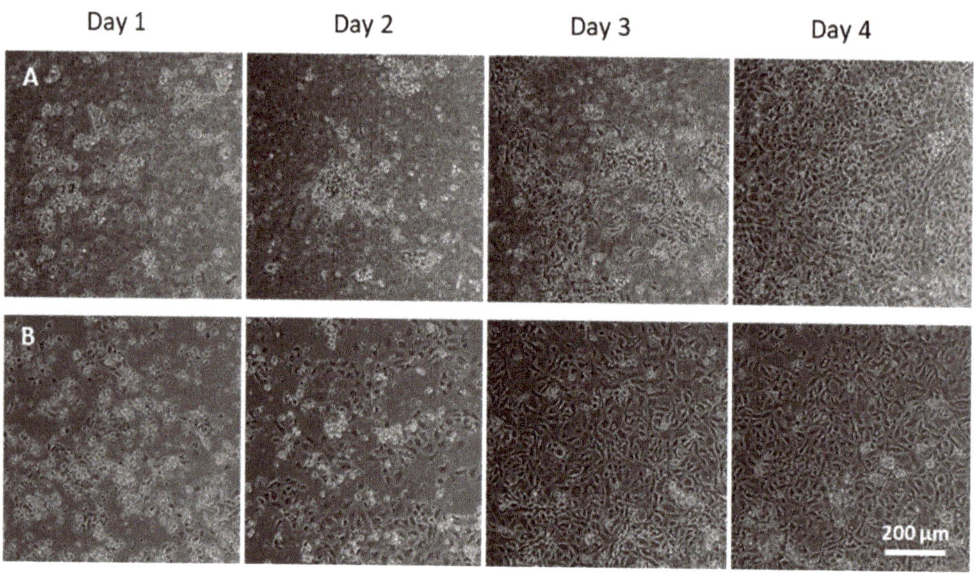


**Figure 1.** Rat primary alveolar cells were cultured either on porcine lung-derived hydrogel (**A**) or on a plate (**B**). Bright field images were taken every 24 h from day 1.

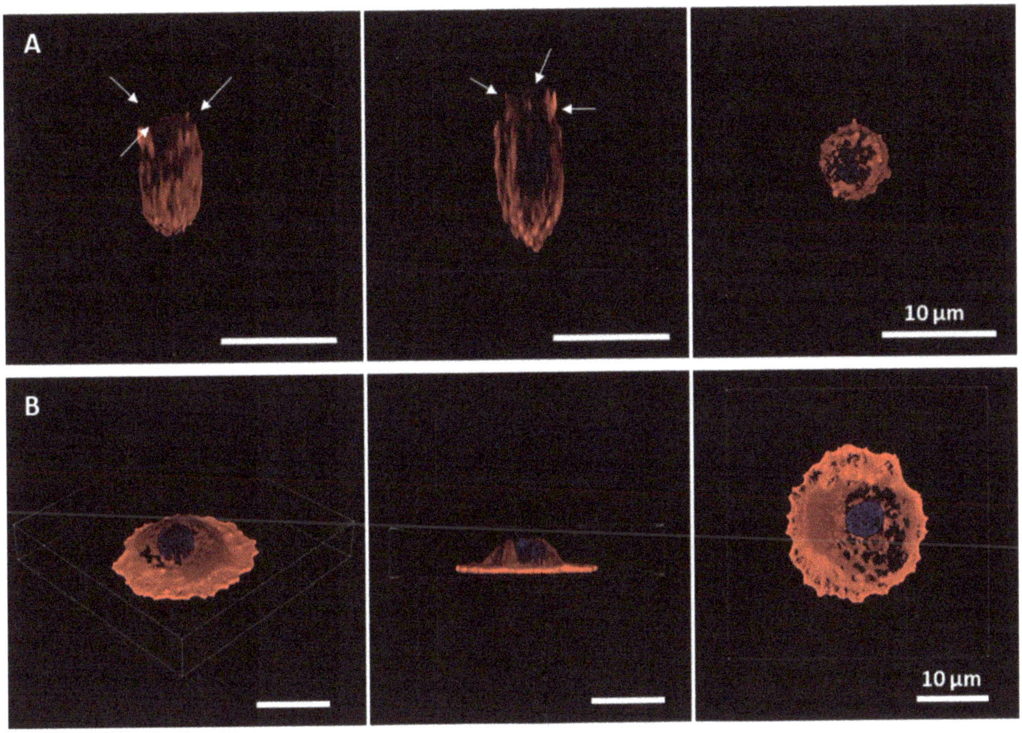


**Figure 2.** Three-dimensional representation of primary alveolar type II cells cultured for three days on hydrogel (**A**) and on a plate (**B**). Nuclei are stained in blue and actin cytoskeleton in red. In the case of the hydrogel-cultured cell, it can be distinguished by a cuboidal morphology and the presence of microvilli, indicated by arrows. In the case of the cell cultured on a plate, a more spread out and wider cytoplasm can be observed showing increased size.

### *2.2. Culture of Alveolar Epithelial Cells on Lung-Derived Hydrogels Preserves the Expression of Type II Markers for Longer Periods*

The results of the genetic expression of ATI and ATII typical markers (from day 2 to 5) are shown in Figure 3A. The gene expression of *sftpc* and *sftpb* decreased over time. There was a significant increase in ATII markers (surfactant proteins B and C) at earlier times (day 2 and day 3) in cells cultured on lung-derived hydrogels. There was no increase in ATI markers over time in cells cultured on hydrogels, in contrast to cells cultured on plates, where the expression of *pdpn* and *aqp5* increased with time (classical ATI markers). These results show that the ATII phenotype and gene expression levels were maintained for longer in lung-derived hydrogels. The results from immunostaining are shown in Figure 3B–E, revealing that the differential expression of surfactant protein C caused by the substrate was noticeable not only at the gene expression level, but also at the protein level. SPC expression was higher in cells cultured on lung hydrogels compared to that of cells cultured on a plate, as shown by the immunostainings. Furthermore, culture time affected cells differently depending on the substrate they were cultured on. Indeed, cells cultured on lung hydrogel were able to maintain the SPC expression over time, whereas cells cultured on a plate rapidly lost this ability.

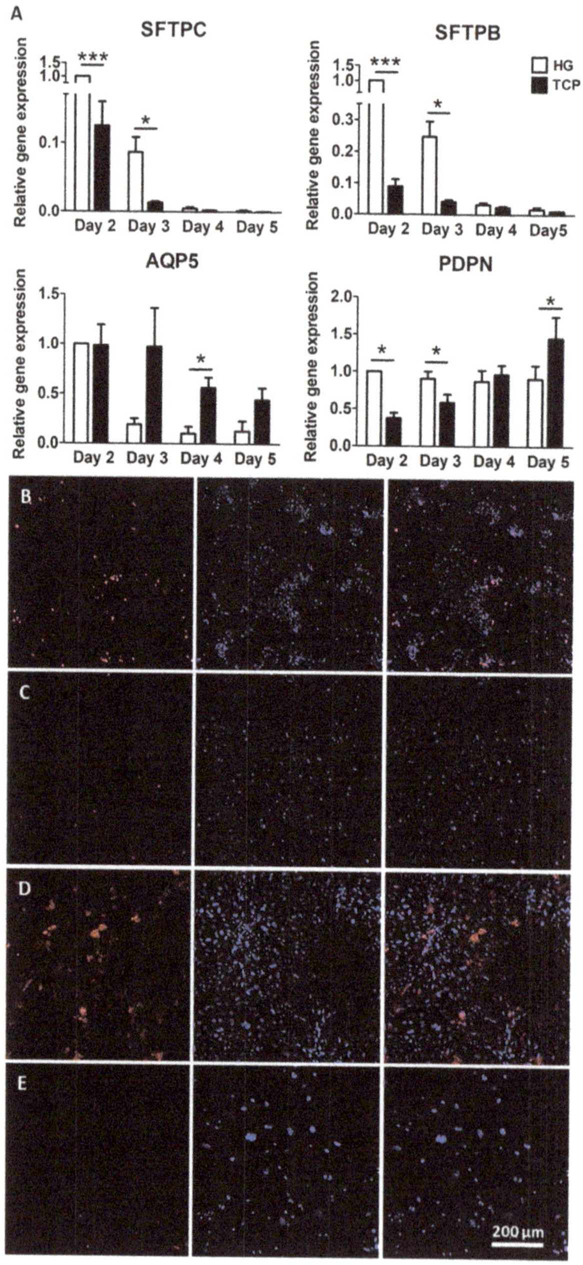


**Figure 3.** ATII cells were cultured either on hydrogel (HG) or a tissue culture plate (TCP). ATII typical marker surfactant proteins C (*sftpc*) and B (*sftpb*) and ATI typical marker aquaporin 5 (*aqp5*), podoplanin (*pdpn*) were studied at different time points (from day 2 to day 5) by qPCR (**A**). Relative gene expression is shown. Expression of surfactant protein C (SPC) was studied using immunofluorescence on ATII cells cultured on hydrogel at day 2 (**B**) and day 4 (**D**) and on a plate at day 2 (**C**) and day 4 (**E**). * $p < 0.05$, *** $p < 0.001$.

*2.3. Culture of Primary Alveolar Epithelial Cells in Lung-Derived Hydrogels Inhibits Type II-to-Type I Transdifferentiation by Altering the Hippo/Rho Pathway*

Since YAP is a key mechanotransduction protein, its expression was studied in cells cultured on hydrogels or plates for three days (Figure 4A,B). In cells cultured on hydrogels, the YAP protein was located mainly in the cytoplasm, which indicates that it was being phosphorylated and subsequently degraded. In contrast, in cells cultured on plates, nuclear active YAP was observed. Specifically, there was a two-fold increase in the amount of nuclear YAP when cells were cultured on a plate compared to that of cells cultured on hydrogels, indicating a higher transcriptional activity in the first group (Figure 4C).

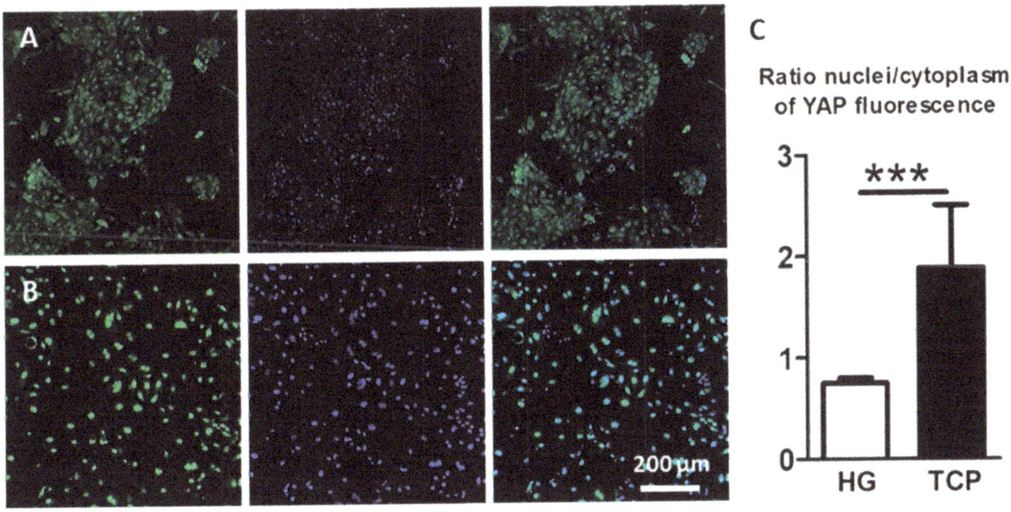


**Figure 4.** Alveolar type II cells were cultured on hydrogel (**A**) or on a plate (**B**) and stained for YAP protein (green). YAP nuclear and cytoplasmic expressions were quantified and expressed as the nuclei/cytoplasm fluorescence ratio in both conditions hydrogel (HG) and tissue culture plate (TCP) (**C**), *** $p < 0.001$.

Owing to the role of focal adhesions (FAs) and the actin cytoskeleton in sensing extracellular matrix cues and transmitting them to the cell, the expressions of actin and paxillin, which is one of the proteins comprising FAs, were studied. Both were reported to be implicated in the hippo pathway, by inhibiting it and promoting the YAP nuclear expression. In cells cultured for three days on lung-derived hydrogels, a poor focal adhesion assembly was observed by the paxillin immunostaining. Moreover, no stress fibers were formed as indicated by the phalloidin staining (Figure 5A). On the contrary, in cells that were cultured for three days on a plate, assembled paxillin and stress fibers were clearly observed (Figure 5B). The role of Rho, which is involved in the maturation of focal adhesions and YAP regulation [13,14] was studied by the use of the ROCK inhibitor (Y27632). The results show that it could play a role in the maintenance of the ATII phenotype, as reflected by an increase in SPC in conventional culture ($p = 0.02$) (Figure 5D) together with a decrease in focal adhesion size ($p = 0.001$) (Figure 5D), suggesting that the use of HGs as a substrate for AECs culture could be inhibiting the Rho pathway, and thus allowing for type II phenotype maintenance.

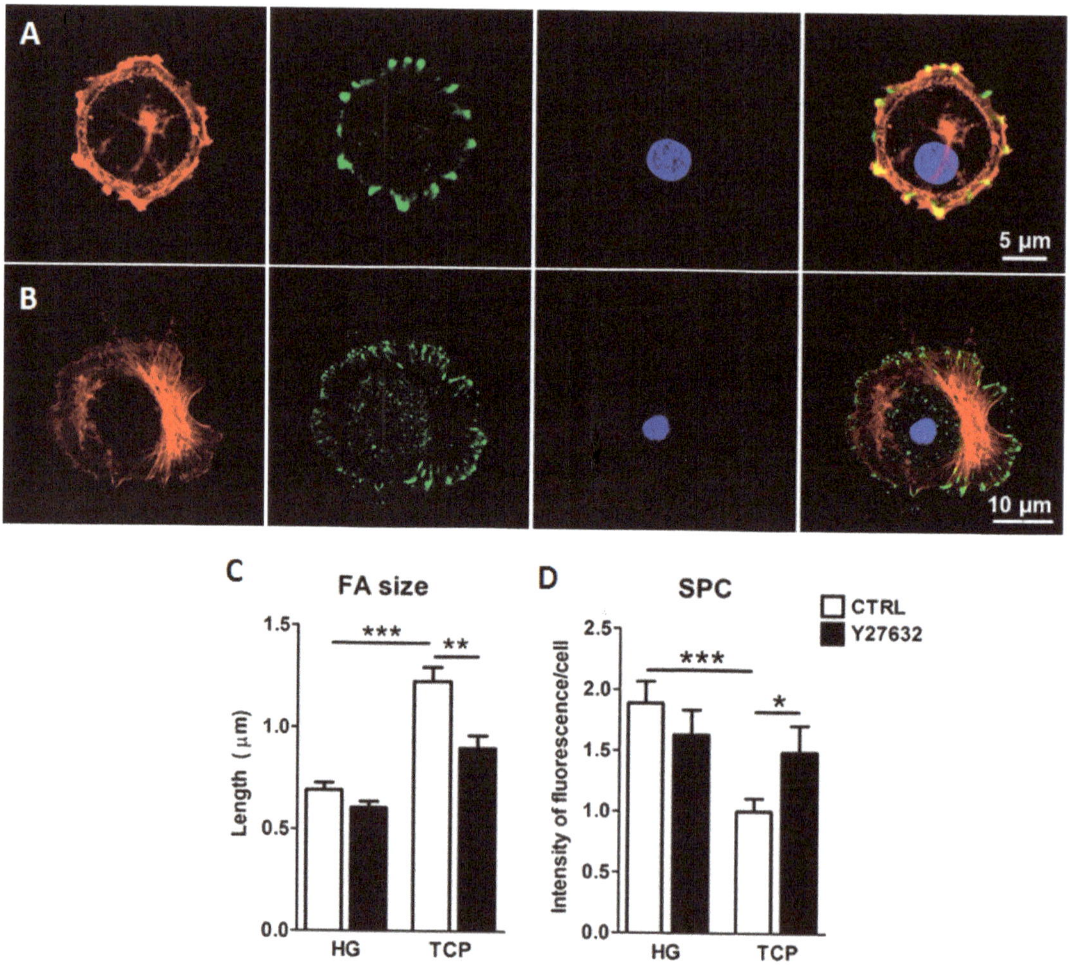


**Figure 5.** Representative images of focal adhesions (FAs) in cells cultured on hydrogels (**A**) and plate (**B**). Red: phalloidin, green: paxillin, blue: nuclei. Quantification of the FA length (**C**) and the intensity of surfactant protein C (SPC) of alveolar type II cultured on hydrogels for 3 days with (Y27632) and without (CTRL) the addition of the ROCK inhibitor (**D**), * $p < 0.05$, ** $p < 0.01$, *** $p < 0.001$.

### 2.4. Primary Alveolar Epithelial Cells Cultured on Lung-Derived Hydrogels Can Be Subcultured

The potential ability to subculture primary ATII cells was studied. Cells were cultured for 3 days on lung hydrogels and, after that time, lung-derived hydrogels were digested with collagenase, and cells were seeded again on plates. Sixteen hours later, they were immunostained for typical epithelial (EpCAM) and ATII (SFTPC) markers (Figure 6). Subcultured cells were positive for both, EpCAM and SFTPC markers. These results indicate that lung-derived hydrogels allow for primary AECs subculture.

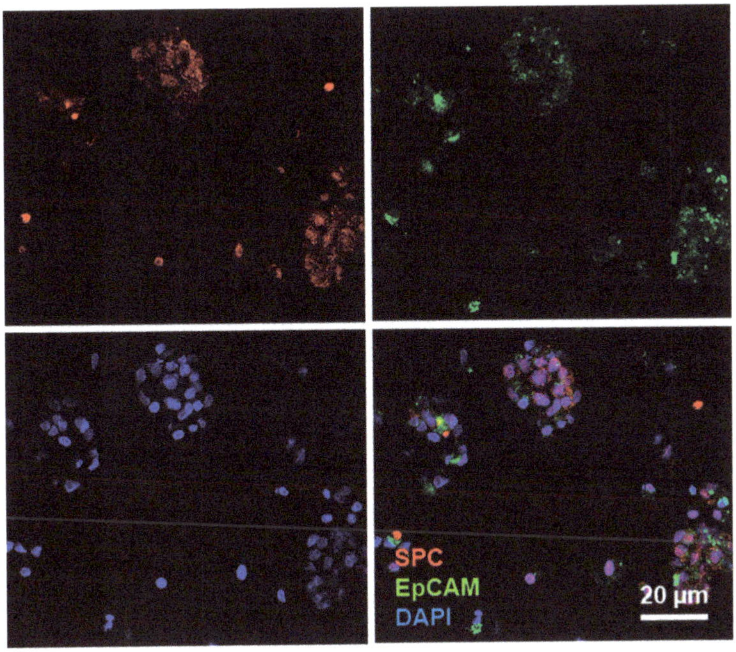


**Figure 6.** Alveolar type II cells were cultured on hydrogels for three days. After that time, they were subcultured on plates and stained for surfactant protein C (red) and EpCAM (green).

## 3. Discussion

We have provided evidence that using lung-derived ECM hydrogels as a culture substrate allows the maintenance of the type II phenotype in primary AECs to be enhanced. Moreover, our results for YAP and when using ROCK inhibitors on cells cultured on lung hydrogels, which preserve several proteins from their native organs [7], suggest that the biophysical properties of the hydrogel have a high impact on cell mechanosensing pathways, thereby playing a role in the maintenance of type II phenotype. This first-time study culturing cells on lung ECM-derived hydrogels, which was carried out in rat primary AECs as a proof of concept, opens the door for further research in primary human AECs, with a potential future impact on cell therapies for diseases such as pulmonary fibrosis and acute lung injury [15,16].

Similar results to those obtained in the present study were provided by Shannon and coworkers [6], who conducted the first study showing the importance of the protein content of the substrate in the maintenance of ATII morphology and phenotype. The main limitation of that first study was the inability of the cells to form a monolayer, which is easily accomplished in our lung-derived hydrogel. Since then, efforts have been made to improve the ability to maintain the ATII phenotype, which has only been achieved by media supplementation [3,17–19]. Both the biochemical and physical properties of the substrate are involved in cell behavior. It is worth noting that cells formed a monolayer slower in hydrogels than on plates, which could be due to the fact that cells proliferate quicker on the plate or that the ATII-to-ATI transdifferentiation allows for a faster substrate coverage, as ATI cells are much larger in size. In view of the differing stiffness of the hydrogels when compared to the culture plate and the cuboidal morphology of the cells, differences in their cytoskeleton distribution and the formation of focal adhesions, which are key points in the crosstalk of cell-ECM, were expected. Cells cultured on hydrogels showed shorter focal adhesions that correlated with a decreased YAP translocation to the nuclei. YAP is an important transcription factor that is implied in the regulation of several genes, and it is reported to be associated with inflammatory and epithelial damage situations in vivo where the ATII-to-ATI transdifferentiation is required [20,21]. In this case, YAP nuclear location of cells cultured on a plate could be triggering the differentiation in the ATI cell population. As Rho is implicated in the maturation of focal adhesions [14], and previous authors have pointed to it as a key factor in the transdifferentiation process [4,22], its role was studied. The determination of the individual contribution of different Rho proteins would be highly interesting. However, as it would be complex [14], we studied this effect through its inhibition using Y27632. The obtained results support the implication of Rho in the ATII phenotype, as its inhibition produced an increase in SPC and decrease in FAs only in cells cultured on a plate. The origin of this inhibition of the Rho pathway, although out of the scope of the present work, is probably related to the complex molecular composition of lung-derived hydrogels and their biomechanical properties.

Primary ATII cells have previously been subcultured [4] by coculturing them with fibroblasts as feeder cells and with the addition of the Rho inhibitors. The role of fibroblasts in maintaining the ATII phenotype is not exactly known but it could be that they work as a source of the keratinocyte growth factor (KGF). The addition of KGF has been used by other authors to increase the phenotypic features of ATII cells [3,19], and in turn, to inhibit Rho kinases, which are shown to play a key role in the ATII-to-ATI transdifferentiation [22]. In our case, the use of lung-derived hydrogels as substrates for culturing allows these cells to be subcultured without the of use of additional chemicals or cocultures. Therefore, culturing primary alveolar epithelial cells on lung ECM-derived hydrogels may facilitate the prolonged culturing of these cells, and thus help in the research of lung diseases. However, more research should be carried out regarding the number of possible passages, especially considering that factors different from the ATI-to-ATII transdifferentiation would limit the time that these primary cells could be maintained in vitro. Further research is also needed to determine which components of ECM-derived hydrogels have more impact on the maintenance of the type II phenotype in vitro. As the extracellular matrix used to produce hydrogels in the present work was obtained from lung tissue, it is expected that some key biochemical and biophysical factors specific to the lung are being preserved in the process of hydrogel production. The identification of these factors, although out of the scope of the present work, would allow the development of novel optimized scaffolds for the in vitro culture of pulmonary cells.

## 4. Materials and Methods

Unless otherwise specified, all reagents were purchased from ThermoFisher Scientific (Waltham, MA, USA) or Sigma-Aldrich (Saint Louis, MO, USA).

### *4.1. Decellularized Lung Extracellular Matrix Hydrogels Preparation*

Lung-derived hydrogels were prepared by following a previously described protocol [23]. Briefly, porcine lungs were decellularized by consecutive perfusion of the following reagents through the vasculature and the airways: 0.1% Triton X-100 and 2% SDC for 24 h at 4 °C, and NaCl 1 M and DNase solution for 1 h at 4 °C. Three washes of miliQ water were performed between consecutive reagent perfusion, and a last wash of PBS 1X was carried out. Decellularized lungs were cut into small pieces, freeze-dried (Telstar Lyoquest-55 Plus, Terrassa, Spain) and milled in liquid $N_2$ (SPEX SamplePrep, Metuchen, NJ, USA). The obtained powder was resuspended at 20 mg/mL in 0.01 M HCl and pepsin digested at a 1/10 proportion under magnetic stirring at room temperature for 16 h. To produce hydrogels, the digested solution was pH-adjusted to 7.4 ± 0.4 by using 0.1 M NaOH and incubated at 37 °C for 20 min.

### *4.2. Primary Alveolar Epithelial Cells Isolation*

The procedure was approved by the Ethical Board for Animal Research of the University of Barcelona, in compliance with regional, national and European regulations. Rat lungs were obtained from 180–250 g Sprague Dawley male rats. The animals were intraperitoneally anesthetized with 1 g/kg urethane and euthanized by exsanguination. AECs were isolated following a protocol described in [24]. Briefly, lungs were perfused with saline through the pulmonary artery and were resected en bloc. Three bronchioalveolar lavages were performed with 10 mL of PBS 1X to remove alveolar macrophages. Then, lungs were digested with 50 mL of 0.25% of trypsin through the airways for 30 min, cut into small pieces, and digested in a 100 units/mL DNase solution. The resulting suspension was filtered through a 100 µm and a 40 µm mesh and centrifuged through a Percoll (GE HealthCare, Chicago, IL, USA) gradient at 500× *g* for 20 min. The band containing ATII cells was recovered and digested using 20 units/mL of DNase. The resulting solution was centrifuged for 15 min at 500× *g*, and the pellet was resuspended in DCCM-1 (Biological Industries, Kibbutz Beit Haemek, Israel) medium and cultured for 1 h. Subsequently, medium containing ATII non-adherent cells was recovered and centrifuged for 10 min at 800× *g*. Cells were finally counted and seeded either on 24-well culture plates or on top of lung-ECM hydrogels.

### 4.3. Rho Kinase Inhibition Assay

For the inhibition of the Rho pathway, Y27638 (10 µM) was added to the medium for 24 h.

### 4.4. Reverse Transcription qPCR

For studying the expression of ATI and ATII markers, cells were cultured either on plastic or lung-derived hydrogels for different times, and RNA was subsequently extracted from samples by employing the RNeasy kit (Qiagen, Hilden, Germany). The cDNA was obtained by a reverse transcription-polymerase chain reaction (TaqMan Reverse Transcription Reagents, Invitrogen, Waltman, MA, USA) according to the manufacturer's instructions. The expression level of surfactant protein C (*sftpc*), surfactant protein B (*sftpb*), aquaporin 5 (*aqp5*) and podoplanin (*pdpn*) was studied using the Taqman Fast Advanced Master Mix and the TaqMan Gene Expression Assays in a StepOnePlus thermocycler (Applied Biosystems, Waltham, MA, USA). The expression level of genes was normalized to the constitutively expressed gene PPIA and calculated using the $2^{-\Delta\Delta Ct}$ method [25].

### 4.5. Immunohistochemistry and Image Processing

For immunohistochemistry experiments, cells were fixed with 4% paraformaldehyde for 30 min. Primary antibodies were incubated overnight, and secondary antibodies were incubated for 2 h at 37 °C. Nuclei were stained with Hoechst 33342. To avoid unspecific binding, especially in the hydrogel samples, a blocking buffer consisting of 2% BSA (Thermo Fisher, MA, USA) diluted in PBS 1X (Gibco, MA, USA) was employed for 40 min. Primary antibodies employed were rabbit anti-SFTPC (Invitrogen, Waltman, MA, USA), mouse anti-YAP (Santa Cruz Biotechnology, Dallas, TX, USA) and EpCAM (Miltenyi, Bergisch Gladbach, Germany). Secondary antibodies used were goat anti-rabbit cy5 (Abcam, Cambridge, UK) and goat anti-mouse Alexa Fluor 488 (Abcam, Cambridge, UK). Images were acquired with a Nikon Confocal Eclipse Ti microscope using a 20 × Plan Fluor Multi-immersion objective (0.75 NA) in the case of the SFTPC, EpCAM and YAP staining with a 10× objective (0.3 NA) for the bright field images, and a 100× objective (1.45 NA) for the paxillin and actin images. Nuclear images were obtained at 450 nm when illuminating the sample at 408 nm. Samples were excited at 488 nm and acquired at 515 nm for YAP, EpCAM and Paxillin images, and excited at 543.5 nm and acquired at 605 nm for SFTPC and actin stains.

For the analysis of YAP images, five images per condition were randomly selected and analyzed using a blind procedure with ImageJ Software. Quantification of the ratio nuclear fluorescence/cytoplasmic fluorescence was assessed following a previously described procedure [26] with slight modifications. To calculate the total cell fluorescence, a triangle threshold was employed, and the integrated fluorescence was calculated in the YAP channel. For calculating the YAP nuclear fluorescence, the perimeter of the nuclei was delimited by the Huang threshold in the DAPI channel. After that, the resulting mask was redirected to the YAP channel and the integrated intensity contained in the nuclear perimeter was obtained. To calculate the cytoplasmic fluorescence, the differences in the intensities in the total cell and in the nuclei were calculated.

For focal adhesion length measurements, five representative adhesions per cell in paxillin stainings at the cell edge were manually quantified with Image J Software.

### *4.6. Statistical Analysis*

Data are expressed as mean ± SE unless stated otherwise. Statistical analysis was performed with Graphpad Prism software. Differences in ATI/ATII markers gene expression and YAP nuclear/cytoplasmic expression were analyzed using paired *t*-test. A $p$-value < 0.05 was considered significant.

**Author Contributions:** Conceptualization, E.M., I.A. and J.O.; Formal analysis, E.M., A.V., A.A., R.F., I.A. and J.O.; Funding acquisition, R.F., I.A., N.G. and J.O..; Investigation, E.M., A.V., I.A. and J.O.; Methodology, E.M., A.V., M.L.N., M.C.-R., I.A. and J.O.; Project administration, R.F., I.A. and J.O.; Resources, N.G., A.A., R.F., N.G., I.A. and J.O.; Software, E.M., M.L.N. and N.G.; Supervision, I.A. and J.O.; Writing of original draft, E.M. and J.O. All authors have read and agreed to the published version of the manuscript.

**Funding:** This research was funded by the Spanish Ministry of Science, Innovation and Universities, grants numbers SAF2017-85574-R, DPI2017-83721-P, PID2019-108958RB-I00, and PGC2018-097323-A-I00, and by European Union's Horizon 2020, Marie Sklodowska-Curie, grant 821772.

**Institutional Review Board Statement:** Animal experiments were approved by the Ethical Committee for Animal Research of the University of Barcelona.

**Data Availability Statement:** Data supporting the findings of this study are available from the corresponding authors upon reasonable request.

**Acknowledgments:** The authors wish to thank Miguel A. Rodríguez from the Unit of Biophysics and Bioengineering for his excellent technical assistance.

**Conflicts of Interest:** The authors declare no conflict of interest.

## References

1. Sun, T.; Huang, Z.; Zhang, H.; Posner, C.; Jia, G.; Ramalingam, T.R.; Xu, M.; Brightbill, H.; Egen, J.G.; Dey, A.; et al. TAZ is required for lung alveolar epithelial cell differentiation after injury. *JCI Insight* **2019**, *5*, 128674. [CrossRef] [PubMed]
2. Dobbs, L.G. Isolation and culture of alveolar type II cells. *Am. J. Physiol.* **1990**, *258 Pt 1*, L134–L147. [CrossRef] [PubMed]
3. Rice, W.R.; Conkright, J.J.; Na, C.L.; Ikegami, M.; Shannon, J.M.; Weaver, T.E. Maintenance of the mouse type II cell phenotype in vitro. *Am. J. Physiol. Lung Cell Mol. Physiol.* **2002**, *283*, L256–L264. [CrossRef] [PubMed]
4. Bove, P.F.; Dang, H.; Cheluvaraju, C.; Jones, L.C.; Liu, X.; O'Neal, W.K.; Randell, S.H.; Schlegel, R.; Boucher, R.C. Breaking the in vitro alveolar type II cell proliferation barrier while retaining ion transport properties. *Am. J. Respir. Cell Mol. Biol.* **2014**, *50*, 767–776. [CrossRef] [PubMed]
5. Seo, J.; Kim, J. Regulation of Hippo signaling by actin remodeling. *BMB Rep.* **2018**, *51*, 151–156. [CrossRef]
6. Shannon, J.M.; Emrie, P.A.; Fisher, J.H.; Kuroki, Y.; Jennings, S.D.; Mason, R.J. Effect of a reconstituted basement membrane on expression of surfactant apoproteins in cultured adult rat alveolar type II cells. *Am. J. Respir. Cell Mol. Biol.* **1990**, *2*, 183–192. [CrossRef]
7. Pouliot, R.A.; Young, B.M.; Link, P.A.; Park, H.E.; Kahn, A.R.; Shankar, K.; Schneck, M.B.; Weiss, D.J.; Heise, R.L. Porcine Lung-Derived Extracellular Matrix Hydrogel Properties Are Dependent on Pepsin Digestion Time. *Tissue Eng. Part C Methods* **2020**, *26*, 332–346. [CrossRef]
8. Falcones, B.; Sanz-Fraile, H.; Marhuenda, E.; Mendizábal, I.; Cabrera-Aguilera, I.; Malandain, N.; Uriarte, J.J.; Almendros, I.; Navajas, D.; Weiss, D.J.; et al. Bioprintable Lung Extracellular Matrix Hydrogel Scaffolds for 3D Culture of Mesenchymal Stromal Cells. *Polymers* **2021**, *13*, 2350. [CrossRef]
9. Das, A.; Fischer, R.S.; Pan, D.; Waterman, C.M. YAP Nuclear Localization in the Absence of Cell-Cell Contact Is Mediated by a Filamentous Actin-dependent, Myosin II- and Phospho-YAP-independent Pathway during Extracellular Matrix Mechanosensing. *J. Biol. Chem.* **2016**, *291*, 6096–6110. [CrossRef]
10. Chen, Q.; Rehman, J.; Chan, M.; Fu, P.; Dudek, S.M.; Natarajan, V.; Malik, A.B.; Liu, Y. Angiocrine Sphingosine-1-Phosphate Activation of S1PR2-YAP Signaling Axis in Alveolar Type II Cells Is Essential for Lung Repair. *Cell Rep.* **2020**, *31*, 107828. [CrossRef]

11. Kim, N.G.; Gumbiner, B.M. Adhesion to fibronectin regulates Hippo signaling via the FAK-Src-PI3K pathway. *J. Cell Biol.* **2015**, *210*, 503–515. [CrossRef] [PubMed]
12. Ito, H.; Uchida, T.; Makita, K. Interactions between rat alveolar epithelial cells and bone marrow-derived mesenchymal stem cells: An in vitro co-culture model. *Intensive Care Med. Exp.* **2015**, *3*, 53. [CrossRef] [PubMed]
13. Pennarossa, G.; Paffoni, A.; Ragni, G.; Gandolfi, F.; Brevini, T.A.L. Rho Signaling-Directed YAP/TAZ Regulation Encourages 3D Spheroid Colony Formation and Boosts Plasticity of Parthenogenetic Stem Cells. *Adv. Exp. Med. Biol.* **2020**, *1237*, 49–60. [CrossRef]
14. Warner, H.; Wilson, B.J.; Caswell, P.T. Control of adhesion and protrusion in cell migration by Rho GTPases. *Curr. Opin. Cell Biol.* **2019**, *56*, 64–70. [CrossRef] [PubMed]
15. Lopez-Rodriguez, E.; Gay-Jordi, G.; Knudsen, L.; Ochs, M.; Serrano-Mollar, A. Improved Alveolar Dynamics and Structure After Alveolar Epithelial Type II Cell Transplantation in Bleomycin Induced Lung Fibrosis. *Front. Med.* **2021**, *8*, 640020. [CrossRef]
16. Serrano-Mollar, A.; Gay-Jordi, G.; Guillamat-Prats, R.; Closa, D.; Hernandez-Gonzalez, F.; Marin, P.; Burgos, F.; Martorell, J.; Sánchez, M.; Arguis, P.; et al. Safety and Tolerability of Alveolar Type II Cell Transplantation in Idiopathic Pulmonary Fibrosis. *Chest* **2016**, *150*, 533–543. [CrossRef]
17. Griffin, M.; Bhandari, R.; Hamilton, G.; Chan, Y.C.; Powell, J.T. Alveolar type II cell-fibroblast interactions, synthesis and secretion of surfactant and type I collagen. *J. Cell Sci.* **1993**, *105 Pt 2*, 423–432. [CrossRef]
18. Buckley, S.; Driscoll, B.; Anderson, K.D.; Warburton, D. Cell cycle in alveolar epithelial type II cells: Integration of Matrigel and KGF. *Am. J. Physiol.* **1997**, *273 Pt 1*, L572–L580. [CrossRef]
19. Wang, J.; Edeen, K.; Manzer, R.; Chang, Y.; Wang, S.; Chen, X.; Funk, C.J.; Cosgrove, G.P.; Fang, X.; Mason, R.J. Differentiated human alveolar epithelial cells and reversibility of their phenotype in vitro. *Am. J. Respir. Cell Mol. Biol.* **2007**, *36*, 661–668. [CrossRef]
20. Chen, W.; Zhuo, Y.; Duan, D.; Lu, M. Effects of Hypoxia on Differentiation of Mesenchymal Stem Cells. *Curr. Stem. Cell Res. Ther.* **2020**, *15*, 332–339. [CrossRef]
21. LaCanna, R.; Liccardo, D.; Zhang, P.; Tragesser, L.; Wang, Y.; Cao, T.; Chapman, H.A.; Morrisey, E.E.; Shen, H.; Koch, W.J.; et al. Yap/Taz regulate alveolar regeneration and resolution of lung inflammation. *J. Clin. Investig.* **2019**, *129*, 2107–2122. [CrossRef] [PubMed]
22. Foster, C.D.; Varghese, L.S.; Gonzales, L.W.; Margulies, S.S.; Guttentag, S.H. The Rho pathway mediates transition to an alveolar type I cell phenotype during static stretch of alveolar type II cells. *Pediatr. Res.* **2010**, *67*, 585–590. [CrossRef] [PubMed]
23. Pouliot, R.A.; Link, P.A.; Mikhaiel, N.S.; Schneck, M.B.; Valentine, M.S.; Kamga Gninzeko, F.J.; Herbert, J.A.; Sakagami, M.; Heise, R.L. Development and characterization of a naturally derived lung extracellular matrix hydrogel. *J. Biomed. Mater. Res. A* **2016**, *104*, 1922–1935. [CrossRef] [PubMed]
24. Guillamat-Prats, R.; Camprubí-Rimblas, M.; Puig, F.; Herrero, R.; Tantinyà, N.; Serrano-Mollar, A.; Artigas, A. Alveolar Type II Cells or Mesenchymal Stem Cells: Comparison of Two Different Cell Therapies for the Treatment of Acute Lung Injury in Rats. *Cells* **2020**, *9*, 1816. [CrossRef]
25. Livak, K.J.; Schmittgen, T.D. Analysis of Relative Gene Expression Data Using Real-Time Quantitative PCR and the $2^{-\Delta\Delta CT}$ Method. *Methods* **2001**, *25*, 402–408. [CrossRef]
26. Marhuenda, E.; Campillo, N.; Gabasa, M.; Martínez-García, M.A.; Campos-Rodríguez, F.; Gozal, D.; Navajas, D.; Alcaraz, J.; Farré, R.; Almendros, I. Effects of Sustained and Intermittent Hypoxia on Human Lung Cancer Cells. *Am. J. Respir. Cell Mol. Biol.* **2019**, *61*, 540–544. [CrossRef]

International Journal of
*Molecular Sciences*

MDPI

*Article*

# microRNA Expression of Renal Proximal Tubular Epithelial Cells and Their Extracellular Vesicles in an Inflammatory Microenvironment In Vitro

**Patrick C. Baer [1,*], Ann-Kathrin Neuhoff [2] and Ralf Schubert [2,*]**

1 Department of Internal Medicine 4, Nephrology, University Hospital, Goethe-University, 60596 Frankfurt/M., Germany
2 Division of Allergology, Pneumology and Cystic Fibrosis, Department for Children and Adolescents, University Hospital, Goethe-University, 60596 Frankfurt/M., Germany; patrick.baer@kgu.de
* Correspondence: p.baer@em.uni-frankfurt.de or pcbaer@arcor.de (P.C.B.); ralf.schubert@kgu.de (R.S.); Tel.: +49-69-6301-83611 (R.S.); Fax: +49-69-6301-83349 (R.S.)

**Abstract:** Renal proximal tubular epithelial cells (PTCs) are central players during renal inflammation. In response to inflammatory signals, PTCs not only self-express altered mRNAs, microRNAs (miRNAs), proteins, and lipids, but also release altered extracellular vesicles (EVs). These EVs also carry inflammation-specific cargo molecules and are key players in cell–cell-communication. Understanding the precise molecular and cellular mechanisms that lead to inflammation in the kidney is the most important way to identify early targets for the prevention or treatment of acute kidney injury. Therefore, highly purified human PTCs were used as an in vitro model to study the cellular response to an inflammatory microenvironment. A cytokine-induced inflammatory system was established to analyze different miRNA expression in cells and their EVs. In detail, we characterized the altered miR expression of PTCs and their released EVs during induced inflammation and showed that 12 miRNAs were significantly regulated in PTCs (6 upregulated and 6 downregulated) and 9 miRNAs in EVs (8 upregulated and 1 downregulated). We also showed that only three of the miRNAs were found to overlap between cells and EVs. As shown by the KEGG pathway analysis, these three miRNAs (miR-146a-5p, miR-147b, and miR-155-5p) are functionally involved in the regulation of the Toll-like receptor signaling pathway and significantly correlated with the inflammatory mediators IL6 and ICAM1 released by stimulated PTCs. Especially with regard to a possible clinical use of miRs as new biomarkers, an accurate characterization of the miR expression altered during inflammatory processes is of enormous importance.

**Keywords:** renal tubular cells; epithelial cells; proximal tubule; extracellular vesicles; cytokines; inflammation; inflammatory microenvironment; kidney

**Citation:** Baer, P.C.; Neuhoff, A.-K.; Schubert, R. microRNA Expression of Renal Proximal Tubular Epithelial Cells and Their Extracellular Vesicles in an Inflammatory Microenvironment In Vitro. *Int. J. Mol. Sci.* **2023**, *24*, 11069. https://doi.org/10.3390/ijms241311069

Academic Editor: Manoocher Soleimani

Received: 15 June 2023
Revised: 29 June 2023
Accepted: 3 July 2023
Published: 4 July 2023



## 1. Introduction

Renal proximal tubular epithelial cells (PTCs) are not only an integral part of the nephron, which is responsible for the reabsorption of water, electrolytes, and other substances from the filtrate produced by the glomerulus, but they are also central players during renal inflammation and mediate the response to inflammation during kidney diseases [1]. Inflammation in the proximal tubular segment is often associated with an immune response to injury or infection [2]. The inflammatory process involves the activation of immune cells, the release of pro-inflammatory molecules, the recruitment of inflammatory cells to the affected area, and tissue damage. This inflammation can lead to various renal diseases, including acute and chronic renal injury, tubulointerstitial nephritis, and finally, fibrosis [3,4].

During the inflammatory process, renal tubular epithelial cells are exposed to various stimuli, such as cytokines, chemokines, and damage-associated molecular patterns. In response to these inflammatory signals, PTC release extracellular vesicles (EVs) that carry inflammation-specific cargo molecules [5]. These EVs are than taken up by neighboring cells or transported to distant sites, where they can influence the immune response and contribute to the progression or resolution of inflammation. EVs are small membrane-bound particles released by nearly all cell types, including renal tubular epithelial cells, and they play crucial roles in cell-to-cell communication. There are different types of known EVs, leading to their classification according to their sub-cellular origin, release pathways, and size [6]. On the one hand, there are EVs released by budding from the cells' plasma membrane, known as microvesicles, ectosomes, or microparticles. On the other hand, there are EVs which are generated inside multivesicular bodies and are secreted by the fusion of the body with the plasma membrane, known as exosomes. Both subtypes are mainly involved in intercellular communication in health and disease. Besides these types, there are also apoptotic bodies belonging to the class of EVS, which, however, have a different task compared to the other subgroups. Apoptotic vesicles are released from the plasma membrane during programmed cell death and therefore transport substances from dying cells [7].

All EVs exhibit some diversity, particularly in terms of their surface molecules, which allows them to be targeted to recipient cells. It has been shown that when EVs attach to a target cell, they trigger signaling pathways via receptor–ligand interactions or internalization via endocytosis or phagocytosis or even by fusion with the membrane of the target cell [8]. It has also been shown that EVs from parent cells can interact in a specific manner with recipient cells, e.g., in the renal nephron, and it has further been suggested that a specific signaling pathway exists from proximal to distal tubular epithelial cells [9]. The signaling pathways and communication via EVs are mostly accomplished by nucleic acids, lipids, proteins, and carbohydrates included in the EVs. By delivering their content into the cytosol of a target cell, EVs can modify the physiological state of the recipient cell [8]. In particular, small RNA molecules, so-called microRNAs (miRNAs), were proven to play a central role as epigenetic regulators in this context, since they regulate gene expression via post-transcriptional mechanisms. Therefore, they are touted as the next generation of biomarkers for different human diseases and biological states [9]. They were, inter alia, shown to prevent the uncontrollable progression of inflammation but were also shown to be frequently dysregulated in the development of renal fibrosis and other diseases [10,11]. Since free RNA included in serum or medium is relatively instable, EVs could serve as a protector of RNA from degradation and therefore contribute to the status of a potential biomarker [12].

In our present study, highly purified human renal proximal tubular epithelial cells were used as an in vitro model to study the cellular response to an inflammatory microenvironment. Therefore, a cytokine-induced inflammatory system was established to analyze different miRNA expression levels in cells and their EVs. Specifically, we describe and discuss the altered miRNA expression of PTC and their released EVs during induced inflammation. Especially with regard to a possible clinical use of miRs as new biomarkers, the accurate characterization of the miR expression altered during inflammatory processes is of enormous importance.

## 2. Results

### 2.1. Characterization of PTCs and Their EVs

Cultured PTCs displayed an epithelial morphology with a highly compact cell monolayer (Figure 1A). The expression of the characteristic markers CD13 and CD26 was shown via flow cytometry. In addition, the expression of the epithelial cell adhesion molecule (Ep-CAM) and the marker CD63, a characteristic marker of exosomes, was shown (Figure 1B). After incubation in an inflammatory microenvironment, the cells retained their epithelial morphology but showed smaller gaps at some sites of the cell monolayer.

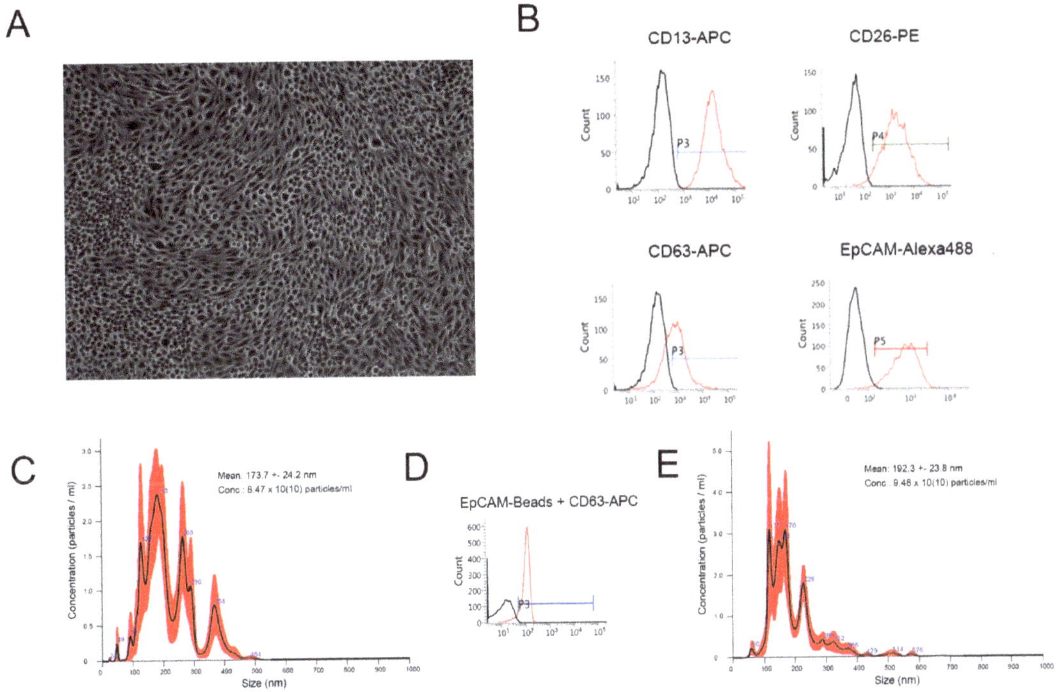


**Figure 1.** Characterization of PTCs (**A**,**B**) and EVs isolated from PTC supernatant (**C**,**D**). (**A**) Characteristic phase contrast microscopy of confluent PTCs cultured in standard cell culture. (**B**) Representative flow cytometric overlay histograms of characteristic PTC marker expression (CD13, CD26, and EpCAM), and of CD63, a characteristic marker of exosomes. The black histograms represent isotype controls. (**C**) Representative nanoparticle tracking analysis (NTA) of EVs isolated from unstimulated PTC after standard cell culture for 48 h. (**D**) Representative flow cytometric overlay histogram of PTC-EVs isolated using EpCAM-beads and immunostained with CD63-APC. The black histogram represents an unstained control. (**E**) Representative nanoparticle tracking analysis (NTA) of EVs isolated from PTCs after culture in an inflammatory microenvironment for 48 h.

SEC-isolated PTC-EVs were characterized via nanoparticle tracking analysis (NTA) (Figure 1C,E) and flow cytometry (Figure 1D). EVs isolated from unstimulated controls showed a mean size of 173.7 ± 24.2 nm (distribution between 24 and 484 nm), with an average concentration of $8.47 \times 10^{10}$ particles/mL ($n$ = 3, Figure 1C). EVs isolated from stimulated cells were slightly larger in their average mean size (192.3 ± 23.8 nm, distribution between 60 and 576 nm), and with a slightly increased concentration ($9.48 \times 10^{10}$ particles/mL, $n$ = 3, Figure 1E). Nevertheless, these differences were not significant. Flow cytometric analyses further showed CD63 expression on magnetic-bead-purified PTC-EVs (using specific EpCAM-marked magnetic beads) (Figure 1D).

Additional qPCR analyses to validate the effect of an inflammatory microenvironment on the cells showed significantly upregulated mRNA expression levels of interleukin (IL) 6 (IL6), IL1β, and intercellular adhesion molecule 1 (ICAM1) (Figure 2A–C). For these experiments, we used a total of six PTC samples: two identical purifications that were also used for miR sequencing and four additional cell cultures.

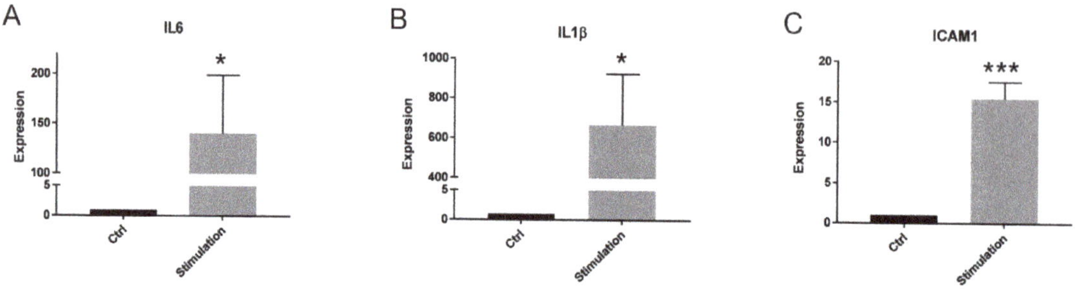


**Figure 2.** Validation of inflammatory stimulation of PTCs. mRNA expression of selected factors under an inflammatory culture condition for 48 h: (**A**) IL6, (**B**) IL1β, and (**C**) ICAM1. The expression levels in each experiment were normalized to a housekeeping gene (β-actin) and expressed relative to the control using the ΔΔCT method (Expression). A mixture of pro-inflammatory cytokines (stimulation) significantly increased the mRNA expression of all three readouts after 48 h of incubation. Mean ± SEM, * $p < 0.05$ (A, B), *** $p < 0.001$ (C), $n = 6$.

### 2.2. Effect of Inflammation on the miR Expression in PTC and Their EVs

To sequence all samples, the high-throughput Illumina sequencing technology together with the Miseq instrument were used, resulting in the generation of raw DNA sequence reads. In this study, a total of around 9,000,000 raw reads were obtained from miRNAs and the slightly longer so-called piwi-interacting RNAs (piRNAs) of PTCs or of EVs released from PTCs, respectively. Nevertheless, the actual number of reads was approximately 1,500,000 reads for samples obtained from the PTCs and 50,000 reads for samples obtained from the EVs.

To finally determine the identity and quantity of all reads included in each sample, the raw data of the sequencing were uploaded to the Qiagen database for further analysis. Therefore, the specific miRNAs and piRNAs that corresponded to the raw reads generated from Illumina sequencing were evaluated. Hierarchical clustering in the form of a heatmap should give an overview about the differences in the quantities of the different miRNAs and piRNAs between the unstimulated controls and the stimulated samples. Those miRNAs and piRNAs that had a minimum log fold change ($log_2FC$) of $>\pm0.75$ and an adjusted $p$ value (p-adj) of less than 0.1 were each shown in a heatmap as up- or downregulated (Figure 3A,B, Supplementary Tables S1 and S2). Of the detected small RNAs from PTCs, there were 13 upregulated and 10 downregulated (Figure 3A, Supplementary Table S1). Of the detected miRNAs and piRNAs from the EV samples, there were 16 upregulated and 34 downregulated (Figure 3B, Supplementary Table S2).

For further analyses, we only used highly significant regulated miRNAs (p-adj < 0.05) with a minimum $log_2FC$ of $>\pm1$, and no piRNAs (Figures 4A,B and 5). We found 12 miRNAs in the PTC samples, of which 6 (miR-146A-5p, miR-147b, miR-146a-3p, miR-155-5p, miR-99b-5p, and miR-100-3p) were upregulated and 6 were (miR-210-3p, miR-128-3p, miR-186-5p, miR-335-5p, miR-140-3p, and miR-296-3) downregulated (Figure 4A), and 9 miRNAs in the EVs from PTCs, of which 8 were upregulated (miR-146a-5p, miR-155-5p, miR-141-3p, miR-221-3p, miR-23b-3p, miR-147b, miR-320c, and miR-3613-5p) and 1 was downregulated (miR-3687) (Figure 4B).

The Venn diagram shows the overlap of the highly significant miRNAs found in PTCs (12 miRNAs) and their EVs (9 miRNAs) (Figure 5A). These three miRNAs (miR-146a-5p ($log_2FC$ 3.384 (PTC) and 1.944 (EVs)), miR-147b ($log_2FC$ 4.244 (PTC) and 3.007 (EVs)), and miR-155-5p ($log_2FC$ 2.438 (PTC) and 1.232 (EVs)), also shown in Tables S1 and S2) were then used for network analyses to identify the contribution of these miRNAs to biological processes and target interactions (using miRNet 2.0, miRTarBase v8.0) (Figure 5B,C). In total, the three miRNAs were predicted to be involved in 55 pathways, including the Toll-like receptor signaling pathway (p-adj < 0.0000305), the T cell receptor signaling pathway (p-adj < 0.0000404), apoptosis (p-adj < 0.00142), and the B cell receptor signaling pathway (p-adj < 0.00677), and were predicted to target 3775 genes (Figure 5B,C and Supplementary Figure S1). Shared targets in the Toll-like receptor signaling pathway for miR-146a-5p and miR-155-5p were IL6, ICAM1, NF*k*B1, FADD, RHOA, RAC1, CXCL8, STAT1, and FOS and COL4A2 for miR-147b and miR-155-5, respectively.

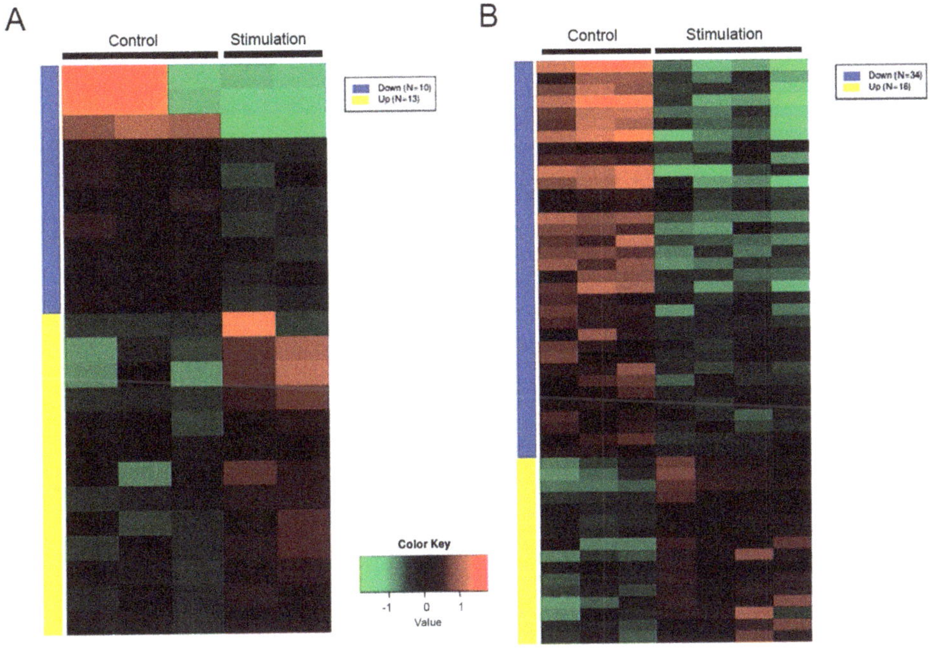


**Figure 3.** RNAseq analysis of miRNA extracted from PTCs (**A**) and EVs (**B**). Heatmap showing hierarchical clustering of analyzed miRNAs and piRNA from stimulated PTCs or their EVs and unstimulated controls (minimum fold change ±1.5, p-adj < 0.1). The groups are represented in columns and the specific miRNAs in rows. Each colored cell on the map corresponds to an expression value.

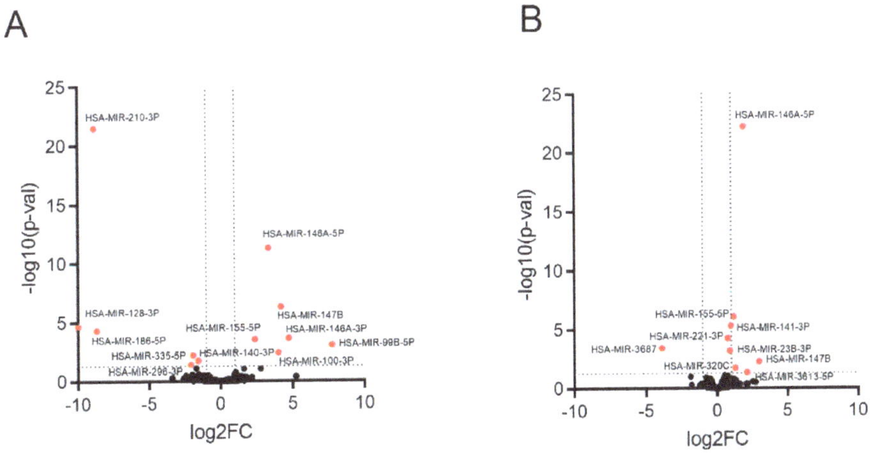


**Figure 4.** RNAseq analysis of miRNA extracted from (**A**) PTCs and (**B**) their EVs. Volcano Plot combining log fold change ($log_2FC$) analysis. Fold changes with a highly significant upregulation or downregulation are highlighted ($log_2FC > \pm 1$; p-adj < 0.05).

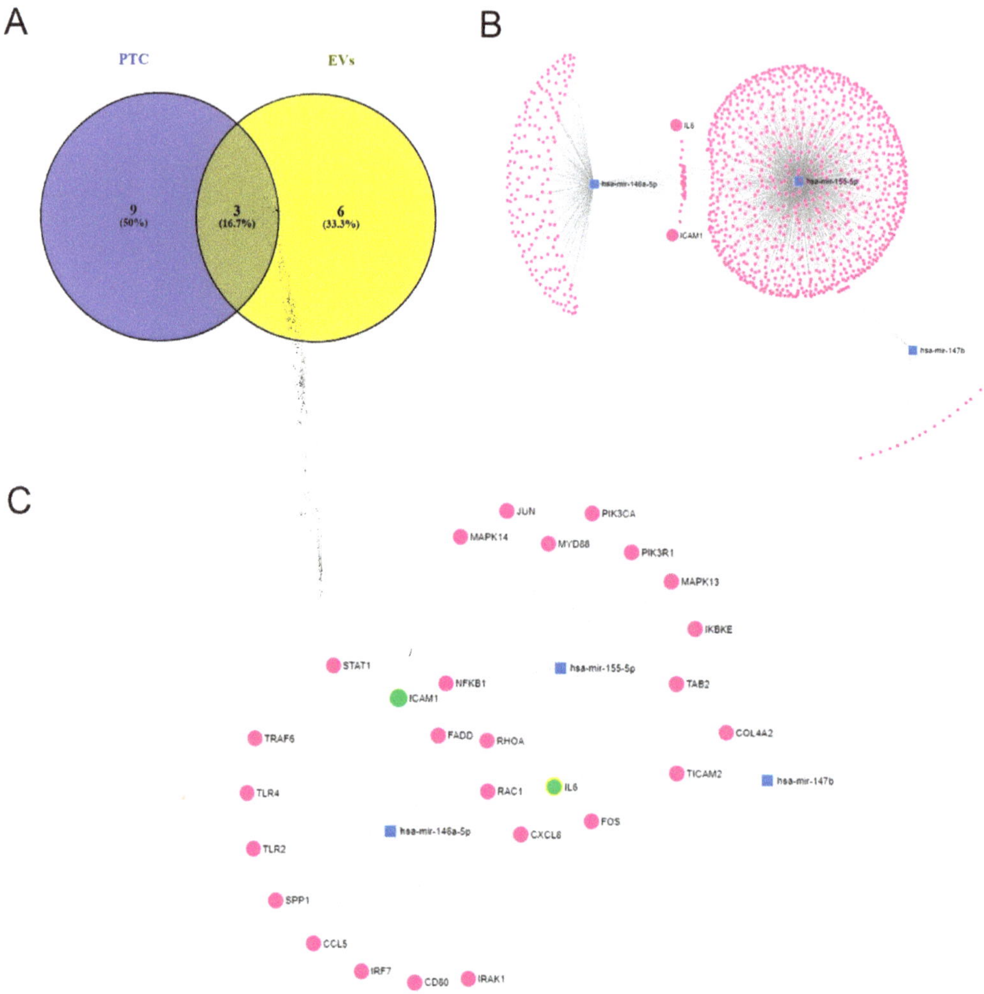


**Figure 5.** Venn diagram and network analysis. (**A**) Venn diagram to show the overlapping highly significant regulated miRNA from PTCs and their EVs. Only three miRNAs (miR-146a-5p, miR-147b, and miR-155-5p) overlapped between the two groups (Venny 2.1.0). (**B**,**C**) Network analyses showing miRNA target interactions. Network visualization: miR (blue squares) and the potentially interacting gene network (magenta and green dots) created with miRNet 2.0 (miRTarBase). (**B**) Dissected miRNA–target interactions and functional associations through network-based visual analysis (miRNet). miR-146a-5p, miR-155-5p, and miR-147b regulate 1092 genes, including ICAM1 and IL6. (**C**) Interaction and targets involved in Toll-like receptor signaling pathway.

The expression of the three miRNAs found (Figure 5A) was then further validated using the purchasable miRCURY system from Qiagen. Using this system, we confirmed the highly significant upregulation of all three miRNAs after the incubation of PTCs in the inflammatory microenvironment (Figure 6). Nevertheless, due to a lack of samples, we did not validate the expression of the miRNAs in EVs.

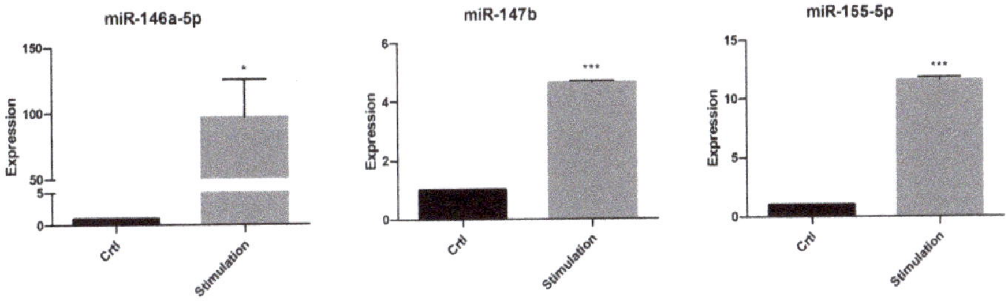


**Figure 6.** Validation of miR-146a-5p, miR-147b, and miR-155-5p expression in unstimulated PTCs (controls) and after stimulation. Values were normalized using the UniSP6 spike-in control primer set. Data are mean ± SEM of three independent experiments. Significance was calculated using Student's *t*-test. * $p < 0.05$; *** $p < 0.001$.

Finally, we examined whether the three miRNAs found correlated with the levels of the three inflammatory molecules from the PCR of the associated samples (see Figure 2). This showed that both IL6 and ICAM correlated well with the reads of the three miRNAs (for all correlations: $r > 0.89$, $p < 0.05$) (Figure 7A,B). Nevertheless, this was not the case for IL1β (*p* not significant for all the correlations).

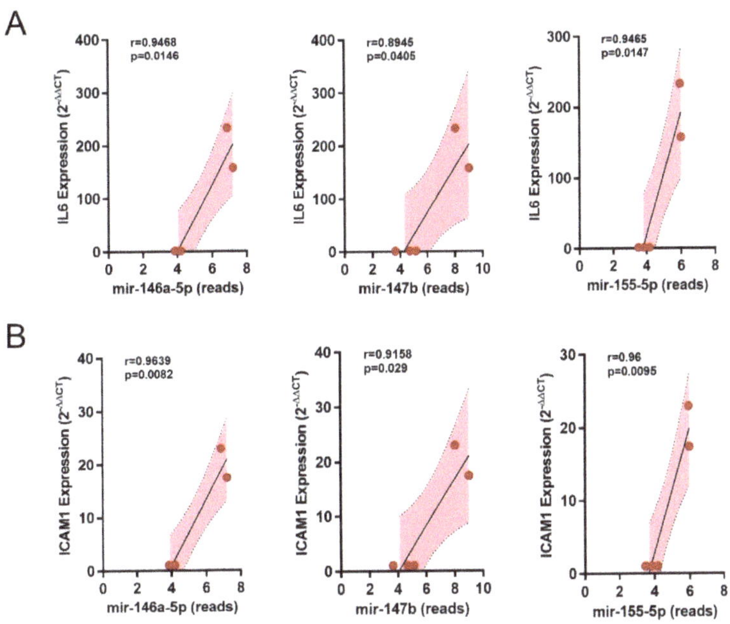


**Figure 7.** Correlations. Correlation of overlapping miRNAs (reads of miR-146a-5p, miR-147b, and miR-155-5p) with (**A**) IL6- and (**B**) ICAM1-mRNA expression.

## 3. Discussion

Understanding the precise molecular and cellular mechanisms that lead to inflammation in the kidney is the most important way to identify early targets for the prevention or treatment of acute kidney injury. Inflammation is a physiological process to protect the body from acute injury induced, for example, by toxins or pathogens. It is mediated by inflammatory cytokines, such as IL1β, IFNγ, and TNFα, released by immune cells, but also by resident tissue cells like PTCs. Thus, PTCs are involved in physiological and pathophysiological processes. The transmission of intercellular signal molecules is of great significance, including endocrine, paracrine, and autocrine biochemical pathways. Following injury or the stimulation of tubular epithelial cells, various cascades of mediator systems may be activated, leading to an increased local release of cytokines, chemokines, and other pro-inflammatory molecules. All these locally produced mediators may subsequently lead to injury enhancement, either directly or indirectly through increased the immigration of proinflammatory cells such as macrophages and lymphocytes [13]. In addition, the expression of various intracellular non-coding RNAs (miRNAs, long non-coding RNAs, and piRNAs) is altered, and extracellular vesicles as mediators of cell–cell communication are released with a cargo of inflammation-specific molecules [5,14]. Although the mechanisms regulating inflammation and the subsequent defense mechanism have not been fully elucidated, recent evidence suggests that non-coding RNAs, particularly miRNAs, play a critical role in the generation and control of the inflammatory response. These non-coding RNAs are critical for the stability and maintenance of mRNA at a post-transcriptional level. Several miRNAs, such as miR-146 and miR-155, have been shown to be important regulators of inflammation-related mediators [15]. Alexander and co-workers showed that miR-146a and miR-155 are present in exosomes and pass between immune cells in vivo and demonstrated that exosomal miR-146a inhibits while miR-155 promotes endotoxin-induced inflammation in mice [15].

There are currently limited data on altered miRNA expression in renal epithelial cells and their EVs during episodes of renal inflammation in general. Nevertheless, several studies investigated the involvement of miRNA during injury and repair in the kidney [16]. Recent studies have shown that miRNAs play a critical role in the development of acute renal failure. The overexpression of miR-150 in mice generated renal cell injury due to the downregulation of a growth factor receptor and the consequent enhancement of inflammation and the apoptosis of interstitial cells [17]. The upregulation of miR-687 in renal tubular epithelial cells both in vitro and in vivo induced renal injury by suppressing the expression of specific molecules, the activation of the cell cycle, and the induction of apoptosis [18]. On the other hand, the upregulation of miR-21 has been shown to ameliorate induced kidney injury by inhibiting inflammation and cell apoptosis [19]. Joo and co-workers demonstrated that miR-125b expression was enhanced in injury and protected the kidney from cisplatin-induced kidney injury [20]. Another study suggested that miR-155 had a protective role in cisplatin-induced nephrotoxicity [21]. The induction of miR-26a has been shown to increase tubular cell viability through the modulation of a growth and a transcription factor during kidney injury in an in vivo model [22]. Several other studies showed the influence of specific miR on cell viability and proliferation during renal regeneration (reviewed in [16]).

EVs released by (renal) cells are important mediators of cell–cell-communication and signal transduction. Urinary EVs have been intensely investigated and characterized for use as diagnostic non-invasive biomarkers of inflammatory renal diseases [23]. However, their specific cellular origin is unclear, as they could originate from any of several cell types from the kidney and bladder. For this reason, the accurate characterization of altered miRNA expression in inflammatory processes in a well-defined system of highly purified and differentiated PTCs and their released EVs is necessary. For example, Wang and co-workers identified differential expression levels of miRs and proteins from PTC-derived exosomes, a specific subpopulation of EVs, under different disease-culture conditions and described different readouts linked to renal pathological processes [14]. They identified different miRs differently expressed in exosomes between the control and the inflammatory group, such as miR-200a, miR-222, and miR-204, which were previously attributed a role within the renal disease pathways [14]. Another study by the same authors used a Transwell system to examine the molecular content and function of EVs released from the apical and the basolateral surface of polarized human primary tubular epithelial cells under inflammatory diseased conditions [5]. They used multiomic analysis to characterize the distinct molecular profiles of miRs, proteins, and lipids released from EVs of the apical versus basolateral cell membranes. The study demonstrated that EVs released, particularly basolaterally, play a central role in modulating tubulointerstitial inflammatory responses observed in immune-mediated renal disease.

Our data revealed that miR 146a, miR-155, and miR-147b were significantly upregulated in both PTCs and EVs after inflammatory stimulation. All three miRNAs have been implicated in inflammatory processes and are directly involved in the Toll-like receptor signaling pathway [24,25]. Following this, the expression of the three microRNAs has been found increased in patients with bacterial sepsis [26]. Whilst it has been proposed that the overexpression of miR-147b inhibited LPS-induced inflammation via the inhibition of the p38 MAPK signaling pathway and alleviates inflammation and apoptosis in acute lung injury, miR-146a and miR-155 have emerged as important transcriptional regulators in the regulation of NF-κB-driven inflammation [27,28]. LPS-induced inflammation in human microvascular endothelial cells showed increased levels of IL6 and ICAM1 together with an upregulation of miR-146a [29]. The overexpression of miR-146a resulted in the downregulation of IL6, as well as ICAM1, while the inhibition of miR-146a led to opposite effects. In this regard, Zhou and co-workers reported a positive feedback loop between miR-155-5p, STAT3, and NF-κB in A. cantonensis infection and the miR-155-5p-dependent expression of eosinophil inflammatory cytokines such as CCL6/C10, ICAM1, and MMP9 [30]. In inflammatory responses, the NF-κB-miR-155 axis and the NF-κB-miR-146a axis regulate the intensity and duration of inflammation [31]. As one of the first miRNAs identified as an immune system regulator, miR-146 is rapidly expressed upon NF-κB activation and downregulates IRAK1 and TRAF6 to reduce the activity of NF-κB by a negative regulatory loop mechanism [32]. miR-146a is involved in the pathogenesis and progression of several immune inflammatory diseases such as rheumatoid arthritis, inflammatory bowel disease, multiple sclerosis, psoriasis, atherosclerosis, hepatitis, or chronic obstructive pulmonary disease.

Like miR-146, miR-155 also plays a critical role in inflammatory diseases such as inflammatory autoimmune diseases (e.g., rheumatoid arthritis, systemic lupus erythematosus, multiple sclerosis, type 1 diabetes, and systemic sclerosis) and is abnormally expressed in inflammatory bowel disease, colitis ulcerosa, and neuroinflammation [33]. miR-155 is rapidly upregulated by NF-κB within the inflammatory response. Once activated, miR-155 controls the expression of both IKKβ and IKKε, forming a positive feedback loop, which leads to the repression of NF-κB activation [31]. Thus, miR-155 and miR-146 regulate the inflammatory response in a combinatorial manner [34]. Clinical and preclinical studies have shown that renal miR-155 expression increases significantly in acute renal failure [35]. An in vitro study has shown that the overexpression of miR-155 promoted cellular apoptosis and suppressed proliferation, while the inhibition of miR-155 expression exerted opposite effects [36]. For this reason, we additionally examined the correlation of the miRs found with the molecules IL6, IL1β, and ICAM1, which are increased by inflammation, and we were able to show that all three miRs correlate with the expression of IL6 and ICAM1, but not with that of IL1β.

miR-146a is also involved in the regulation of inflammation in the kidney. Fu and co-workers reported that miR-146a reduced lupus-erythematosus-induced renal injury in MRL/lpr mice by downregulating the NF-κB pathway [37]. In addition, a protective effect of miR-146 against kidney injury in diabetic nephropathy rats through mediating the NF-κB signaling pathway has also been shown [38]. It is tempting to speculate about the role of miR-147b in renal inflammation. Interestingly, a recent publication described renal injury due to the induction of miR-147 during cold-storage-associated transplantation [39]. miR-147 suppressed NDUFA4, resulting in mitochondrial damage and renal tubule cell death.

In summary, our current study provides further insight into the inflammatory response of renal proximal tubular epithelial cells. We characterized their altered expression of small RNAs in response to cytokine stimulation and compared it with their altered cargo to released EVs. We found three characteristic miRNAs, miR-146a, miR-147b, and miR-155, which were significantly upregulated in cells and in EVs, and thus may play a critical role in regulating the inflammatory response in the kidney.

## 4. Materials and Methods

### 4.1. Isolation and Culture of Human Renal Proximal Tubular Epithelial Cells

Human renal proximal tubular epithelial cells (PTCs) were separated as described previously [40]. Briefly, PTCs were isolated after nephrectomies from tissue not involved in renal cell carcinoma. Kidney tissue was minced with crossed blades, digested with collagenase/dispase, and forced through a sieve (106 μm). The tissue slurry was then incubated with collagenase IV, DNase, and MgCl2 and centrifuged over a Percoll density gradient. PTCs were then isolated in high purity using a mAb against aminopeptidase M (CD13) and the Mini-MACS system (Miltenyi, Bergisch Gladbach, Germany). Cells were characterized as previously described [40,41], seeded in cell culture plates, and cultured in standard culture medium (Medium 199 (M4530, Sigma, Taufkirchen, Germany) with a physiologic glucose concentration (100 mg/dL) and with 10% fetal bovine serum (FBS; Biochrom, Berlin, Germany)). The medium was replaced every three to four days, and confluent cells were passaged via trypsinization. Cells isolated from four different patients were used. Cells between passages 3 and 6 were used for the experiments. Selected PTCs were used for further characterizations via flow cytometry, as previously described [40].

### 4.2. Preconditioning in an Inflammatory Microenvironment

Cells were grown in culture flasks (75cm$^2$) in standard cell culture medium until confluence. The cells were then washed and kept in serum-free medium 199 for 48 h, either in standard medium (controls) or in an inflammatory microenvironment. The inflammatory microenvironment was induced by culturing PTCs in medium with a mixture of cytokines (CM) containing γ-interferon (200 U/mL), interleukin-1β (25 U/mL), and TNF-α (10 ng/mL). Then, PTCs were used for direct miR isolation (see Section 4.5), and EVs were isolated from the supernatant via size exclusion chromatography (SEC) (see Section 4.3).

In addition, the effect of the inflammatory microenvironment was checked via qPCR analysis. The complete method was described earlier [42]. The primers used were IL6 (150 bp, forward AAA GAT GGC TGA AAA AGA TGG ATG C, reverse ACA GCT CTG GCT TGT TCC TCA CTA C), IL1β (83 bp, forward AGC TGA TGG CCC TAA ACA GA, reverse AGA TTC GTA GCT GGA TGC CG), ICAM1 (135 bp, forward CAA CCT CAG CCT CGC TAT GG, reverse CGG GGC AGG ATG ACT TTT GA).

### 4.3. Isolation of Extracellular Vesicles

The preconditioned supernatant was used to isolate EVs from cells cultured under normal and inflammatory pretreatment for 48 h (growth area of 150 cm$^2$ with 16 mL of serum-free DMEM). After 48 h of incubation, the medium was centrifuged for 10 min at 600× *g* in order to remove cell debris, and then filtered using a 0.45 μm PVDF filter and concentrated approximately 40-fold via centrifugation at 2800× *g* for 20 min using a Vivaspin® Turbo 30 MWCO Centrifugal Filter. The concentrated medium was then used for the isolation of EVs via SEC using Sepharose CL-2B columns [43,44]. The concentrated medium was applied onto a Sepharose CL-2B column, which was washed and equilibrated with PBS. As elution buffer, PBS was applied to the column until 18 flow-through fractions with a respective volume of 500 μL were collected. The fractions 7–12 containing extracellular vesicles were pooled and subsequently concentrated (to approximately 400 μL EV solution) using a 3 kDa molecular weight cut-off Amicon filter by centrifuging for 25 min at 2800× *g*. EV samples were used immediately for characterization or miR isolation. Selected EV isolations were used for further characterizations.

### 4.4. Characterization of Extracellular Vesicles

We used nanoparticle tracking analysis (NTA) to determine the size distribution and concentration of the isolated EVs, as previously described [42]. In brief, 10 μL of concentrated EV solution was diluted 1:100 and immediately analyzed using a NanoSight NS500 (Malvern Panalytical, Malvern, UK) according to the manufacturer's instructions (camera settings: level 14, camera gain 1.5, and temperature 28 °C). Videos were analyzed with the Nano Sight NTA 3.2 software (threshold 14, gain 1.5).

To further analyze PTC-EVs via flow cytometry, a commercially available Flow Detection Magnetic Bead Reagent Kit was used (Invitrogen, Carlsbad, CA, USA, No. 10618D (anti-EpCAM)). In brief, a 40 µL anti-EpCAM magnetic bead solution was added to 1 mL of isolation buffer (PBS, 2% FCS, 1 mM EDTA, 0.1% Sodium azide) and placed on a magnetic stand for 2 min. Then, the supernatant was removed, and the bound beads were washed and re-suspended in 90 µL of isolation buffer. Then, a 20 µL SEC-isolated PTC-EV solution in PBS was added and incubated overnight at 4 °C. The sample was then washed twice with 1 mL of isolation buffer before the beads were re-suspended in isolation buffer (300 µL). The bead-EV solution (100 µL) was then stained with 20 µL of APC-labeled anti-CD63 detection antibody (ImmunoTools, Friesoythe, Germany) and incubated under rotation at RT in the dark for 45 min and washed and re-suspended in a 300 µL isolation buffer. The samples were measured using a flow cytometer (BD Biosciences, Heidelberg, Germany) and analyzed on the instrument until 10,000 events were detected. As a control for the fluorescent measurements, a 100 µL bead-EV solution without detection antibody was mixed with 200 µL of isolation buffer and measured. In addition, we used commercially available Dynabeads M-280 (Fisher Scientific, Schwerte, Germany, No. 11205D) to calibrate forward and sideward scatter in the flow cytometric measurement.

*4.5. Isolation of miR from PTC*

For miR extraction from cultured PTC, the NucleoSpin® miRNA kit (Macherey-Nagel, Düren, Germany) was used in accordance with the manufactures protocol. First, cells were harvested by adding 1.0 mL of Nucleozol to each flask (75 $cm^2$) and then scratching all cells from the bottom. The mixture was then transferred into a new 1.5 mL tube, and 400 µL of RNase free water was added to each sample and mixed by shaking the tube vigorously multiple times. Afterwards, the mixture was incubated for 5 min at RT. The samples were centrifugated at 12,000× *g* for 15 min at RT. Then, 300 µL of each supernatant was transferred to a new tube, and 100 µL of 100% ethanol was added and incubated for 5 min at RT. Afterwards, a NucleoSpin RNA Column was placed into a 2 mL tube and the supernatants were loaded onto the column. The supernatant was collected for further miRNA extraction. Next, 350 µL of buffer was added to the column and the tubes were centrifugated for 1 min at 11,000× *g*. The flow-through was discarded, and 100 µL of rDNase was added to the column to digest DNA. The samples were incubated for around 15 min at RT. In the meantime, the flow-through containing small RNAs was used for adjusting binding conditions for small RNAs by adding 300 µL of Buffer MP to each flow-through. After vortexing the samples for 5 s, the tubes were centrifuged for 3 min at 11,000× *g* to pellet the protein. To remove the protein precipitate, the samples were loaded onto a NucleoSpin Protein Removal Column in a 2 mL tube. After a centrifugation step for 1 min at 11,000× *g*, the column was discarded, 800 µL of buffer was added to each flow-through, and the samples were vortexed for 5 s. To bind small RNA, for each sample, a new NucleoSpin RNA Column was placed in a 2 mL tube, and 725 µL of the sample was loaded onto the column. The tubes were centrifuged for 30 s at 11,000× *g*, and the flow-through was discarded. This step was repeated until the whole sample was loaded onto the column. In the next step, the columns were washed by adding 600 µL of buffer. After a centrifugation step for 30 s at 11,000× *g*, the flow-through was discarded, and afterwards, 700 µL of buffer was added to each sample. Again, the tube was centrifugated at 11,000× *g* for 30 s and, 250 µL of buffer was added and the tube was centrifugated for 2 min at 11,000× *g* to remove all residuals from the membrane. The column was transferred to a new collection tube, and 15–30 µL of nuclease-free water was added on the center of the membrane. After incubation for 1 min at RT, a centrifugation step followed for 1 min at 11,000× *g* to recover the miR.

*4.6. Isolation of miR from EVs*

Whole miR was isolated from EVs using the NuceloSpin® miRNA Plasma Kit (Macherey-Nagel, Düren, Germany) according to the manufacturer's instructions. Therefore, 90 µL of buffer was added to 300 µL of the sample and mixed by vortexing the tube for 5 s. The mixture was incubated for 3 min at RT, and in connection, 30 µL of buffer was added. The sample was mixed by vortexing it for 5 s, and it was incubated for 1 min at RT. Afterwards, the tube was centrifugated at 11,000× *g* for 3 min at RT. The clear supernatant was transferred to a new 2 mL tube and mixed with 400 µL of isopropanol. The mixture was then transferred onto an miRNA column which was placed into a collection tube and centrifugated at 11,000× *g* for 2 min at RT. The flow-through was discarded, and 100 µL of buffer was applied on the column. Again, the tube was centrifugated at 11,000× *g* for 30 s at RT, and 700 µL of buffer was added. After another centrifugation step at 11,000× *g* for 30 s at RT, the flow-through was discarded, and the washing step was repeated with 250 µL of buffer, and the tube was centrifugated for 2 min at 11,000× *g* to remove all residuals from the membrane. The column was transferred to a new collection tube, and 30 µL of nuclease-free water was added on the center of the membrane. After a centrifugation step for 1 min at 11,000× *g* to recover the miR, the tube was incubated for 1 min at RT.

*4.7. miR Sequencing*

The concentration of miR was analyzed using the HS RNA Kit for TapeStation 4150 (Agilent, Waldbronn, Germany). The miR library was prepared using the QIAseq miRNA Library Kit according to the manufacturer's instructions. After library preparation, the DNA concentration was analyzed by using a Qubit dsDNA Assay Kit in Qubit 3.0 Fluorometer (ThermoFisher Scientific, Darmstadt, Germany). In addition, the DNA quality was checked using an HS DNA Kit of the Bioanalyzer 2100 (Agilent, Waldbronn, Germany). Next-generation sequencing (NGS) was performed using the MiSeq Reagent Kit v3 (Illumina, San Diego, CA, USA), PhiX Sequencing Control v3, and MiSeqTM Desktop Sequencer. For this, samples were pre-diluted (1:5 or 1:10) and prepared for sequencing according to the manufacturer's instructions. Samples were loaded into the MiSeqTM Desktop Sequencer in a sequencing cassette, and sequencing was performed.

Coverage files were converted into raw count matrices using the Qiagen pipeline. A priori filtering for sparse read counts was applied. The bioinformatics tool iDEP.96 was used to statistically analyze and display the miR sequencing data [45]. We performed a hierarchical cluster analysis of raw count data using the normalized read counts of the DESeq2 package. Kyoto encyclopedia of genes and genomes (KEGG) pathway analysis based on microRNA signature was performed with miRNet (miRTarBase v8.0) [46] for all the identified miRNAs. The Venn diagram was created in Venny 2.1.0 [47].

*4.8. Validation of miR Expression*

For the validation of selected miRNAs, cDNA synthesis was performed using the miRCURY LNA miR Assay Kit (339340, Qiagen, Venlo, The Netherlands) according to the manufacturer's protocol. In brief, a master mix was prepared on ice containing 5 µL 2× miRCURY SYBR® Green Master Mix, 0.5 µL ROX Reference Dye, 1 µL PCR primer (miR-146a-5p, No. YP00204688; miR-147b, No. YP00204368; miR-155-5p, No. YP02104687; all from Qiagen) and 1 µL of RNase free water per sample. After vortexing, 7 µL of the Master mix and 3 µL of the cDNA template were loaded onto a 96-well PCR plate. For the no-template controls, 3 µL of RNase free water was used. Quantitative PCR was carried out under the following conditions: two minutes at 95 °C for enzyme activation, then 10 s at 95 °C and 60 s at 56 °C for 40 cycles. For the quantification of the PCR fragments, we used the ABI Prism® 7900HT Fast Real-Time PCR System with a Sequence Detection System SDS 2.4.1 (Thermo Fisher Scientific, Darmstadt, Germany). Relative quantification was carried out with the ΔΔCT method using spike-in control UniSP6 (Qiagen, No. YP00203954) for normalization [48], and the level of target gene expression was calculated using $2^{-\Delta\Delta Ct}$.

### 4.9. Statistical Analysis

For the analysis of qPCR data, we used Student's *t*-test. The data are expressed as mean ± SEM. $p$ values < 0.05 were considered significant. The statistical analysis and Pearson r value correlation of the measured data as well as their graphic representation was performed with the software GraphPad Prism 7.04.

**Supplementary Materials:** The supporting information can be downloaded at: https://www.mdpi.com/article/10.3390/ijms241311069/s1.

**Author Contributions:** Investigation, A.-K.N. and P.C.B.; conceptualization, P.C.B. and R.S.; formal analysis, P.C.B. and R.S.; writing, review, and editing, P.C.B. and R.S. All authors have read and agreed to the published version of the manuscript.

**Funding:** This research received no external funding.

**Institutional Review Board Statement:** The study was conducted in accordance with the Declaration of Helsinki and approved by the ethics committee of the clinic of the Goethe University, Frankfurt (UGO 03/10, Amendment, 5 December 2014).

**Informed Consent Statement:** Informed consent was obtained from all subjects involved in the study.

**Data Availability Statement:** Not applicable.

**Acknowledgments:** We thank Rita Schmitt-Prokopp and Michael Lein, SANA, Klinik für Urologie und Kinderurologie, Offenbach, Germany, for providing us with human renal tissue.

**Conflicts of Interest:** The authors declare no conflict of interest.

## References


1. Schlondorff, D.O. Overview of factors contributing to the pathophysiology of progressive renal disease. *Kidney Int.* **2008**, *74*, 860–866. [CrossRef] [PubMed]
2. McWilliam, S.J.; Wright, R.D.; Welsh, G.I.; Tuffin, J.; Budge, K.L.; Swan, L.; Wilm, T.; Martinas, I.-R.; Littlewood, J.; Oni, L. The complex interplay between kidney injury and inflammation. *Clin. Kidney J.* **2021**, *14*, 780–788. [CrossRef] [PubMed]
3. Vallon, V. The proximal tubule in the pathophysiology of the diabetic kidney. *Am. J. Physiol. Regul. Integr. Comp. Physiol.* **2011**, *300*, R1009–R1022. [CrossRef] [PubMed]
4. Andrade-Oliveira, V.; Foresto-Neto, O.; Watanabe, I.K.M.; Zatz, R.; Câmara, N.O.S. Inflammation in Renal Diseases: New and Old Players. *Front. Pharmacol.* **2019**, *10*, 1192. [CrossRef] [PubMed]
5. Wang, X.; Wilkinson, R.; Kildey, K.; Ungerer, J.P.J.; Hill, M.M.; Shah, A.K.; Mohamed, A.; Dutt, M.; Molendijk, J.; Healy, H.; et al. Molecular and functional profiling of apical versus basolateral small extracellular vesicles derived from primary human proximal tubular epithelial cells under inflammatory conditions. *J. Extracell. Vesicles* **2021**, *10*, e12064. [CrossRef] [PubMed]
6. Colombo, M.; Raposo, G.; Théry, C. Biogenesis, secretion, and intercellular interactions of exosomes and other extracellular vesicles. *Annu. Rev. Cell Dev. Biol.* **2014**, *30*, 255–289. [CrossRef]
7. Battistelli, M.; Falcieri, E. Apoptotic Bodies: Particular Extracellular Vesicles Involved in Intercellular Communication. *Biology* **2020**, *9*, 21. [CrossRef]
8. Tkach, M.; Théry, C. Communication by Extracellular Vesicles: Where We Are and Where We Need to Go. *Cell* **2016**, *164*, 1226–1232. [CrossRef] [PubMed]
9. Jeon, J.S.; Kim, E.; Bae, Y.-U.; Yang, W.M.; Lee, H.; Kim, H.; Noh, H.; Han, D.C.; Ryu, S.; Kwon, S.H. microRNA in Extracellular Vesicles Released by Damaged Podocytes Promote Apoptosis of Renal Tubular Epithelial Cells. *Cells* **2020**, *9*, 1409. [CrossRef]
10. Chung, A.C.-K.; Lan, H.Y. MicroRNAs in renal fibrosis. *Front. Physiol.* **2015**, *6*, 50. [CrossRef]
11. Khoury, S.; Tran, N. Circulating microRNAs: Potential biomarkers for common malignancies. *Biomark. Med.* **2015**, *9*, 131–151. [CrossRef] [PubMed]
12. Yang, Y.; Wang, Y.; Wei, S.; Zhou, C.; Yu, J.; Wang, G.; Wang, W.; Zhao, L. Extracellular vesicles isolated by size-exclusion chromatography present suitability for RNomics analysis in plasma. *J. Transl. Med.* **2021**, *19*, 104. [CrossRef] [PubMed]
13. Daha, M.R.; van Kooten, C. Is the proximal tubular cell a proinflammatory cell? *Nephrol. Dial. Transpl.* **2000**, *15* (Suppl. S6), 41–43. [CrossRef]
14. Wang, X.; Wilkinson, R.; Kildey, K.; Potriquet, J.; Mulvenna, J.; Lobb, R.J.; Möller, A.; Cloonan, N.; Mukhopadhyay, P.; Kassianos, A.J.; et al. Unique molecular profile of exosomes derived from primary human proximal tubular epithelial cells under diseased conditions. *J. Extracell. Vesicles* **2017**, *6*, 1314073. [CrossRef]
15. Alexander, M.; Hu, R.; Runtsch, M.C.; Kagele, D.A.; Mosbruger, T.L.; Tolmachova, T.; Seabra, M.C.; Round, J.L.; Ward, D.M.; O'Connell, R.M. Exosome-delivered microRNAs modulate the inflammatory response to endotoxin. *Nat. Commun.* **2015**, *6*, 7321. [CrossRef] [PubMed]

16. Liu, Z.; Wang, Y.; Shu, S.; Cai, J.; Tang, C.; Dong, Z. Non-coding RNAs in kidney injury and repair. *Am. J. Physiol. Cell Physiol.* **2019**, *317*, C177–C188. [CrossRef]
17. Ranganathan, P.; Jayakumar, C.; Tang, Y.; Park, K.; Teoh, J.; Su, H.; Li, J.; Kim, I.; Ramesh, G. MicroRNA-150 deletion in mice protects kidney from myocardial infarction-induced acute kidney injury. *Am. J. Physiol. Renal Physiol.* **2015**, *309*, F551–F558. [CrossRef]
18. Bhatt, K.; Wei, Q.; Pabla, N.; Dong, G.; Mi, Q.-S.; Liang, M.; Mei, C.; Dong, Z. MicroRNA-687 Induced by Hypoxia-Inducible Factor-1 Targets Phosphatase and Tensin Homolog in Renal Ischemia-Reperfusion Injury. *J. Am. Soc. Nephrol.* **2015**, *26*, 1588–1596. [CrossRef]
19. Zhang, W.; Shu, L. Upregulation of miR-21 by Ghrelin Ameliorates Ischemia/Reperfusion-Induced Acute Kidney Injury by Inhibiting Inflammation and Cell Apoptosis. *DNA Cell Biol.* **2016**, *35*, 417–425. [CrossRef]
20. Joo, M.S.; Lee, C.G.; Koo, J.H.; Kim, S.G. miR-125b transcriptionally increased by Nrf2 inhibits AhR repressor, which protects kidney from cisplatin-induced injury. *Cell Death Dis.* **2013**, *4*, e899. [CrossRef]
21. Pellegrini, K.L.; Han, T.; Bijol, V.; Saikumar, J.; Craciun, F.L.; Chen, W.W.; Fuscoe, J.C.; Vaidya, V.S. MicroRNA-155 deficient mice experience heightened kidney toxicity when dosed with cisplatin. *Toxicol. Sci.* **2014**, *141*, 484–492. [CrossRef] [PubMed]
22. Gattai, P.P.; Maquigussa, E.; Da Silva Novaes, A.; Da Silva Ribeiro, R.; Varela, V.A.; Ormanji, M.S.; Boim, M.A. miR-26a modulates HGF and STAT3 effects on the kidney repair process in a glycerol-induced AKI model in rats. *J. Cell. Biochem.* **2018**, *119*, 7757–7766. [CrossRef] [PubMed]
23. Lv, L.-L.; Feng, Y.; Tang, T.-T.; Liu, B.-C. New insight into the role of extracellular vesicles in kidney disease. *J. Cell. Mol. Med.* **2019**, *23*, 731–739. [CrossRef]
24. Lin, L.; Hu, K. MiR-147: Functions and Implications in Inflammation and Diseases. *Microrna* **2021**, *10*, 91–96. [CrossRef] [PubMed]
25. Nahid, M.A.; Satoh, M.; Chan, E.K. MicroRNA in TLR signaling and endotoxin tolerance. *Cell. Mol. Immunol.* **2011**, *8*, 388–403. [CrossRef] [PubMed]
26. Trung, N.T.; Lien, T.T.; Sang, V.V.; Hoan, N.X.; Manh, N.D.; Thau, N.S.; Quyen, D.T.; Hien, T.T.T.; Hoan, P.Q.; Bang, M.H.; et al. Circulating miR-147b as a diagnostic marker for patients with bacterial sepsis and septic shock. *PLoS ONE* **2021**, *16*, e0261228. [CrossRef]
27. Xu, Q.; Huang, G.-D.; Duan, G.-C.; Qin, H.-J. MicroRNA-147b alleviates inflammation and apoptosis in acute lung injury via inhibition of p38 MAPK signaling pathway. *Eur. Rev. Med. Pharmacol. Sci.* **2021**, *25*, 1974–1981. [CrossRef]
28. Ma, X.; Becker Buscaglia, L.E.; Barker, J.R.; Li, Y. MicroRNAs in NF-kappaB signaling. *J. Mol. Cell Biol.* **2011**, *3*, 159–166. [CrossRef]
29. Feng, L.-L.; Xin, W.-N.; Tian, X.-L. MALAT1 modulates miR-146's protection of microvascular endothelial cells against LPS-induced NF-κB activation and inflammatory injury. *Innate Immun.* **2019**, *25*, 433–443. [CrossRef]
30. Zhou, X.; Zhang, J.; Liu, J.; Guo, J.; Wei, Y.; Li, J.; He, P.; Lan, T.; Peng, L.; Li, H. MicroRNA miR-155-5p knockdown attenuates Angiostrongylus cantonensis-induced eosinophilic meningitis by downregulating MMP9 and TSLP proteins. *Int. J. Parasitol.* **2021**, *51*, 13–22. [CrossRef]
31. Mahesh, G.; Biswas, R. MicroRNA-155: A Master Regulator of Inflammation. *J. Interferon Cytokine Res.* **2019**, *39*, 321–330. [CrossRef]
32. Mortazavi-Jahromi, S.S.; Aslani, M.; Mirshafiey, A. A comprehensive review on miR-146a molecular mechanisms in a wide spectrum of immune and non-immune inflammatory diseases. *Immunol. Lett.* **2020**, *227*, 8–27. [CrossRef] [PubMed]
33. Hu, J.; Huang, S.; Liu, X.; Zhang, Y.; Wei, S.; Hu, X. miR-155: An Important Role in Inflammation Response. *J. Immunol. Res.* **2022**, *2022*, 7437281. [CrossRef]
34. Mann, M.; Mehta, A.; Zhao, J.L.; Lee, K.; Marinov, G.K.; Garcia-Flores, Y.; Lu, L.-F.; Rudensky, A.Y.; Baltimore, D. An NF-κB-microRNA regulatory network tunes macrophage inflammatory responses. *Nat. Commun.* **2017**, *8*, 851. [CrossRef] [PubMed]
35. Saikumar, J.; Hoffmann, D.; Kim, T.-M.; Gonzalez, V.R.; Zhang, Q.; Goering, P.L.; Brown, R.P.; Bijol, V.; Park, P.J.; Waikar, S.S.; et al. Expression, circulation, and excretion profile of microRNA-21, -155, and -18a following acute kidney injury. *Toxicol. Sci.* **2012**, *129*, 256–267. [CrossRef] [PubMed]
36. Zhang, X.-B.; Chen, X.; Li, D.-J.; Qi, G.-N.; Dai, Y.-Q.; Gu, J.; Chen, M.-Q.; Hu, S.; Liu, Z.-Y.; Yang, Z.-M. Inhibition of miR-155 Ameliorates Acute Kidney Injury by Apoptosis Involving the Regulation on TCF4/Wnt/β-Catenin Pathway. *Nephron* **2019**, *143*, 135–147. [CrossRef]
37. Fu, H.-X.; Fan, X.-P.; Li, M.; Liu, M.-J.; Sun, Q.-L. MiR-146a relieves kidney injury in mice with systemic lupus erythematosus through regulating NF-κB pathway. *Eur. Rev. Med. Pharmacol. Sci.* **2019**, *23*, 7024–7032. [CrossRef]
38. Yu, H.-Y.; Meng, L.-F.; Lu, X.-H.; Liu, L.-H.; Ci, X.; Zhuo, Z. Protective effect of miR-146 against kidney injury in diabetic nephropathy rats through mediating the NF-κB signaling pathway. *Eur. Rev. Med. Pharmacol. Sci.* **2021**, *25*, 6445. [CrossRef]
39. Zhu, J.; Xiang, X.; Hu, X.; Li, C.; Song, Z.; Dong, Z. miR-147 Represses NDUFA4, Inducing Mitochondrial Dysfunction and Tubular Damage in Cold-storage Kidney Transplantation. *J. Am. Soc. Nephrol.* 2023; *Online ahead of print*. [CrossRef]
40. Baer, P.C.; Nockher, W.A.; Haase, W.; Scherberich, J.E. Isolation of proximal and distal tubule cells from human kidney by immunomagnetic separation. Technical note. *Kidney Int.* **1997**, *52*, 1321–1331. [CrossRef] [PubMed]
41. Baer, P.C.; Bereiter-Hahn, J.; Schubert, R.; Geiger, H. Differentiation status of human renal proximal and distal tubular epithelial cells in vitro: Differential expression of characteristic markers. *Cells Tissues Organs (Print)* **2006**, *184*, 16–22. [CrossRef]
42. Koch, B.; Geßner, A.; Farmand, S.; Fuhrmann, D.C.; Chiocchetti, A.G.; Schubert, R.; Baer, P.C. Effects of Hypoxia on RNA Cargo in Extracellular Vesicles from Human Adipose-Derived Stromal/Stem Cells. *Int. J. Mol. Sci.* **2022**, *23*, 7384. [CrossRef] [PubMed]

43. Mol, E.A.; Goumans, M.-J.; Doevendans, P.A.; Sluijter, J.P.G.; Vader, P. Higher functionality of extracellular vesicles isolated using size-exclusion chromatography compared to ultracentrifugation. *Nanomedicine* **2017**, *13*, 2061–2065. [CrossRef]
44. Böing, A.N.; van der Pol, E.; Grootemaat, A.E.; Coumans, F.A.W.; Sturk, A.; Nieuwland, R. Single-step isolation of extracellular vesicles by size-exclusion chromatography. *J. Extracell. Vesicles* **2014**, *3*, 23430. [CrossRef] [PubMed]
45. Ge, X. iDEP Web Application for RNA-Seq Data Analysis. *Methods Mol. Biol.* **2021**, *2284*, 417–443. [CrossRef]
46. Chang, L.; Zhou, G.; Soufan, O.; Xia, J. miRNet 2.0: Network-based visual analytics for miRNA functional analysis and systems biology. *Nucleic Acids Res.* **2020**, *48*, W244–W251. [CrossRef] [PubMed]
47. Oliveros, J.C. Venny. An Interactive Tool for Comparing Lists with Venn's Diagrams. Available online: https://bioinfogp.cnb.csic.es/tools/venny/index.html (accessed on 22 May 2023).
48. Faraldi, M.; Gomarasca, M.; Sansoni, V.; Perego, S.; Banfi, G.; Lombardi, G. Normalization strategies differently affect circulating miRNA profile associated with the training status. *Sci. Rep.* **2019**, *9*, 1584. [CrossRef]

International Journal of *Molecular Sciences*

MDPI

*Article*

# Determination of Krogh Coefficient for Oxygen Consumption Measurement from Thin Slices of Rodent Cortical Tissue Using a Fick's Law Model of Diffusion

**D. Alistair Steyn-Ross [1,*], Moira L. Steyn-Ross [1], Jamie W. Sleigh [2] and Logan J. Voss [3]**

1 School of Engineering, University of Waikato, Hamilton 3240, New Zealand; msr@waikato.ac.nz
2 Anaesthesia Department, Waikato District Health Board, Hamilton 3204, New Zealand; jamie.sleigh@waikatodhb.health.nz
3 Department of Anaesthesia, Waikato Clinical Campus, University of Auckland, Hamilton 3204, New Zealand; logan.voss@waikatodhb.health.nz
* Correspondence: asr@waikato.ac.nz

**Abstract:** To investigate the impact of experimental interventions on living biological tissue, ex vivo rodent brain slices are often used as a more controllable alternative to a live animal model. However, for meaningful results, the biological sample must be known to be healthy and viable. One of the gold-standard approaches to identifying tissue viability status is to measure the rate of tissue oxygen consumption under specific controlled conditions. Here, we work with thin (400 μm) slices of mouse cortical brain tissue which are sustained by a steady flow of oxygenated artificial cerebralspinal fluid (aCSF) at room temperature. To quantify tissue oxygen consumption ($Q$), we measure oxygen partial pressure ($pO_2$) as a function of probe depth. The curvature of the obtained parabolic (or parabola-like) $pO_2$ profiles can be used to extract $Q$, providing one knows the Krogh coefficient $K_t$, for the tissue. The oxygen trends are well described by a Fick's law diffusion–consumption model developed by Ivanova and Simeonov, and expressed in terms of ratio $(Q/K)$, being the rate of oxygen consumption in tissue divided by the Krogh coefficient (oxygen diffusivity × oxygen solubility) for tissue. If the fluid immediately adjacent to the tissue can be assumed to be stationary (i.e., nonflowing), one may invoke conservation of oxygen flux $K \cdot (\partial P/\partial x)$ across the interface to deduce $(K_t/K_f)$, the ratio of Krogh coefficients for tissue and fluid. Using published interpolation formulas for the effect of salt content and temperature on oxygen diffusivity and solubility for pure water, we estimate $K_f$, the Krogh coefficient for aCSF, and hence deduce the $K_t$ coefficient for tissue. We distinguish experimental uncertainty from natural biological variability by using pairs of repeated profiles at the same tissue location. We report a dimensionless Krogh ratio $(K_t/K_f) = 0.562 \pm 0.088$ (mean ± SD), corresponding to a Krogh coefficient $K_t = (1.29 \pm 0.21) \times 10^{-14}$ mol/(m s Pa) for mouse cortical tissue at room temperature, but acknowledge the experimental limitation of being unable to verify that the fluid boundary layer is truly stationary. We compare our results with those reported in the literature, and comment on the challenges and ambiguities caused by the extensive use of 'biologically convenient' non-SI units for tissue Krogh coefficient.

**Keywords:** Krogh coefficient; mouse cortical slice; diffusion–consumption equation; oxygen flux in brain tissue; thin-slice metabolism

**Citation:** Steyn-Ross, D.A.; Steyn-Ross, M.L.; Sleigh, J.W.; Voss, L.J. Determination of Krogh Coefficient for Oxygen Consumption Measurement from Thin Slices of Rodent Cortical Tissue Using a Fick's Law Model of Diffusion. *Int. J. Mol. Sci.* **2023**, *24*, 6450. https://doi.org/10.3390/ijms24076450

Academic Editors: Patrick C. Baer and Ralf Schubert

Received: 30 December 2022
Revised: 13 March 2023
Accepted: 24 March 2023
Published: 29 March 2023



## 1. Introduction

The acute ex vivo brain slice technique [1] is widely used by electrophysiologists as an experimental tool for probing neurophysiological health and disease. Because brain slices lack a blood flow, a continuous supply of glucose and oxygen must enter the tissue by diffusion from the perfusing liquid that serves as an artificial cerebral spinal fluid (aCSF) to maintain cell vitality. This slow diffusion of essential elements, combined with inevitable mechanical damage by the sectioning procedure, compromises the viability of the bulk tissue. As a result, for slice experimentalists in general (and particularly those wishing to follow the time course of tissue repair and regeneration), it is essential to have an objective and validated method for quantifying the health of test slices. Electrophysiological methods are commonly used to assess slice viability [2,3]. However, functional electrophysiological output, while helpful, does not provide a direct or complete readout of tissue status [4]. A cleaner approach would be to measure tissue oxygen consumption—healthier tissue consumes more oxygen [5–7]—providing a direct signal of cell viability.

Oxygen consumption can be quantified from thin sections of living tissue by profiling oxygen tension (partial pressure) as a function of tissue depth, probing from the upper surface to the lower surface. The curvature of the pressure vs. depth profile allows extraction of $Q$, the rate of oxygen consumption per unit volume of tissue, provided one knows the Krogh coefficient, $K_t$, for the tissue. The Krogh coefficient is a lumped constant that describes oxygen permeability, being the product of oxygen diffusion coefficient ($D$) and oxygen solubility ($S$). Because oxygen solubility is difficult to measure in metabolising tissue [8], the $K_t$ lumped constant is usually reported for active tissue experiments. A wide range of values of Krogh coefficient for biological tissue appear in the literature [8–15], so it is not immediately clear which value to choose. In addition, most reported values assume a bath temperature of 37 °C, while many experiments (including those reported here) are run at room temperature (20–25 °C). The fluid Krogh coefficients $K_f$ for pure water and saline are known to increase with temperature, so it is very probable that a temperature correction for $K_t$ would also be required. For these reasons, we sought to derive a tissue $K_t$ value specific to our experimental conditions and, in so doing, provide a unified theoretical and experimental framework for deriving $K_t$ for other experimental models.

A complicating factor when attempting to compare metabolic studies is the fact that at least six different compound units for Krogh coefficient are in common use (see Table 1 for a survey), and while these are all metric combinations, none adhere to the SI standard. This unfortunate state of affairs makes study comparisons challenging since multiple unit interconversions are required, leading to potential ambiguity and error. In this paper, our measurements for oxygen partial pressure [mmHg] and probe depth [µm] are dictated by the instrumentation at hand (Clark-style oxygen electrode and micromanipulator, respectively), but we ensure that our final Krogh results are either quoted in ratio form $(K_t/K_f)$ [tissue:fluid Krogh ratio, dimensionless] or using standard SI [mol/(m·s·Pa)].

**Table 1.** Selection of non-SI units for oxygen Krogh coefficient appearing in the physiology literature. The numerator represents quantity of gas, e.g., (mmol $O_2$), (mL $O_2$), (cm$^3$ $O_2$), etc. Bracketed rows of the table indicate equivalent units.

| | Unit | Reference | |
|---|---|---|---|
| 1 | $\frac{\text{mol}}{\text{m}\cdot\text{s}\cdot\text{Pa}}$ | SI | |
| 2 | $\frac{\text{mmol}}{\text{cm}\cdot\text{min}\cdot\text{mmHg}}$ | Ivanova & Simeonov (2012) [10] | } |
| 3 | $\frac{\text{mmol}}{\text{cm}\cdot\text{min}\cdot\text{torr}}$ | Kawashiro & Scheid (1976) [12] | } |
| 4 | $\frac{\text{nM}\cdot\text{mm}^3}{\text{mm}\cdot\text{s}\cdot\text{mmHg}}$ | van der Laarse et al. (2005) [13] | |
| 5 | $\frac{\text{mL}}{\text{cm}\cdot\text{min}\cdot\text{atm}}$ | Ganfield et al. (1970) [8] | } |
| 6 | $\frac{\text{cm}^3}{\text{cm}\cdot\text{min}\cdot\text{atm}}$ | Chen & Liew (1975) [11]; Poole et al. (2020) [15] | } |
| 7 | $\frac{\text{mL}}{\text{cm}\cdot\text{s}\cdot\text{mmHg}}$ | Sasaki et al. (2012) [14] | |

The goals in this paper are twofold. Building on earlier work by Ivanova & Simeonov (2012) [10], our first goal is to provide a unified mathematical foundation for the experimental determination of the Krogh coefficient in a thin slice of metabolically active tissue sustained by a continuous flow of oxygenated fluid. Our second is to demonstrate application of the theory to slices of mouse cortex in order to quantify error bounds on the

Krogh coefficient, and to identify limitations in the theory. Our overarching motivation is to provide a strong theoretical and experimental basis for quantifying oxygen consumption in mouse cortical slices, thereby allowing clear and unambiguous classification of tissue viability status.

The paper is structured as follows. In Section 2.1, we present the classical 1D Fick's law partial differential equation (PDE) model to describe the concentration-driven diffusion of oxygen through a thin slab of metabolically active tissue. At steady-state, the oxygen concentration and gas pressure (tension) at a given depth $x$ will be unchanging; so, the PDE reduces to a second-order differential equation in oxygen tension $P$ whose curvature is proportional to $Q$, the rate of oxygen consumption per unit volume of tissue. By selecting an idealised piecewise-linear model for consumption rate (Section 2.2), the differential equation can be solved exactly. In Section 2.3, we list the Ivanova & Simeonov [10] solutions for three distinct boundary conditions, but choose here to work with dimensioned quantities, and to show explicitly the mirror symmetry of the $P$ vs. $x$ solutions about the slice central axis at $x = 0$. In Section 2.4, we demonstrate the validity of these solutions via optimised curve fits to four representative oxygen profiles obtained from cortical slices sustained in our perfusion bath (described later in Section 3.1).

Section 2.5 surveys the surprisingly wide range of non-SI but 'biologically convenient' units used for Krogh coefficient in the literature, and comments on the potential ambiguity that can arise for unit conversions involving $V_m$, the molar volume of oxygen. In Section 2.6, we use literature sources to investigate the temperature dependence of the Krogh coefficient of pure water, then consider the effect on $K_f$ of adding salts and glucose to create aCSF, the artificial cerebrospinal fluid. We conclude the theoretical discussion in Section 2.7 by describing how the tissue Krogh coefficient $K_t$ can be determined via flux conservation across the tissue–fluid boundary.

Materials and Methods (Section 3) details how the slices of brain tissue are prepared and sustained with a steady flow of oxygenated aCSF in the perfusion bath (Section 3.1). Section 3.2 describes the experimental setup that allows dual-hemisphere recording of both LFP (local-field potential) electrical activity and oxygen-tension variation with depth using a pair of oxygen probes co-located with the LFP wire electrodes. Our standard protocol (Section 3.3) is to measure oxygen tension as a vertical profile, with soundings taken every 50 μm through the fluid and tissue. The resulting $pO_2$ pressure profiles are processed using custom-written MATLAB software (Section 3.4) to locate the slice centre, then extract the $(Q/K_t)$ curvature via iterative curve fitting. If a stationary fluid layer is detected, then the Krogh ratio $(K_t/K_f)$ can be determined, and hence an estimate for $K_t$ for a given $pO_2$ profile.

We present our results in Section 4 for a range of fluid flow rates. The statistics for our $K_t$ determinations show a large scatter about the mean value. By using pairs of repeated profiles at the same tissue location, we are able to distinguish natural biological variability from true experimental uncertainty. Our results suggest that, when comparing profiles from different locations within the same slice, natural variability in tissue Krogh coefficient is about three times larger than experimental errors arising from measurement uncertainty, curve fitting, and boundary gradient calculations. In Section 5, we discuss our findings, and compare our results with Krogh coefficient values reported by other workers. We acknowledge limitations in our experimental approach, and make suggestions for future work.

## 2. Theoretical Background: Model Equations and Solutions

### 2.1. Diffusion Equations

Krogh's foundation paper of 1919 [16] examined oxygen flow from a blood capillary source to surrounding metabolically active muscle tissue. He assumed an ideal two-dimensional (2D) geometry in which a cylindrical capillary of radius $r$ is enclosed within a concentric outer cylinder of tissue of radius $R > r$, then solved the 2D Fick's law of diffusion to give an expression for oxygen tension as a function of distance from the centre of the capillary.

For our ex vivo slice preparation, there is no blood supply. Instead, a steady flow of oxygenated perfusion fluid passing above and below the slice provides the tissue with oxygen via diffusion from the fluid to the tissue. The slice of brain material can be treated as a horizontal plane sheet of homogeneous tissue sufficiently thin that all of the diffusing oxygen enters through the top and bottom plane faces with a negligible amount through the edges (see Figure 1 for geometry). Oxygen concentration is then a function of vertical displacement from the centre of the slice, and the problem becomes one-dimensional.

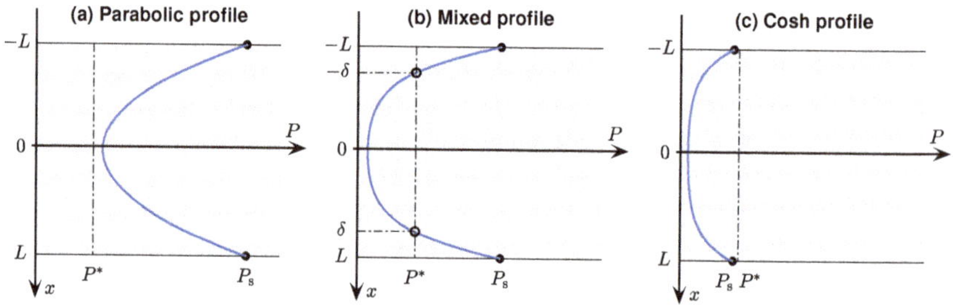


**Figure 1.** Representative $P(x)$ solutions of Ivanova & Simeonov diffusion–consumption Equations (4) and (5), assuming a piecewise linear model (6) for consumption rate (see blue curve in Figure 2). Shading represents thin slab of tissue which extends from $-L$ to $L$. Here, $P_s$ = surface pressure; $P^*$ = critical pressure below which oxygen consumption is restricted; $\delta$ = depth at which pressure reaches critical value. (**a**) If minimum tension exceeds $P^*$, pressure profile is a simple parabolic function of depth. (**b**) If the pressure profile crosses the $P^*$ boundary, the central portion forms a flattened hyperbolic-cosine (cosh) 'basin' that smoothly merges with parabolic 'wings' for $P > P^*$. (**c**) If the surface tension falls below the critical value, the cosh basin extends to the tissue boundaries.

We wish to derive an expression describing the steady-state condition in which the local diffusive supply of oxygen exactly balances the local metabolic demand. Our derivation of Equation (4) follows Ganfield et al. (1970) [8] but uses updated naming conventions for symbols.

The concentration $C$ of oxygen [SI units: mol/m$^3$] in the tissue is modelled as a 1D Fick's law diffusion–consumption process,

$$\frac{\partial C}{\partial t} = D\frac{\partial^2 C}{\partial x^2} - Q(C) \tag{1}$$

where $D$ is the diffusion coefficient [m$^2$/s] of oxygen in tissue, and $Q$ is the rate of oxygen consumption [mol/(m$^3\cdot$ s)]. From Henry's gas law, oxygen concentration $C$ in the tissue is proportional to local oxygen partial pressure (tension) $P$ [Pa] with proportionality constant $S$ [mol/(m$^3\cdot$ Pa)] being the gas solubility (also known as Henry's law constant $H$) in tissue, with

$$C = PS \tag{2}$$

allowing (1) to be rewritten in terms of oxygen tension $P$,

$$S\frac{\partial P}{\partial t} = DS\frac{\partial^2 P}{\partial x^2} - Q(P) \tag{3}$$

At steady state, the oxygen concentration (and pressure) is unchanging. Setting the left-hand side of (3) to zero gives the steady-state condition that diffusive oxygen supply in the tissue matches local metabolic demand,

$$DS\,\frac{\partial^2 P}{\partial x^2} = Q(P)$$

which we rewrite as,

$$K\,\frac{\partial^2 P}{\partial x^2} = Q(P) \tag{4}$$

where $K \equiv DS$ is the Krogh diffusion coefficient [mol/(m·s·Pa)], a lumped constant giving the permeability of oxygen in tissue.

Following Ivanova & Simeonov [10], we define oxygen flux $q$ (rate of oxygen flow per unit area of tissue) [mol/(m$^2$· s)],

$$q(x) = K\,\frac{\partial P}{\partial x} \tag{5}$$

(Strictly speaking, the right-hand side of (5) should carry a minus sign indicating that oxygen flows in the direction in which pressure *decreases*, but for simplicity we follow the Ivanova & Simeonov convention, and treat flux as an unsigned quantity).

Assume the tissue slice has thickness $2L$. The goal is to write down analytic expressions for oxygen consumption $Q(x)$ and flux $q(x)$ as a function of tissue depth $x$ (with $-L \leq x \leq L$), where $x = 0$ locates the centre of the slice.

Ivanova & Simeonov [10] have solved differential Equations (4) and (5) by assuming an idealised form (Figure 2) for the $Q(P)$ consumption vs. tension profile, then applying appropriate boundary conditions.

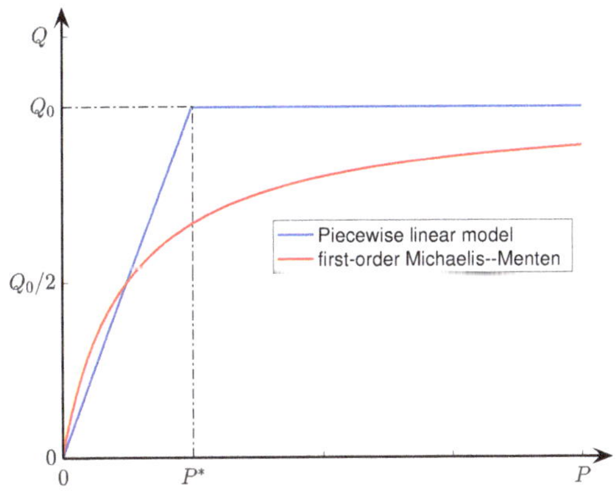


**Figure 2.** Two models for oxygen consumption rate in tissue as a function of oxygen tension. Ivanova and Simeonov's piecewise-linear model of Equation (6) (blue curve) approximates the first-order Michaelis–Menton form $Q(P) = Q_0\,P/(K_M + P)$ with Michaelis constant $K_M = P^*/2$ (red curve).

Their solutions take the form of pressure–depth ($P$ vs. $x$) profiles that are symmetric about $x = 0$, and take one of three forms:

(a) simple parabolic if minimum tension $P_{\text{min}}$ exceeds a critical tension value $P^*$;
(b) mixed parabolic/hyperbolic-cosine (cosh) if the $P(x)$ profile crosses the $P^*$ boundary;
(c) hyperbolic-cosine if the tension at the $x = L$ tissue surface, $P_s$, is less than $P^*$.

Once the theoretical pressure profile has been established, the oxygen consumption rate can be determined—provided the Krogh coefficient $K$ (tissue oxygen permeability) is known or can be assumed.

Because their solutions provide the foundation for the metabolic analyses reported here, we summarise the Ivanova and Simeonov (I & S) model and their mathematical solutions in the following two subsections.

### 2.2. Model for Oxygen Consumption vs. Oxygen Partial Pressure

Ivanova and Simeonov [10] treat respiring tissue as a continuous medium with distributed oxygen consumption that exhibits two behaviours: (i) If the local oxygen tension exceeds a critical pressure $P^*$, then consumption is maximal with $Q = Q_0$, a constant. In this regime, oxygen supply matches metabolic demand. (ii) If, however, oxygen tension falls below the critical value, $P < P^*$, then consumption drops off linearly with pressure such that $Q = 0$ at $P = 0$. This gives the I & S piecewise-linear consumption–pressure model,

$$Q(P) = \begin{cases} (P/P^*)\, Q_0, & 0 \le P \le P^* \\ Q_0, & P > P^* \end{cases} \tag{6}$$

illustrated in Figure 2. This simple piecewise form provides an elegant approximation to more complicated curvilinear relationships such as Michaelis–Menten and its generalisations,

$$Q(P) = Q_0 \frac{P^n}{K_M^n + P^n} \tag{7}$$

illustrated for the $n = 1$ case in Figure 2. The Michaelis constant $K_M$ is the oxygen pressure corresponding to half-maximum oxygen consumption: $Q(K_M) = Q_0/2$.

The significant advantage of the Ivanova and Simeonov piecewise form (6) is that it permits exact analytic solutions of the diffusion and flux differential Equations (4) and (5). In addition, pleasingly, we find that the I & S solutions provide a good match with the pressure profiles we measured within thin (400 μm) slices of mouse cortical tissue.

### 2.3. Solution of Ivanova & Simeonov Diffusion Equations

Oxygen flux is directed from the oxygenated aCSF (artificial cerebral spinal fluid) into the tissue depth through the surface boundary ($x = \pm L$) marking the upper and lower interfaces between tissue and the surrounding perfusion fluid. Oxygen diffusion and consumption within the tissue are controlled by the oxygen tension $P_s$ at the tissue surface. Because of the compound nature of the I & S oxygen consumption model (6), there are three distinct cases to consider:

(a) if the surface pressure $P_s$ is sufficiently high such that tissue pressure everywhere exceeds the critical value, i.e., $P(x) > P^*$, then oxygen consumption rate is constant throughout the tissue: $Q(x) = Q_0$;

(b) if the surface oxygen tension is insufficiently high, then the declining internal tissue pressure may reach the critical value at some depth $x = \pm\delta$, i.e., $P(|\delta|) = P^*$, in which case we separate the tissue into 'outer' layers ($\delta < x \le L$) and ($-L \le x < -\delta$) which are well supplied with oxygen so that $Q(x) = Q_0$, and an 'inner' layer ($-\delta \le x \le \delta$) with linearly restricted oxygen supply: $Q(P) = (P/P^*)Q_0$;

(c) if the surface pressure lies *below* the critical value, $P_s < P^*$, then $P(x) < P^*$ everywhere in the slice, and consequently consumption rate scales linearly with pressure: $Q(P) = (P/P^*)Q_0$.

These three cases result in three different tension–depth profiles as illustrated in Figure 1. Note that, because of the assumed slab symmetry of the 1D slice, the pressure curves are all symmetric about the slice centre at $x = 0$. The three sets of I & S solutions are itemised below:

**Case (a):** $P(x) > P^* \implies Q(x) = Q_0 = \text{const}$

$$P(x) = P_s - \frac{Q_0}{2K}\left(L^2 - x^2\right), \qquad -L \le x \le L \tag{8a}$$

$$q(x) = Q_0 x \tag{8b}$$

This is the simplest case. The tension profile is parabolic throughout, and the flux profile is linear with zero flux at the slice centre.

**Case (b):** Profile crosses critical value at depth $x = \pm\delta$: i.e., $P(|\delta|) = P^*$
For inner layer $(-\delta \le x \le \delta)$:

$$P(x) = P^* \frac{\cosh(\alpha x)}{\cosh(\alpha\,\delta)}, \qquad \text{with } \alpha \equiv \sqrt{\frac{Q_0}{P^* K}} \tag{9a}$$

$$q(x) = K \cdot \left(\alpha P^* \frac{\sinh(\alpha x)}{\cosh(\alpha\,\delta)}\right) \tag{9b}$$

For outer layer $(\delta < x \le L)$:

$$P(x) = P^* + [\alpha P^* \tanh(\alpha\,\delta)](x - \delta) + \frac{Q_0}{2K}(x - \delta)^2 \tag{10a}$$

$$q(x) = K \cdot \left(\alpha P^* \tanh(\alpha\,\delta) + \frac{Q_0}{K}(x - \delta)\right) \tag{10b}$$

These equations for pressure and flux also apply to the mirror-image outer layer $(-L \le x < -\delta)$ after making the substitution $(\delta \to -\delta)$ in Equation (10). Note that the value of $\delta$ in Equations (9) and (10) is unknown in advance, and must be determined numerically by solving the nonlinear equation,

$$\tanh(\alpha\,\delta) = \frac{1 - P_s/P^*}{\alpha(\delta - L)} + \tfrac{1}{2}\alpha(\delta - L) \tag{11}$$

**Case (c):** $P_s < P^* \implies$ entire profile lies below critical oxygen tension

$$P(x) = P_s \frac{\cosh(\alpha x)}{\cosh(\alpha L)}, \qquad -L \le x \le L \tag{12a}$$

$$q(x) = K \cdot \left(\alpha P_s \frac{\sinh(\alpha x)}{\cosh(\alpha L)}\right) \tag{12b}$$

Note that Ivanova & Simeonov present their theoretical solutions in terms of *dimensionless* lengths ($x_n \equiv \alpha x$, $L_n \equiv \alpha L$, $\delta_n \equiv \alpha\delta$), whereas we have chosen to work with dimensioned length quantities ($x$, $L$, $\delta$). With this change in notation, our Equations (8)–(12) are fully equivalent to their solutions (15)–(25).

A significant feature of these equations is that consumption rate $Q_0$ and Krogh coefficient $K$ always appear as a lumped ratio, $(Q_0/K)$; this is an inevitable consequence of the way diffusion equation (4) has been set up. This means that, using oxygen pressure measurements taken entirely within the slice, it is not possible to determine $Q_0$ unless the $K$ is known, and vice versa. However, if there is a stationary boundary layer of fluid immediately adjacent to the tissue surface, then we can assume conservation of oxygen flux across the surface (from fluid to tissue at $x = L$), and use the known value of Krogh coefficient of water to determine the Krogh coefficient for tissue. This idea is developed further in Section 2.7.

The ratio $(Q_0/K)$ [SI: $Pa/m^2$] has a physical meaning. For the pure parabolic profile (8) of Figure 1a, $(Q_0/K)$ is the *curvature* of the pressure parabola at the $x = 0$ midpoint of the tissue slab. Suppose the pressure curve is fitted to the polynomial $P(x) = ax^2 + c$, then the quadratic coefficient, doubled, gives the curvature: $(Q_0/K) = 2a$. More generally, $(Q_0/K)$ can be thought of as the *metabolic rate per unit permeability* for oxygen in tissue.

## 2.4. Representative Oxygen Tension Profiles

Figure 3 shows four representative sets of oxygen partial-pressure measurements obtained from 400-μm slices of mouse cortical tissue using the methods described in Section 3; note that, compared with the exemplar traces of Figure 1, the graph axes here have been rotated 90° counterclockwise.

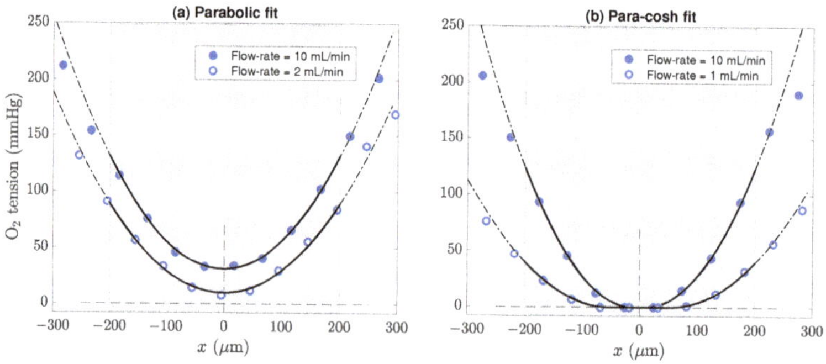


**Figure 3.** Parabolic and para–cosh curve fits for four representative tension vs. depth profiles measured from healthy slices of mouse brain tissue at room temperature (~20 °C). The grey shading indicates the 400-μm extent of the slice for $-200 \leq x/\mu m \leq 200$, with $x = 0$ μm marking the central plane of symmetry of the slice. (**a**) Parabolic profiles showing no restriction of metabolic rate. (**b**) Insufficient oxygen tension leads to flattened profiles with a cosh-modulated central basin joined to parabolic wings to the left and right.

Each dataset has been fitted to either the parabolic profile of Equation (8) as shown in Figure 3a, or to the mixed parabolic-cosh-parabolic profile of Equations (9) and (10) (Figure 3b). Oxygen tension reaches its minimum value at the centre of the slice. Typically, when the flow rate of the perfusing aCSF is reduced, oxygen tension at the tissue/fluid boundary is reduced, and the entire profile is displaced towards the zero-pressure axis. The flattened curves in panel (b) are an indicator of insufficient oxygen tension to satisfy local metabolic demand.

## 2.5. Choice of Units: SI vs. 'Biological'

As described in Section 2.1, the Krogh coefficient quantifies the permeability of oxygen in tissue, with SI units mol/(m·s·Pa) representing (molar amount of gas) per (unit distance in tissue × unit time × unit pressure). In the physiology literature, we find that it is common for authors to use a wide range of 'biologically convenient' (non-SI) units for the Krogh coefficient. Table 1 illustrates the dimensional diversity for choice of units with length (cm, mm); time (min, s); pressure (atm, mmHg, torr); gas amount (mmol, nM·mm$^3$, cm$^3$, mL).

While it is straightforward to convert between units of length, time, and pressure, e.g.,

$$1 \text{ atm} = 101.325 \text{ kPa} = 760 \text{ mmHg} = 760 \text{ torr}$$

the conversions between the various units for gas amount (mmol, cm$^3$, mL) require that we know the value of the gas molar volume at a specified temperature and pressure. If the author has not stated the temperature and pressure conditions for the metabolic measurements, then the gas molar volume is unknown. One might consider assuming IUPAC *standard* temperature and pressure (STP) conditions, or, alternatively, the NIST definition for *normal* temperature and pressure (NTP), and then apply ideal gas theory to compute $V_m$, the oxygen volume per mole,

$$pV = nRT \implies V_m = \frac{RT}{p}$$

where $R = 8.314462$ J/(mol·K) is the universal gas constant. The resulting STP and NTP values for $V_m$ for an ideal gas are listed in Table 2; note that the IUPAC definition for STP changed in 1982. We observe that the molar volume at NTP is ~10% larger than that for either definition of STP.

**Table 2.** Ideal-gas molar volumes (L/mol) for STP and NTP temperature and pressure definitions. Bracketed value in first row is the actual (non-ideal) molar volume for $O_2$ at STP (pre-1982).

| Standard | Temperature, $T$ | Pressure, $p$ | Molar Volume, $V_m$ |
|---|---|---|---|
| STP (IUPAC to 1982) | 273.15 K (0 °C) | 1 atm = 101.325 kPa | 22.414 (22.392) |
| STP (IUPAC after 1982) | 273.15 K (0 °C) | 100 kPa | 22.711 |
| NTP (NIST) | 293.15 K (20 °C) | 1 atm | 24.055 |

Of course, real gases are only approximately ideal. The measured molar volume for oxygen (at $T = 273.15$ K, $p = 1$ atm) is 22.392 L/mol ([17] [p. 329, Table 6]), about 0.1% below the ideal gas value of 22.414 L/mol.

### 2.6. Krogh Coefficient for Oxygen in Water

In our slice experiments (see Section 3 for details), oxygen is delivered to the tissue-slice via diffusion from the surrounding no-Magnesium (no-Mg) aCSF perfusate. Table 3 lists its chemical composition. The no-Mg fluid contains glucose; three chloride salts (NaCl, KCl, $CaCl_2$); and pH buffering agents ($NaHCO_3$, HEPES). We assume the buffering agents have negligible impact on oxygen transport and solubility.

**Table 3.** Chemical composition of Normal and no-Magnesium (no-Mg) artificial cerebrospinal fluids (aCSF) used in mouse cortical thin-slice experiments. Last column lists chloride concentrations for the no-Mg fluid. Total chloride concentration $\Sigma = 4.928$ g/L corresponds to chlorinity $[m_{Cl}] = 4.904$ g/kg.

| Species | Molar Mass | Normal aCSF | no-Magnesium aCSF | | Chloride Content |
|---|---|---|---|---|---|
| | (g/mol) | (mM) | (mM) | (g/L) | (g/L) |
| NaCl | 58.44 | 130 | 130 | 7.597 | 4.609 |
| KCl | 74.55 | 2.5 | 5 | 0.373 | 0.177 |
| $MgCl_2$ | 95.21 | 1 | — | — | — |
| $CaCl_2$ | 110.98 | 2 | 2 | 0.222 | 0.142 |
| $NaHCO_3$ | 84.01 | 2.5 | 2.5 | 0.210 | |
| NaOH | 40.00 | 3.5 | 3.5 | 0.140 | |
| HEPES | 238.30 | 10 | 10 | 2.383 | |
| D-glucose | 180.16 | 20 | 20 | 3.603 | |
| | | | | | $\Sigma = 4.928$ |

Relative to pure water, the glucose content of aCSF is expected to increase fluid viscosity slightly, thereby reducing gas diffusivity within the fluid. The presence of dissolved salts will reduce oxygen solubility ('salting-out' effect) by an amount proportional to total chlorinity (chloride content expressed in g of $Cl^-$ per kg of solution). In addition to these direct solute effects, both $O_2$ transport and solubility vary with temperature but in opposite directions: $O_2$ diffusivity increases moderately strongly with temperature, while $O_2$ solubility decreases less strongly, so the Krogh coefficient, being the product of diffusion and solubility, is expected to increase weakly as fluid temperature increases.

2.6.1. Diffusion Coefficient for Oxygen in Water

Han and Bartels (1996) [18] give an interpolation formula for the temperature dependence of oxygen diffusion [in $cm^2/s$] in pure water. We convert to SI units [$m^2/s$] by adjusting their base-10 exponential offset from $-4.410 \rightarrow -8.410$, and their formula now reads,

$$\log_{10}\left[\frac{D(T)}{m/s^2}\right] = -8.410 + \frac{773.8}{T} - \left(\frac{506.4}{T}\right)^2 \tag{13}$$

for temperature $T$ in kelvin. The resulting $O_2$ diffusion curve $D(T)$ is plotted in Figure 4a.

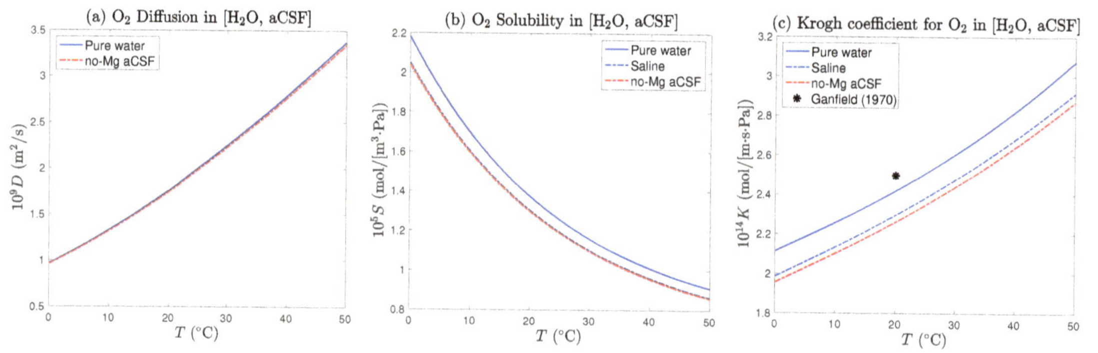


**Figure 4.** Temperature dependence of oxygen diffusion, solubility, and permeability in pure water, and in no-Mg artificial cerebral spinal fluid (aCSF). (**a**) Oxygen diffusion in aCSF (dashed-red trace) is depressed by ~1% relative to pure water (solid blue trace) by small viscosity increase caused by presence of 0.02 mol/L glucose. (**b**) Oxygen solubility in aCSF is reduced by ~5% by 'salting out' effect of chloride salts (dashed-blue trace) with chlorinity $[m_{Cl}] = 4.904$ g/kg; and by a further ~0.5% by glucose effect. (**c**) Oxygen Krogh coefficient in fluid is given by the product of diffusion and solubility, $K = DS$. Glucose effect is assumed to be cumulative, resulting in a ~1.5% depression below the saline curve. Asterisk (*) marks the value ($K^* = 2.5 \times 10^{-14}$ mol/(m·s·Pa)) used by Ganfield et. al. (1970) [8] for Krogh coefficient for fluid at 20 °C.

2.6.2. Solubility of Oxygen in Water

For oxygen solubility in pure water and saline solutions, Green and Carritt (1967) [19] provide an interpolation expression that shows explicit dependence on both fluid temperature and chloride content,

$$\alpha = 10^{-3}\exp\left\{(-7.424 + \frac{4417}{T} - 2.927\ln T + 0.04238\,T) - [m_{Cl}](-0.1288 + \frac{53.44}{T} - 0.04442\ln T + 7.145\times 10^{-4}\,T)\right\} \tag{14}$$

where $\alpha$ is the Bunsen absorption coefficient of oxygen, i.e., the volume of gas ($T = 0$ °C and $p = 1$ atm) dissolved per unit volume of solvent when the partial pressure of oxygen is 1 atm; $[m_{Cl}]$ is the chlorinity, defined as the amount of chloride in grams per 1000 g of solution. Expression (14) has been used by Stroe and Janssen (1993) [20] (but note their Equation (5) has omitted '$T$' in the final term in the chlorinity component), and the chlorinity expression of (14) is recommended by Forstner and Gnaiger (1983) [17].

To convert (14) from (volume ratio per atm) to oxygen solubility $S$ in SI units [$mol/(m^3{\cdot}Pa)$], we write

$$\frac{S(T,[m_{Cl}])}{mol/(m^3{\cdot}Pa)} = \alpha \times \left(\frac{1\text{ atm}}{101\,325\text{ Pa}}\right)\left(\frac{1\text{ mol}}{22.392\text{ L}}\right)\left(\frac{10^3\text{ L}}{1\text{ m}^3}\right) = 4.4075\times 10^{-4}\,\alpha \tag{15}$$

where the (22.392 L) divisor is the (non-ideal) molar volume for $O_2$ at 0 °C, 1 atm. The oxygen solubility curve for pure water and for no-Mg aCSF saline (chlorinity $[m_{\text{Cl}}] = 4.904$ g/kg) are illustrated in Figure 4b.

A recent NASA–JPL report ([21] [Table 5.5, pp. 5–162] on atmospheric data tabulates a list of Henry's law constants [mol/(L·atm)] for pure water. After converting to SI, the entry for oxygen reads,

$$\frac{S(T)}{\text{mol}/(\text{m}^3\cdot\text{Pa})} = 9.8692 \times 10^{-3} \exp\left[-161.6 + \frac{8160}{T} + 22.39 \ln T\right] \tag{16}$$

We compared $S(T)$ of (16) against the zero-chlorinity limit $S(T, 0)$ of (15) for the temperature range $0 \leq T/°\text{C} \leq 50$, finding excellent agreement with an rms difference of ∼0.72%. We choose to use (15) for our aCSF solubility modelling because of its convenient inclusion of salting-out effects.

For completeness, we also list the two-parameter expression for $O_2$ solubility in pure water extracted from the comprehensive online database of Henry's law constants maintained by Sander (2015) [22],

$$\frac{S(T)}{\text{mol}/(\text{m}^3\cdot\text{Pa})} = S_{\text{ref}} \exp\left[1500\left(\frac{1}{T} - \frac{1}{T_{\text{ref}}}\right)\right] \tag{17}$$

where $S_{\text{ref}} = 1.3 \times 10^{-5}$ mol/(m$^3$·Pa) at reference temperature $T_{\text{ref}} = 298.15$ K (25 °C). As expected, this two-parameter, two-sig-fig fit is less accurate than Green and Carritt (15) (rms difference ∼2.6% for [0–50] °C ), but has the advantage of direct interpretation of its numerical coefficients.

#### 2.6.3. Glucose Effects on Oxygen Diffusion and Solubility

Stroe-Biezen et al. (1993) (see Figures 7, 8 of [23]) quantified the effect of dissolved glucose on oxygen transport in water at 25 and 37 °C for glucose concentration $C_g$ ranging from 0 to 1 mol/L. Their $O_2$ diffusion and $O_2$ concentration graphs reveal slow monotonic decays as glucose content is increased. By interpolating for $C_g = 0.02$ mol/L (see Table 3), we are able to estimate the glucose depression effects for our no-Mg perfusion fluid:

- $O_2$ diffusion coefficient: 1% depression
- $O_2$ solubility: 0.5% depression

We assume these effects are temperature independent, but also cumulative when computing the Krogh permeability for oxygen in aCSF, leading to a net 1.5% depression below the saline curve as illustrated in Figure 4c.

Our brain-slice experiments were run at room temperature which typically lay within 20 to 25 °C. Reading off the aCSF curve (dashed-red trace) of Figure 4c, this temperature range corresponds to Krogh permeabilities [2.262 to 2.348] $\times 10^{-14}$ mol/[m·s·Pa], suggesting a reasonable working value for $K_f$, the Krogh coefficient for no-Mg aCSF at room temperature is

$$K_f = (2.30 \pm 0.04) \times 10^{-14} \text{ mol}/[\text{m}\cdot\text{s}\cdot\text{Pa}] \tag{18}$$

### 2.7. *Krogh Coefficient for Tissue via Flux Conservation*

In their oxygen profile measurements for a cat cerebral cortex, Ganfield et al. (see Figure 4 in [8]) observed a linear decrease in $O_2$ tension across a narrow layer (∼40 µm) of solution immediately above the surface of the tissue, indicating a thin boundary layer of nonflowing solution exists above the tissue slice. Equating the oxygen flux $q_f$ crossing this fluid layer to $q_t$, the flux crossing the tissue surface, and the tissue Krogh coefficient $K_t$ can be determined as a fraction of $K_f$, the (assumed) Krogh permeability of the fluid.

We now derive three distinct expressions for $K_t$ corresponding to the three possible $O_2$ tension profiles illustrated in Figure 1: (a) parabolic; (b) parabolic–cosh; (c) cosh.

Assume the perfusion fluid is nonflowing and biologically inert. Because no oxygen is consumed in the fluid, Equation (4) for $O_2$ diffusion within the fluid reads

$$\left(\frac{\partial^2 P}{\partial x^2}\right)_{\mathrm{f}} = 0 \quad \Longrightarrow \quad \left(\frac{\partial P}{\partial x}\right)_{\mathrm{f}} = c_1 \quad \Longrightarrow \quad P_{\mathrm{f}}(x) = c_1 x + c_2$$

with $c_1$, $c_2$ being constants. Thus pressure varies linearly with depth within the nonflowing fluid layer. Let the stationary layer have thickness $\Delta x$, then from definition (5), the $O_2$ flux across the fluid layer will be,

$$q_{\mathrm{f}} = K_{\mathrm{f}}\left(\frac{\partial P}{\partial x}\right)_{\mathrm{f}} \approx K_{\mathrm{f}}\left(\frac{\Delta P}{\Delta x}\right)_{\mathrm{f}} \tag{19}$$

where $\Delta P$ is the pressure change across fluid layer $\Delta x$.

**(a) Parabolic profile**

Define coefficient $a \equiv (Q_0/2K)$ in Equation (8a). Then, $O_2$ tension in tissue is

$$P_{\mathrm{t}}(x) = P_{\mathrm{s}} - a(L^2 - x^2) \tag{20}$$

giving a pressure gradient at the $x = L$ boundary,

$$\left.\left(\frac{\partial P_{\mathrm{t}}}{\partial x}\right)\right|_{x=L} = 2aL \tag{21}$$

so the incoming $O_2$ flux at the fluid–tissue interface will be

$$q_{\mathrm{t}}(L) = K_{\mathrm{t}}\left.\left(\frac{\partial P_{\mathrm{t}}}{\partial x}\right)\right|_{x=L} = 2aLK_{\mathrm{t}}. \tag{22}$$

Equating fluid flux (19) with tissue-boundary flux (22), we obtain the desired expression for Krogh permeability in tissue,

$$K_{\mathrm{t}} = K_{\mathrm{f}}\left(\frac{\Delta P}{\Delta x}\right)_{\mathrm{f}} \cdot \frac{1}{2aL} \tag{23}$$

Note that is a particular case of the general form which applies for all tissue profiles,

$$K_{\mathrm{t}} = K_{\mathrm{f}}\left(\frac{\Delta P}{\Delta x}\right)_{\mathrm{f}} \Big/ \left.\left(\frac{\partial P_{\mathrm{t}}}{\partial x}\right)\right|_{x=L} \tag{24}$$

**(b) Parabolic–cosh mixed profile**

Making the following substitutions in (10b) in favour of coefficient $a$,

$$\frac{Q_0}{2K_{\mathrm{t}}} = a, \qquad \alpha = \sqrt{\frac{Q_0}{P^* K_{\mathrm{t}}}} = \sqrt{\frac{2a}{P^*}}, \qquad \alpha P^* = \sqrt{2aP^*}$$

we obtain the pressure gradient at the $x = L$ boundary,

$$\left.\left(\frac{\partial P_{\mathrm{t}}}{\partial x}\right)\right|_{x=L} = \sqrt{2aP^*}\,\tanh\left(\sqrt{\frac{2a}{P^*}}\,\delta\right) + 2a(L-\delta) \tag{25}$$

Substituting (25) in (24) gives the tissue Krogh coefficient for the mixed pressure profile (hyperbolic-cosine centre with parabolic wings). Note that in the limit $\delta \to 0$, the paracosh boundary gradient (25) collapses to the (23) parabolic profile value of $2aL$, as expected.

**(c) Cosh profile**

From (12b), the pressure gradient at the fluid–tissue boundary $x = L$ is

$$\left(\frac{\partial P_t}{\partial x}\right)\bigg|_{x=L} = \alpha P_s \tanh(\alpha L) = \sqrt{\frac{2a}{P^*}}\, P_s \tanh\left(\sqrt{\frac{2a}{P^*}}\, L\right) \quad (26)$$

giving tissue Krogh permeability on substitution into (24).

## 3. Materials and Methods

In this section, we describe the experimental setup to measure $pO_2$ profiles in thin slabs of mouse cortical tissue, and outline the software techniques used to analyse the profiles to extract $(Q/K_t)$ curvatures and $(K_t/K_f)$ Krogh ratios.

### *3.1. Tissue Preparation*

The brain was rapidly dissected from adult male and female C57 mice anaesthetised with $CO_2$, then submerged in ice-cold 'Normal' artificial cerebrospinal fluid (aCSF); see Table 3 for chemical composition. Coronal 400 µm-thick sections were sliced between +1 to −5 Bregma using a vibrotome (Campden Instruments Ltd., Sileby, Leics, UK). The slices were then immersed at room temperature in no-magnesium (no-Mg) aCSF (see Table 3, columns 4–6). All solutions were made in double-distilled water and pre-oxygenated (95% oxygen, 5% nitrogen) using an oxygen concentrator (Perfecto2, Invacare, Auckland, New Zealand) prior to use.

### *3.2. Data Recording*

At least 60 min after preparing the tissue, the slices were transferred one at a time to a submersion-style perfusion bath based on a design by Thomas [24] (see Figure 5), and continuously perfused with aCSF. This design of the perfusion bath allows each hemisphere of the slice to be treated independently, but in this study both halves of the slice were exposed to identical experimental conditions. The aCSF flow rate was set at 0.5 mL/min for the repeated-profile experiments; flow values used in other experiments were [1, 2, 10] mL/min. Our expectation was that lower aCSF flow rates might favour the formation of a stationary boundary layer, but recognised that lower flows would restrict $O_2$ supply and could compromise tissue function.

Slices were perfused throughout with no-Mg aCSF. Removing magnesium from the aCSF unblocks NMDA receptors [25], resulting in the generation of ongoing spontaneous bursts of paroxysmal neuronal activity known as seizure-like events (SLEs) [26]; these fast voltage transients provide a convenient and sensitive indicator of tissue viability, and allow correlation of neurophysiological activity with tissue-oxygen profile characteristics.

SLEs were detected using a 75 µm diameter Ag/AgCl wire electrode inserted into the 400 µm-thick tissue slice, and measured as an extracellular local-field potential difference developed between active tissue and a common-ground electrode (Ag/AgCl disc) placed some distance from the slice in the perfusion bath. The analog signal was amplified (1000×) and filtered (bandpass 1–100 Hz, notch filter at 50 Hz) (Model 3000 differential amplifier, A-M Systems, Sequim, WA, USA), then digitised at 1000 $s^{-1}$ (PowerLab, ADInstruments, Bella Vista, NSW, Australia) and recorded using LabChart 7 (ADInstruments, Australia).

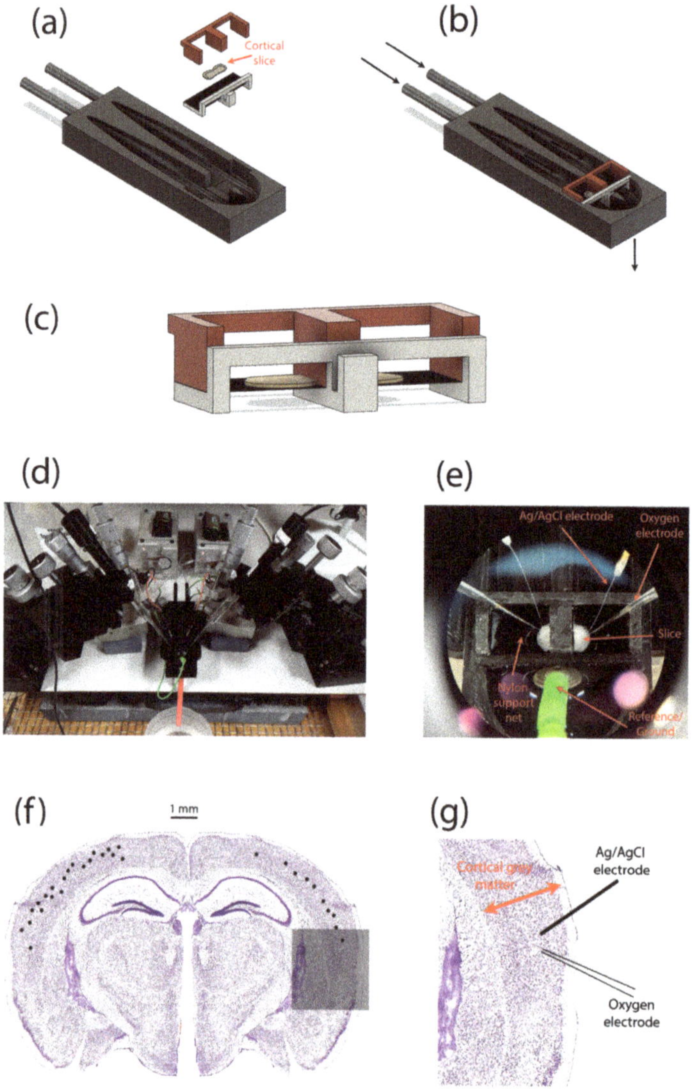


**Figure 5.** Perfusion bath and experimental setup for measuring oxygen-pressure profiles within a thin slice of mouse cortical tissue. (**a**) Exploded CAD model showing the separate components of the dual-compartment slice perfusion apparatus. (**b**) Assembled perfusion bath; arrows indicate direction of solution flow. (**c**) Enlarged view of the brain slice (grey 'pancake' shape) and its support structures; the slice sits on a nylon net (shown as a black-shaded horizontal surface). (**d**) Overview of experimental setup showing left and right pairs of micromanipulators, local-field potential (LFP) electrodes, oxygen probes. (**e**) Zoomed view of in situ cortical slice sitting on nylon net, and probed by LFP (Ag/AgCl wires) and oxygen (glass) electrodes. (**f**) Distribution of recording locations (black dots) within the slice cerebral cortex. For illustrative purposes, locations are shown on a single slice (but slices anterior to the one shown were used in some cases). Repeat profiles captured from some locations are not differentiated. (**g**) Representative arrangement of LFP electrode and oxygen probe for a single recording location, expanded from the shadowed region in (**f**).

Oxygen partial pressures ($pO_2$) were measured using a Clark-style oxygen electrode (50 μm tip diameter, Unisense Ltd., Aarhus, Denmark) inserted into the tissue slice. Prior to experiments, the oxygen electrode was polarised to ensure a stable output signal, then two-point calibrated using (a) aCSF equilibrated with room air ($pO_2$ = 160 mmHg), and (b) a solution of 0.1-M sodium ascorbate (zero-point). The $pO_2$ data were sampled at ~4.8 Hz using SensorTrace (v1.8, Unisense Ltd., Denmark).

### 3.3. Experimental Protocol

Once the slice was established in the perfusion bath, the Ag/AgCl wire electrode was inserted into layer III/IV of the cerebral cortex, with no specific cortical region targeted. If robust SLE activity was detected at that location (indicating viable tissue), the oxygen sensor was positioned using a precision micromanipulator (FX-117, Minitool Inc., Los Gatos, CA, USA) as close as practicable to the Ag/AgCl electrode, placing the tip of the oxygen probe in the bath fluid about 200 μm above the slice surface.

The $pO_2$ vertical profile was recorded every 50 μm by lowering the oxygen probe in 50-μm steps downwards through the upper fluid layer, then through the full depth of the 400-μm tissue slab, continuing until the tip had penetrated ~100 μm beyond the lower slice surface into the lower fluid layer. It was essential to include $O_2$ soundings in the upper and lower fluid layers for two reasons. First, because of lack of visual contrast, it was not possible to accurately discern when the probe entered or exited the tissue, so the locations of the slice centre ($x = 0$) and slice boundaries ($x = \pm L$) could only be determined later, during the curve-fitting stage, by assuming that the curve minimum marked the (zero-flux) centre of symmetry. Second, we were seeking evidence of a nonflowing boundary layer in order to apply the flux-conservation argument of Equation (19); this would manifest as an abrupt change in profile gradient on crossing the tissue–fluid boundary as the $P$ vs. $x$ profile curve transitions from parabolic to linear.

Some profiles exhibited a smooth tissue–fluid boundary crossing similar to those shown in Figure 3, so were not suitable for estimation of tissue Krogh coefficient. In contrast, Figure 6a,b,d display a clear gradient discontinuity at the lower tissue–fluid interface, but no gradient break at the upper interface. We attribute this profile asymmetry to a design characteristic of the slice perfusion system: the finely-woven nylon netting on which the slice sits (see Figure 5c,e) seems to support the formation a locally static fluid layer immediately adjacent to the lower surface of the slice, without impeding the bulk flow of solution below the net. If the fluid is static (i.e., no advection), then oxygen transport through the fluid layer adjacent to the slice is via diffusion only; therefore, diffusive flux through the layer should match that entering the tissue, so the Equation (19) assumption of flux conservation should be applicable here.

The fact that the probe detects a lower-surface stagnant layer in many—but not all—profiles suggests that the emergent location of the probe tip relative to the weave structure of the netting may be significant, but this aspect remains unresolved at present.

### 3.4. Numerical Methods and Curve Fitting

Each pressure profile consists of a vector of ~20 oxygen tension readings [mmHg] recorded at 50-μm intervals during a micromanipulator-controlled vertical descent starting in the fluid above the slice, then stepping down through the fluid into the tissue slice, through the slice, then continuing for several steps beyond the tissue into the fluid beneath. Tissue thickness is $2L = 400$ μm, so either 8 or 9 samples will lie within the tissue. The precise position of the oxygen probe tip relative to the tissue boundary is not easily visualised, so the metabolic centre of the slice is determined mathematically by fitting a parabola to the profile: the $dP/dx = 0$ pressure minimum locates the zero-flux reference depth $x = 0$.

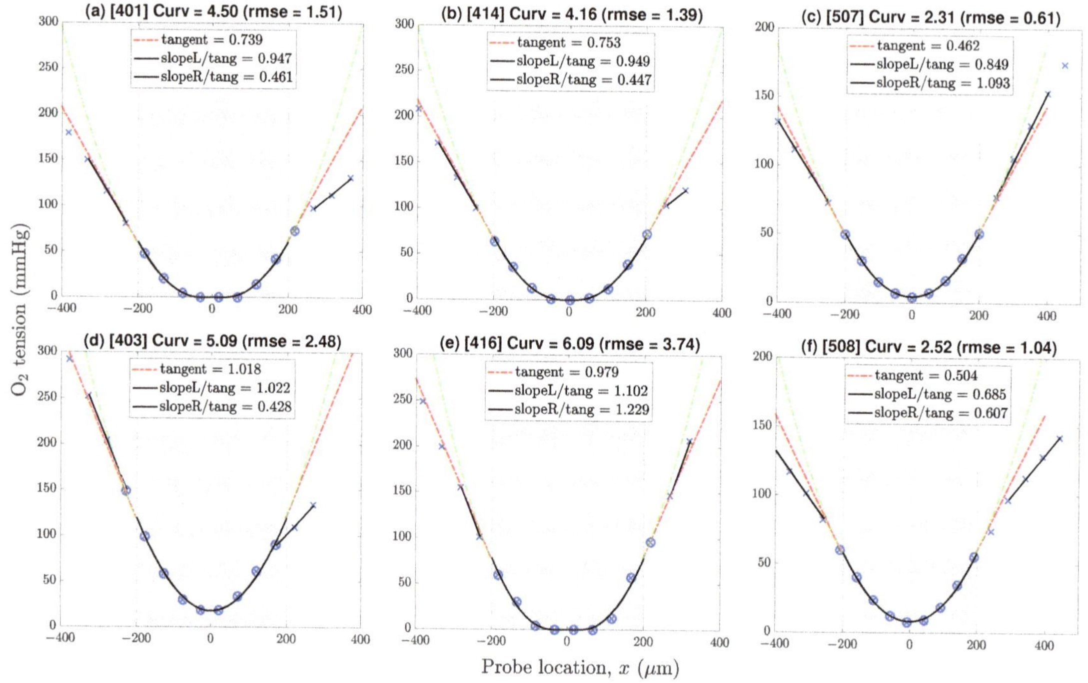


**Figure 6.** Representative $O_2$ pressure profiles drawn from datasets v4 (first two columns) and v5 (third column). Profiles (**a**,**b**,**d**,**f**) show a gradient discontinuity at the lower boundary; profiles (**c**,**e**) do not. Negative/positive locations correspond to points lying above/below the $x = 0$ line of symmetry. Key: × = sample values; ⊗ = samples selected for 9-point in-tissue curve-fit (parabola or paracosh); dashed-green = extrapolation of parabola/paracosh curve into fluid layers above (left) and below (right) the tissue–fluid interface; dashed-red = tangent to curve at boundary with slope $(\partial P/\partial x)|_{x=L}$ [mmHg/μm]; solid-black linear segments identify linear pressure trends in proximal fluid layer; 'Curv' = $10^3 \times$curvature = $10^3 \times (2a)$ [mmHg/μm$^2$]. Gradient ratios 'slopeL/tang', 'slopeR/tang' give estimates for above-slice, below-slice $(K_t/K_f)$ Krogh ratios; ratios exceeding cutoff value 0.725 are rejected.

**Case (a):** If the profile shape is accurately modelled as parabolic, then Equation (20) applies, and the oxygen pressure gradient in tissue is given by

$$\frac{\partial P_t}{\partial x} = 2ax \rightarrow 2aL \quad \text{at the } x = L \text{ boundary}$$

**Case (b):** If the pressure curve is 'flat-bottomed', then Equations (9) and (10) apply, indicating that the profile is a 'paracosh' mixed case with a hyperbolic-cosine (cosh) core for $(-\delta \leq x \leq \delta)$, and parabolic wings for $(|x| > \delta)$. Paracosh fitting proceeds by iterating curvature $(Q/K_t) = 2a$ and critical pressure $P^*$ to maximise agreement between the paracosh curve and the subset of $(x_i, P_i)$ data points lying within the tissue boundaries. Paracosh optimisation makes use of MATLAB function `fminsearchbnd` [D'Errico (2022), www.mathworks.com/matlabcentral/fileexchange/8277-fminsearchbnd-fminsearchcon, (accessed 29 June 2022)] which adds bounded constraints to the standard `fminsearch` optimiser; in our case, we require that both $2a$ and $P^*$ be non-negative. The critical depth $\delta = \delta(P_s, P^*, 2a)$ is obtained via a separate iteration on Equation (11) with surface pressure $P_s$ fixed by linear interpolation of the $(x_i, P_i)$ data pairs bracketing the $x = \pm L$ boundaries. Once the paracosh curve parameters $(2a, P^*, \delta)$ have been established, the pressure gradient at the $x = L$ boundary is given by Equation (25).

**Case (c):** The cosh-only profile ($P_s < P^*$) was never encountered in any of our oxygen tension soundings; nevertheless, the case-(b) curve-fitting algorithm should work equally well here.

If a clear gradient break could be identified at the tissue–fluid boundary, a straight line was fitted to the linear pressure trend in the fluid adjacent to the slice edge. The tissue–fluid Krogh ratio $(K_t/K_f)$ was then obtained using the flux conservation expression of Equation (24).

## 4. Results

We analysed five distinct $pO_2$ profile datasets [labelled v1 to v5] but discarded the first two because of oxygen-probe calibration issues. The remaining datasets are:

**Dataset v3** (10 mL/min): Recorded from 19 cortical locations from 4 slices (1 animal); $n = 19$ profiles

**Dataset v4** (1 and 2 mL/min): Recorded from 8 locations from 4 slices (1 animal), repeated at 1 and 2 mL/min for each location; $n = 16$ profiles

**Dataset v5** (0.5 mL/min): Recorded from 6 cortical locations from 2 slices (1 animal), each profile repeated once, giving 6 profile pairs; $n = 12$ profiles

This gave a total of 47 oxygen tension profiles as summarised in Table A1. Most of the curves (31 of 47) were well-fitted with a simple parabola; the remainder (16 of 47) were fitted with a mixed paracosh function (identified with 'Flat = 1' table entry); none of the profiles exhibited a purely cosh shape. The proportion of flattened curves decreased as aCSF flow rate was reduced: [42%, 31%, 25%] for [v3, v4, v5], respectively. Provided that each curve was correctly classified as parabolic or paracosh, we found no discernible difference between $K_t$ estimates derived from the parabolic set compared with the paracosh set.

### 4.1. Oxygen Tension Profiles in Fluid and Tissue

Figure 6 shows six representative profiles with pressure gradient discontinuities evident at neither, one, or both tissue–fluid boundaries at $x = \pm L$, slice half-thickness being $L = 200$ µm. If the adjacent fluid exhibited a linear $P(x)$ trend, we took this as evidence of a local nonflowing fluid layer which should permit estimation of the $(K_t/K_f)$ Krogh ratio via conservation of $O_2$ flux across the interface (see Equation (24)).

For each $P(x)$ pressure profile, Krogh ratio retrievals proceed via two independent curve-fitting steps. First, the optimal value of profile curvature $(Q/K_t) \equiv 2a$ is determined by minimising the rms difference between measured $(x_i, P_i)$ data points and iterated $(x_i, P(x_i))$ parabola/paracosh predictions for the eight or nine locations $x_i$ lying within the $-L \leq x \leq L$ tissue boundaries. The resulting pressure gradient at the boundary $(\partial P/\partial x)|_{x=L}$ is then calculated using Equation (21) (parabola) or Equation (25) (paracosh): see red 'tangent' extrapolations in Figure 6.

Second, the pressure values in the proximal fluid above and below the slice are inspected for linear trends $\Delta P/\Delta x$, whose slope is markedly *lower* from the tissue tangent extrapolations; a clear gradient break suggests a locally stationary fluid layer. For our experimental setup, these gradient breaks were common at the lower interface (at $x = L$), but rarely occurred at the top interface ($x = -L$), except at the lowest perfusion rates. We attribute this top/bottom—left/right on the graphs—asymmetry to the presence of the fine nylon mesh that supports the underside of the slice (see Figure 5); evidently, the weave of the net can create a 'shadow zone' that shields the slice from longitudinal advective currents.

If the fluid pressure gradient was similar to, or larger than, the extrapolated tissue gradient (i.e., if $(K_t/K_f) \gtrsim 1$), then the candidate ratio was immediately rejected (no stationary layer, therefore flux conservation assumption is invalid). However, a more stringent acceptance criterion is needed since some apparently stationary cases may be 'contaminated' by weak residual advective flows. After inspecting scatter plots of Krogh ratio vs. curvature (see Figures 7 and 8a), and Krogh ratio histograms for $(K_t/K_f) < 1$ (Figure 9), we set the ratio cutoff at

$$(K_t/K_f)^{max} = 0.725$$

This selection was made on the basis that candidate Krogh ratios evidently fall into two clusters:

- candidate ratios larger than unity are biologically disallowed (oxygen permeability in tissue cannot be greater than permeability in fluid);
- candidate ratios larger than 0.73 appear to form part of the $K_t/K_f > 1.0$ 'disallowed' cluster (e.g., see panel (a) of Figure 7);
- the aggregated histogram ratios of Figure 9 suggest a clear break between 'allowed' and 'disallowed' clusters if we set the cutoff at $K_t/K_f = 0.725$.

### 4.2. *Scatter Plots of Krogh Ratio vs. Curvature*

Individual scatter distributions for $(K_t/K_f)$ candidate Krogh ratios vs. profile curvature are displayed as separate panels in Figure 7, then aggregated into a unified cluster-graph in Figure 8a; Figure 8b shows the apparently sigmoidal dependence of profile curvature on perfusion flow rate. The scatterplots of Figure 7 are segregated by interface (upper/lower) and by dataset (v3/v4/v5). Perfusion rates decrease from left to right: 10 mL/min (v3), 1 or 2 mL/min (v4), 0.5 mL/min (v5). We see that, on average, lower flow rates are associated with lower curvatures ('flatter' parabolic/paracosh curves). Since curvature = $2a = (Q/K_t)$, then $O_2$ consumption rate also scales down as $O_2$ supply becomes more restricted (assuming $K_t$ can be taken as a nominal constant). In contrast, flow rate appears to have no influence on *plausible* Krogh ratios (i.e., those that fall below the proposed 0.725 cutoff). These contrary sensitivities to fluid $O_2$ transport rates are summarised in Table 4.

The distribution of candidate Krogh ratios within a given dataset (v3/v4/v5) appears qualitatively rather noisy and erratic, unlike the curvature distributions. This is particularly evident in the paired observations in v4 and v5. For v4, each slice location was profiled twice, first at perfusion rate 1 mL/min, then at the doubled rate 2 mL/min. For every pair, doubling the flow rate raised the curvature by a roughly similar proportion; however, the impact on Krogh ratio was unpredictable and scattered (e.g., see outlier pair 414/416 in Figure 7e), particularly for those candidate Krogh ratios lying in the nominated rejection zone. We attribute this increased scatter to three possible sources:

- Krogh ratio requires separate tissue and fluid curve fits, so variance in the ratio will be the *sum* of the individual curve-fitting variances for tissue and fluid gradients;
- the flux conservation argument used to derive Equation (24) is invalid if the proximal fluid layer is not stationary (hence the need to impose a Krogh ratio cutoff);
- formation of a local stagnant layer is not guaranteed, even within the closely woven structure of the nylon net that supports the slice.

We note that reducing the flow rate increases the proportion of candidate Krogh ratios that lie within our nominal acceptance range (i.e., $(K_t/K_f) < 0.725$), particularly for the lower interface (see bottom row of Figure 7). Evidently, stagnant layer formation is more probable at low flow rates.

**Table 4.** Sensitivity of profile curvature and Krogh ratio to variations in aCSF flow rate. Column headings identify the datasets used for computing statistics; the v4-dataset is partitioned into its high- and low-flow subsets. Perfusion rates decrease from left to right across the columns. Krogh statistics summarise the $x = L$ (lower interface) retrievals, but note that candidate Krogh ratios that exceed the $(K_f/K_f)^{max} = 0.725$ cutoff have been excluded (see Table A1).

| | v3 | v4 (hi) | v4 (lo) | v5 | All |
|---|---|---|---|---|---|
| **Flow (mL/min)** | **10.0** | **2.0** | **1.0** | **0.5** | |
| **Curvature,** $(2a)$ | | | | | |
| $10^3 \times$ mean | 6.55 | 5.02 | 4.23 | 3.14 | |
| $10^3 \times$ stdev | 1.38 | 0.85 | 0.80 | 0.90 | |
| $N$ | 19 | 8 | 8 | 12 | |
| **Krogh ratio,** $(K_t/K_f)$ | | | | | |
| mean | 0.553 | 0.595 | 0.538 | 0.568 | 0.562 |
| stdev | 0.083 | 0.140 | 0.082 | 0.072 | 0.088 |
| $N$ | 11 | 5 | 5 | 9 | 30 |

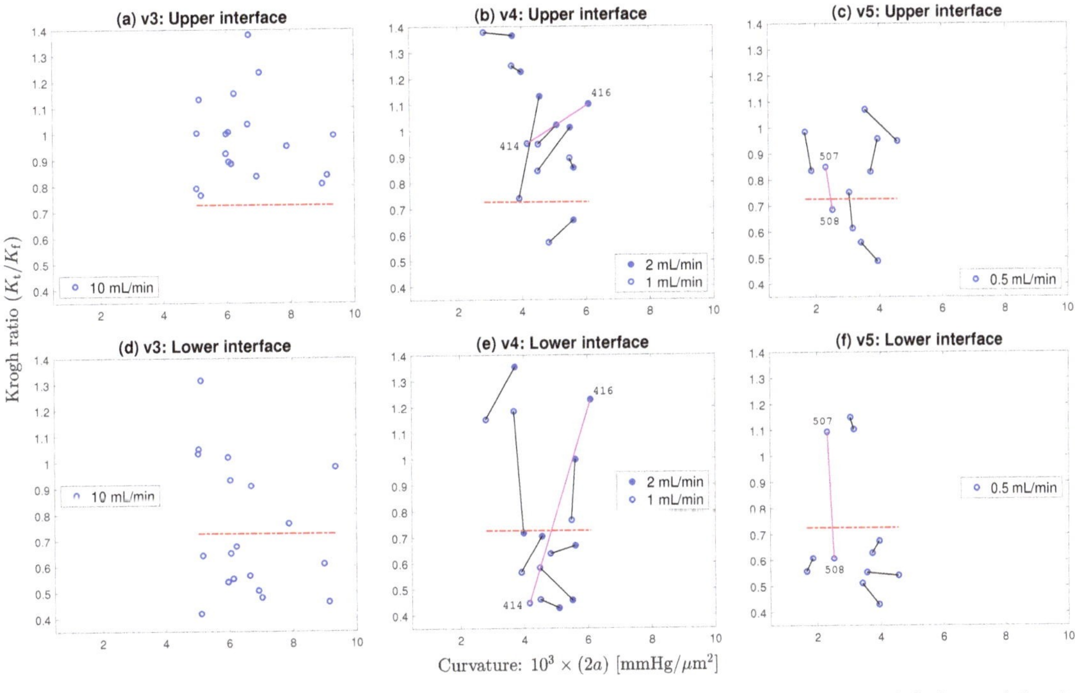


**Figure 7.** Distribution of Krogh ratios as a function of curvature of fitted parabolic/paracosh function. Results are clustered by dataset (three columns: v3/v4/v5) and interface (two rows: upper/lower). Dashed-red horizontal marks the selected cut-off between accepted (below red line) and rejected (above line) Krogh ratios. Scanning from left to right, lower aCSF flow rates are generally associated with reduced curvature values, implying increasingly constrained $O_2$ consumption. Linked pairs show repeated sampling at the same location. For v4, flow rate was set at 1 (open circles) or 2 mL/min (filled circles); for v5, flow rate was fixed at 0.5 mL/min. Outlier pairs 414/416 (v4) and 507/508 (v5) have very discrepant Krogh ratio estimates at the lower interface, possibly due to mechanical disturbance of the slice during withdrawal of $O_2$ probe prior to repeat sounding. See Figure 6 and Table A1.

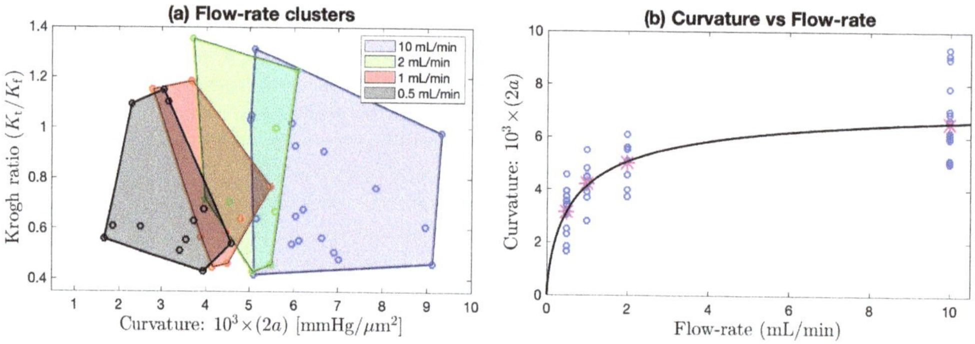


**Figure 8.** Flow rate clustering and curvature dependence aggregated across (lower interface) of Figure 7d–f. (**a**) Aggregated scatterplot of candidate Krogh ratio $(K_t/K_f)$ vs. curvature of fitted parabolic/paracosh function, clustered by flow rate [10, 2, 1, or 0.5] mL/min, as indicated by shaded convex-hull polygons (computed via MATLAB function `convhull`). Qualitatively, the polygon centroids move to the left as flow rate decreases, implying that curvature decreases (profiles become flatter) as perfusion flow rate is reduced. This trend is made quantitative in (**b**) with a sigmoid fit to the Table 4 curvature means (magenta asterisks) at each flow rate. The fitted curve is $y = y^{\max} \cdot x^n/(K^n + x^n)$ with [$y^{\max} = 7.5 \times 10^{-3}$ mmHg/$\mu$m$^2$; $K = 0.75$ mL/min; $n = 0.75$].

Unlike the v4 pairs, the perfusion rate for the v5 paired observations (Figure 7c,f) was maintained at a constant value (0.5 mL/min); consequently, the profile repeats typically have smaller Cartesian separation (the links are smaller) than is the case for v4, and the average link length can be taken as an overall measure of the experimental uncertainty associated with our method. Nevertheless, we still see an outlier pair 507/508 at the lower interface with an abrupt reduction in the candidate Krogh ratio. While the 507/508 profile graphs in Figure 6c,f show consistent in-tissue curvature values, the proximal fluid environment immediately above and below the tissue changes dramatically in the $\sim$1 min between the 507 and 508 soundings: the later profile shows clear evidence of stationary fluid layers at the upper and lower boundaries, while no evidence was apparent in the earlier profile. Perhaps the withdrawal and reinsertion of the $O_2$ probe caused a subtle change in the seating of the tissue slab that favoured the formation of stagnant zones?

### 4.3. Histograms for Krogh Ratio and Profile Curvature

In Figure 9, we present histograms for Krogh ratios aggregated over the [v3, v4, v5] datasets shown in Figure 7, but restrict the analysis to $(K_t/K_f) \leq 1.0$ [allowing $(K_t/K_f) > 1.0$ would imply that $O_2$ permeability in tissue *exceeds* that in fluid; this is physiologically implausible]. Krogh ratios are biased towards larger values at the upper tissue–fluid interface (Figure 9a), and smaller values at the lower interface (panel (b)); this is consistent with our observation that well-defined stagnant layers are a common occurrence at the lower boundary, but rare at the upper boundary. Setting an upper bound of $(K_t/K_f)^{\max} = 0.725$ provides a clean separation between accepted and rejected Krogh ratio candidates.

Figure 10 histograms the profile curvatures across each of the [v3, v4, v5] datasets. It confirms the earlier observation that smaller aCSF flow rates are associated with tissue profiles that have weaker (flatter) curvature, indicating that restricting oxygen flow leads to constrained metabolic activity.

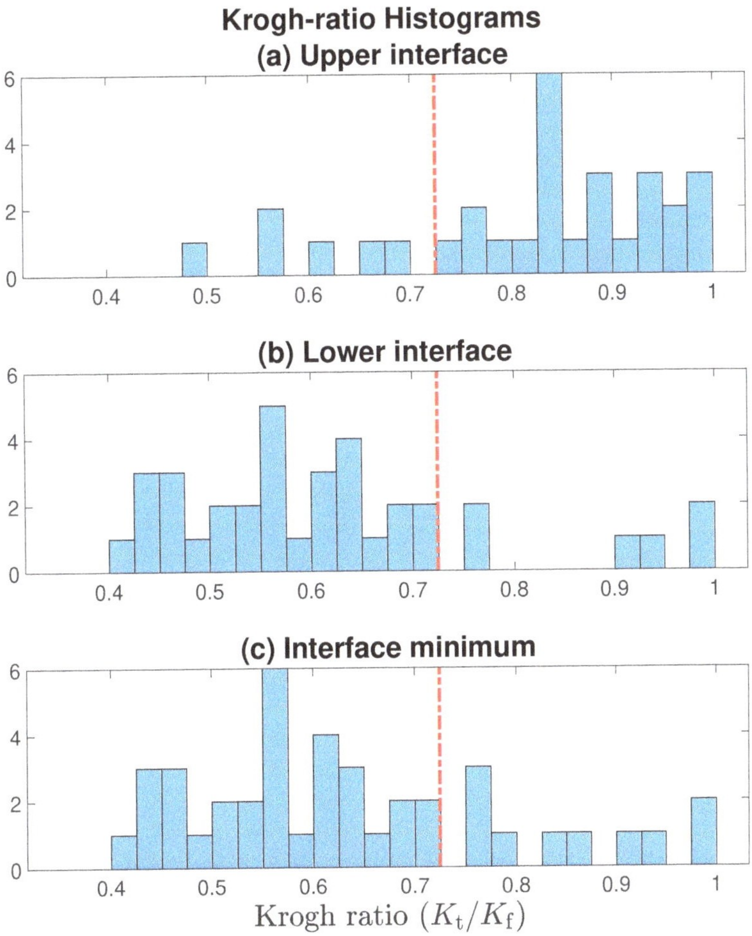


**Figure 9.** Histograms for Krogh ratios aggregated over [v3, v4, v5] datasets illustrated in Figure 7, but restricted to domain $(K_t/K_f) \leq 1.0$. Red-dashed line marks the accept/reject boundary set at 0.725: only Krogh ratios below cutoff are associated with a well-defined stationary layer. Comparing panels (**a**, **b**), the lower tissue–fluid interface is more likely to form a nonflowing boundary layer. In panel (**c**), for each profile, the smaller of the [upper interface, lower interface] Krogh ratio is selected.

### *4.4. Possible Linkage between SLE Activity and Formation of Stationary Boundary Layer*

We postulated that detection of a stagnant fluid layer at the tissue–fluid interface might require the maintenance of at least a minimum level of metabolic activity within the tissue. The 'stationary fluid' idealisation requires that—close to the tissue surface—oxygen diffusion (a molecular random walk with net $O_2$ motion directed towards the face of the tissue) through the fluid strongly dominates any residual advective bulk flow ($O_2$ in fluid moving parallel to the tissue). However, if the tissue is not drawing much oxygen from the fluid, then diffusive flux across the interface will be low, and the $O_2$ advective component may no longer be insignificant, meaning that the flux conservation argument fails because the local fluid layer is insufficiently 'stationary'.

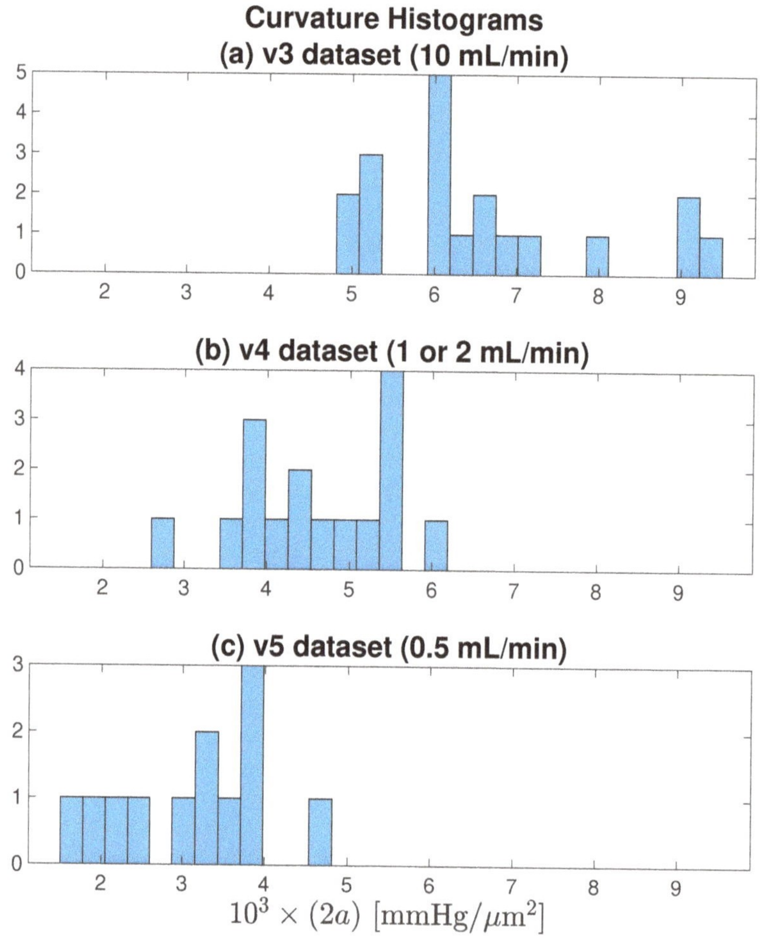


**Figure 10.** Curvature histograms for each of the [v3, v4, v5] datasets illustrated in Figure 7. As flow rate decreases from (**a**) 10 → (**b**) [1 or 2] → (**c**) 0.5 mL/min, average curvature decreases, meaning that the parabolic/paracosh curves become 'flatter' with shallower wings. For fixed $K_t$, a flatter curvature implies reduced metabolism.

We chose to test this hypothesis on the v3-database since this used the fastest aCSF flow rate (10 mL/min), thus maximising both $O_2$ delivery and the potential for seizure-like electrical activity. For each of the nineteen pressure soundings, we identified four consecutive seizure-like events (SLEs) that occurred around the time of profile acquisition, and computed average values for three SLE parameters: peak-to-peak amplitude [µV]; duration [s]; inter-event frequency [$s^{-1}$].

We found that SLE amplitude (but not SLE duration or frequency) correlated *negatively* with $(K_t/K_f)$ Krogh ratio for the under-slice interface ($R^2 = 0.23$, $p = 0.03$). That is, a tissue location generating smaller SLEs was *less* likely to display the pressure-gradient discontinuity associated with a nonflowing fluid layer at the $x = L$ boundary. Conversely, a site generating larger SLEs was more likely to show evidence of a stationary fluid layer.

### 4.5. Estimation of Krogh Coefficient for Cortical Tissue at Room Temperature

We now have sufficient data to derive an estimate for $K_t$, the oxygen Krogh coefficient for mouse cortical tissue at room temperature. In Section 2.6, we computed a theoretical value for $K_f$, the Krogh coefficient for the no-Mg aCSF (no-magnesium artificial cerebrospinal fluid) used to supply oxygen and glucose to the 400 μm-thick slice of brain tissue (18),

$$K_f = (2.30 \pm 0.04) \times 10^{-14} \text{ mol}/[\text{m}\cdot\text{s}\cdot\text{Pa}] \quad (27)$$

This value is based on that for pure water, but adjusted for the presence of dissolved salts and glucose. As displayed in Figure 4, the coefficient is weakly temperature dependent. If the aCSF temperature were to rise from 20 to 25 °C, $K_f$ would increase by ~4% from [2.262 to 2.348] $\times 10^{-14}$ mol/[m·s·Pa]. We have chosen the middle of the range, with an uncertainty of ±2%.

From the Table 4 statistics for the tissue–fluid ratio of Krogh coefficients $(K_t/K_f)$, we learn that the Krogh ratio is completely insensitive to perfusion rate, so we are justified in combining all datasets (last column of Table 4) to give an aggregate (mean ± standard deviation) statistic,

$$(K_t/K_f) = 0.562 \pm 0.088 \quad (\text{dimensionless}; N = 30) \quad (28)$$

The standard deviation in the ratio is ~16% of the mean. This uncertainty estimate is almost an order of magnitude larger than the 2% relative uncertainty in $K_f$ for the perfusion fluid, so the latter can be neglected as source of uncertainty in the Krogh ratio. How much of this uncertainty is due to experimental error (e.g., probe calibration and positioning, pressure measurement, in-tissue and in-fluid curve fitting, tissue movement, unsteady flow rates, degraded stagnant layer, etc.), and how much arises from the natural variability of living cortical tissue?

The [v4, v5] repeated profiles—shown as linked pairs in Figure 7e,f—allow us to apportion the relative contributions of experimental and biological sources of variation. There are eight linked pairs (four in each of [v4, v5]) that lie entirely within the $(K_t/K_f) < 0.725$ acceptance zone. Let $(A, B)$ represent the (first, second) elements of the eight pairs of Krogh ratios retrieved at each repeated location. Define the biological and experimental contributions to the variance (i.e., square of the standard deviation) of the Krogh ratios as $\sigma^2_{\text{biol}}$ and $\sigma^2_{\text{expt}}$, respectively. Then, the following variance identities should apply,

$$\begin{aligned} \text{var}(A - B) &= \sigma^2_{\text{expt}} && \text{(biology cancels)} \\ \text{var}(A + B) &= \sigma^2_{\text{expt}} + \sigma^2_{\text{biol}} && \text{(both contribute)} \\ \implies \quad \sigma^2_{\text{biol}} &= \text{var}(A + B) - \text{var}(A - B) \end{aligned}$$

Our pairwise variance calculations give

$$\text{var}(A + B) = 0.0263, \qquad \text{var}(A - B) = \sigma^2_{\text{expt}} = 0.0069, \qquad \implies \qquad \sigma^2_{\text{biol}} = 0.0194$$

implying that $[\sigma^2_{\text{biol}}/(\sigma^2_{\text{expt}} + \sigma^2_{\text{biol}})] = 74\%$ of the variance in our Krogh ratio determinations is biological in origin, and the remaining 26% is attributable to experimental uncertainty.

Finally, we compute the room temperature Krogh coefficient for mouse cortical tissue by taking the product of (27) and (28),

$$K_t = (1.29 \pm 0.21) \times 10^{-14} \text{ mol}/[\text{m}\cdot\text{s}\cdot\text{Pa}] \quad (29)$$

where we have carried forward the 16% uncertainty from (28).

## 5. Discussion

In this paper we have sought to provide a theoretical and experimental basis for determining oxygen consumption in thin slices of mouse brain tissue. Oxygen consumption within metabolically active tissue can be deduced from the curvature of $pO_2$ oxygen profiles using Fick's law of diffusion, so long as the oxygen diffusion and solubility coefficients of tissue are known. Because oxygen solubility in tissue is difficult to measure, it is standard practice to work with the Krogh coefficient $K_t$, a lumped measure of oxygen permeability given by the product of diffusion and solubility coefficients for oxygen. Early work by Buerk & Saidel [27] identified the Michaelis–Menton kinetics model as the most accurate description of $pO_2$ gradients in tissue slices; its piecewise-linear approximation (see Equation (6)) provides the basis of the analytical solution published by Ivanova & Simeonov [10] and extended here. We followed Ganfield et al. [8] in assuming the existence of a narrow but stationary (i.e., non-flowing) boundary layer of fluid in close proximity to the tissue surface, then invoking conservation of oxygen flux across the fluid/tissue interface in order to deduce a value for $K_t/K_f$, the dimensionless ratio of tissue–fluid Krogh coefficients. By construction, the perfusion fluid is a passive, non-biologically active medium, so the fluid Krogh coefficient $K_f$ can be calculated from well-established interpolation formulas that are functions of both water temperature and saline concentration; this then allows tissue Krogh coefficient $K_t$ to be determined.

There are alternative ways to measure the tissue Krogh coefficient experimentally; however, these are generally not appropriate for soft, delicate tissue such as brain. For example, the original gaseous diffusion method described by Krogh (1919) [16] involved separating two chambers with a stretched membrane, allowing measurement of gas diffusion through the membrane from one chamber to the other. A similar direct measurement method was applied by Sasaki et al. [14], using microscopy techniques to measure the flux of oxygen across thin (10–20 μm) arteriolar walls. Such methods are not easily applied to a slice of brain tissue since it is very easily damaged by physical manipulation.

It is informative to compare our value for tissue Krogh coefficient with those published in the literature. This requires appropriate unit conversions from SI to the various alternative unit systems in use. We select two of the more common metric choices:

| | |
|---|---|
| $(\text{mL O}_2)/(\text{cm}\cdot\text{min}\cdot\text{atm})$: | Ganfield et al. [8] |
| $(\text{mmol O}_2)/(\text{cm}\cdot\text{min}\cdot\text{mmHg})$: | Ivanova & Simeonov [10] |

and list the unit-remapped values for our estimate for $K_t$,

$$\begin{aligned} K_t &= (1.29 \pm 0.21) \times 10^{-14}\ \text{mol}/[\text{m}\cdot\text{s}\cdot\text{Pa}] \\ &= (1.76 \pm 0.29) \times 10^{-5}\ \text{mL}/[\text{cm}\cdot\text{min}\cdot\text{atm}] \\ &= (1.03 \pm 0.17) \times 10^{-9}\ \text{mmol}/[\text{cm}\cdot\text{min}\cdot\text{mmHg}] \end{aligned}$$

Using the second set of units (mL/[cm·min·atm]), Ganfield et al. (1970) derive a Krogh coefficient for cat cortex of $1.29 \times 10^{-5}$, about 27% lower than our $1.76 \times 10^{-5}$ estimate. Their result is unexpectedly low, given that they were working at 37 °C, while our value was derived at room temperature.

Ivanova & Simeonov (2012) tabulate a range of Krogh coefficients at 37 °C for a variety of different tissues drawn from the work of several authors. Using the third set of units (mmol/[cm·min·mmHg]), their quoted values for $10^9 \times K_t$ covered the range

0.59 (kidney), 0.65 (liver), 1.35 (brain), 1.44 (heart),

so our value of $1.03 \times 10^{-9}$ for mouse cortex is certainly plausible, given the temperature difference.

We have focused on calculating tissue/fluid Krogh ratio $(K_t/K_f)$ in the first instance. The advantage of working with a dimensionless ratio is that it should aid direct comparison between studies since it eliminates the need for non-SI unit conversions. Although the biological significance of expressing Krogh permeability as a ratio is uncertain, it is plausible that the $K_t/K_f$ ratio may be less sensitive to temperature than either component. If so, this would be advantageous when attempting to compare the findings of different research groups working with a range of perfusion temperatures. This idea remains to be tested experimentally.

Our method for computing the Krogh ratio is dependent of the formation of a stationary boundary layer at the tissue–fluid interface. As discussed previously, a nonflowing fluid layer was often detected at the lower interface, but rarely at the upper interface. This asymmetrical behaviour is probably a favourable artifact created by the support netting on which the slice sits: it seems that the tight weave of the netting can provide shielding from the bulk advective flow. This likely explains some of the variability in our $K_t$ estimate, because precise positioning of the oxygen electrode within this stationary layer could not be guaranteed from one recording location to the next.

Importantly, when (repeat) profiles were collected from the *same* location, the variance in the $K_t/K_f$ ratio was substantially lower (contributing only 26% of the total variation), indicating that the majority of the variability was of biological origin. This hints at the possibility that the oxygen permeability characteristics of brain tissue are not uniform across and between slices, with potential influences from variation in cortical layer structure, variation in regional cortical anatomy, and spatial differences in tissue viability. In these experiments, we did not attempt to control for specific cortical anatomical location or for non-uniformity in slice viability.

In summary, we have found that the Ivanova & Simeonov diffusion–consumption model provides an excellent description of oxygen-tension distribution within a thin slice of active tissue. We have extended the model to include the effects of a stationary fluid layer at the boundary, and have shown how to compute the ratio of tissue and fluid Krogh coefficients via a flux conservation argument. Being dimensionless, the Krogh ratio allows unambiguous and direct comparisons between studies by different researchers, since it obviates the need for unit conversions. The mapping to dimensioned Krogh coefficients can be delayed until the choice of units for the fluid Krogh coefficient has been made.

**Author Contributions:** Experiment design and data acquisition, L.J.V.; data analysis, L.J.V. and D.A.S.-R.; mathematical development, D.A.S.-R. and M.L.S.-R.; software, D.A.S.-R.; original draft preparation, D.A.S.-R. and L.J.V.; manuscript review and editing, D.A.S.-R., M.L.S.-R., J.W.S. and L.J.V. All authors have read and agreed to the published version of the manuscript.

**Funding:** This research received no external funding.

**Institutional Review Board Statement:** The tissue recovery methods in this study were approved by the Animal Ethics Committee at the University of Waikato, New Zealand.

**Informed Consent Statement:** Not applicable.

**Data Availability Statement:** The MATLAB `paracosh` curve-fitting codes plus associated sample $pO_2$ depth profiles will be supplied on request to the corresponding author (D.A.S.-R.).

**Conflicts of Interest:** The authors declare no conflict of interest for this work.

## Abbreviations

| | |
|---|---|
| 1D | one-dimensional |
| aCSF | artificial cerebral spinal fluid |
| Ag/AgCl | silver/silver chloride (electrode) |
| I & S | Ivanova and Simeonov (2012) |
| IUPAC | International Union of Pure and Applied Chemistry |
| LFP | local-field potential |
| NIST | National Institute of Standards and Technology |
| NMDA | N-methyl-D-aspartate |
| no-Mg | no magnesium |
| PDE | partial differential equation |
| $pO_2$ | partial pressure of oxygen |
| rmse | root-mean-square error |
| SD | standard deviation |
| SLE | seizure-like event |

## Appendix A. Curve-Fitting Summary

**Table A1.** Summary table for 47 $pO_2$ pressure profiles. 'Ref' = vnn where v = [3, 4, 5] identifies dataset, nn = profile index; 'Flow' = aCSF flow rate [mL/min]; 'Flat' = flag indicating paracosh (1) or parabolic (0) fit; 'rmse' = rms error for curve fit [mmHg]; 'Curvature' = $10^3 \times (2a)$ [mmHg/μm$^2$]; $P_s$ = fitted surface pressure at $x = L$ boundary [mmHg]; $P_{min}$ = fitted pressure at $x = 0$ [mmHg]; 'Tangent' = $(\partial P/\partial x)|_{x=L}$ [mmHg/μm]; 'Ratio (top)', 'Ratio (bot)' = Krogh ratio ($K_t/K_f$) at upper, lower interface. NB: Values in [brackets] are to be disregarded (existence of stationary layer is implausible).

| Ref | Flow | Flat? | rmse | Curvature | $P_s$ | $P_{min}$ | Tangent | Ratio (top) | Ratio (bot) |
|---|---|---|---|---|---|---|---|---|---|
| 301 | 10.0 | 0 | 4.62 | 5.97 | 120.2 | 0.8 | 1.194 | [0.999] | [1.018] |
| 302 | 10.0 | 1 | 1.15 | 6.92 | 124.2 | 0.0 | 1.311 | [0.835] | 0.505 |
| 303 | 10.0 | 1 | 3.20 | 9.34 | 83.3 | 0.0 | 1.247 | [0.993] | [0.981] |
| 304 | 10.0 | 0 | 1.20 | 5.04 | 131.7 | 31.0 | 1.007 | [0.789] | [1.032] |
| 305 | 10.0 | 0 | 3.40 | 6.04 | 165.8 | 45.0 | 1.208 | [1.006] | [0.930] |
| 306 | 10.0 | 0 | 3.26 | 5.17 | 170.5 | 67.0 | 1.035 | [0.763] | 0.640 |
| 307 | 10.0 | 0 | 5.86 | 8.97 | 205.6 | 26.2 | 1.795 | [0.806] | 0.608 |
| 308 | 10.0 | 0 | 3.65 | 6.05 | 154.1 | 33.1 | 1.210 | [0.892] | 0.649 |
| 309 | 10.0 | 0 | 3.18 | 5.96 | 134.4 | 15.2 | 1.193 | [0.923] | 0.538 |
| 310 | 10.0 | 1 | 2.38 | 9.13 | 134.9 | 0.0 | 1.570 | [0.839] | 0.462 |
| 311 | 10.0 | 1 | 4.17 | 6.65 | 121.4 | 0.0 | 1.270 | [1.037] | 0.563 |
| 312 | 10.0 | 0 | 3.31 | 6.69 | 191.1 | 57.3 | 1.338 | [1.379] | [0.908] |
| 313 | 10.0 | 1 | 1.64 | 5.05 | 93.1 | 1.3 | 0.962 | [1.001] | [1.049] |
| 314 | 10.0 | 0 | 5.13 | 7.87 | 187.2 | 29.8 | 1.575 | [0.951] | [0.764] |
| 315 | 10.0 | 1 | 2.46 | 5.13 | 47.5 | 0.0 | 0.698 | [1.132] | [1.314] |
| 316 | 10.0 | 0 | 2.13 | 6.22 | 133.4 | 8.9 | 1.245 | [1.154] | 0.675 |
| 317 | 10.0 | 1 | 2.31 | 7.03 | 95.0 | 0.0 | 1.155 | [1.235] | 0.479 |
| 318 | 10.0 | 0 | 2.68 | 6.12 | 135.1 | 12.6 | 1.225 | [0.884] | 0.551 |
| 319 | 10.0 | 1 | 1.63 | 5.12 | 49.2 | 0.0 | 0.710 | [1.640] | 0.417 |
| 401 | 1.0 | 1 | 1.51 | 4.50 | 60.6 | 0.0 | 0.739 | [0.947] | 0.461 |
| 402 | 1.0 | 1 | 2.15 | 3.90 | 73.6 | 0.0 | 0.758 | [0.740] | 0.565 |
| 403 | 2.0 | 0 | 2.48 | 5.09 | 119.8 | 18.0 | 1.018 | [1.022] | 0.428 |
| 404 | 2.0 | 0 | 1.47 | 4.56 | 115.0 | 23.8 | 0.912 | [1.132] | 0.704 |
| 405 | 1.0 | 0 | 1.36 | 4.49 | 93.6 | 3.8 | 0.897 | [0.845] | 0.582 |
| 406 | 1.0 | 0 | 5.27 | 5.49 | 128.3 | 18.5 | 1.098 | [0.894] | [0.766] |
| 407 | 2.0 | 0 | 4.18 | 5.51 | 145.0 | 34.8 | 1.102 | [1.013] | 0.458 |
| 408 | 2.0 | 0 | 3.78 | 5.62 | 171.3 | 59.0 | 1.123 | [0.856] | [0.998] |
| 409 | 1.0 | 0 | 1.43 | 2.81 | 58.3 | 2.1 | 0.562 | [1.378] | [1.151] |
| 410 | 1.0 | 1 | 0.90 | 3.69 | 40.0 | 0.0 | 0.543 | [1.250] | [1.185] |
| 411 | 2.0 | 0 | 2.77 | 3.72 | 89.6 | 15.1 | 0.745 | [1.365] | [1.355] |
| 412 | 2.0 | 0 | 2.00 | 3.98 | 89.3 | 9.7 | 0.796 | [1.226] | 0.717 |
| 413 | 1.0 | 0 | 2.85 | 4.82 | 99.1 | 2.7 | 0.965 | 0.570 | 0.637 |
| 414 | 1.0 | 1 | 1.39 | 4.16 | 68.2 | 0.0 | 0.753 | [0.949] | 0.447 |
| 415 | 2.0 | 0 | 3.83 | 5.61 | 140.7 | 28.6 | 1.121 | 0.656 | 0.668 |
| 416 | 2.0 | 1 | 3.74 | 6.09 | 78.7 | 0.0 | 0.979 | [1.102] | [1.229] |
| 501 | 0.5 | 1 | 0.58 | 4.57 | 49.9 | 0.0 | 0.675 | [0.949] | 0.541 |
| 502 | 0.5 | 1 | 1.33 | 3.55 | 50.0 | 0.0 | 0.596 | [1.070] | 0.553 |
| 503 | 0.5 | 0 | 1.64 | 3.05 | 92.3 | 31.3 | 0.610 | [0.752] | [1.150] |
| 504 | 0.5 | 0 | 1.11 | 3.16 | 90.8 | 27.7 | 0.631 | 0.614 | [1.103] |
| 505 | 0.5 | 1 | 2.82 | 3.95 | 71.8 | 3.7 | 0.730 | [0.958] | 0.675 |
| 506 | 0.5 | 0 | 4.69 | 3.72 | 79.4 | 4.9 | 0.745 | [0.830] | 0.629 |
| 507 | 0.5 | 0 | 0.61 | 2.31 | 50.7 | 4.5 | 0.462 | [0.849] | [1.093] |
| 508 | 0.5 | 0 | 1.04 | 2.52 | 58.6 | 8.2 | 0.504 | 0.685 | 0.607 |
| 509 | 0.5 | 0 | 3.77 | 3.94 | 99.7 | 21.0 | 0.788 | 0.488 | 0.430 |
| 510 | 0.5 | 0 | 2.78 | 3.41 | 110.2 | 41.9 | 0.683 | 0.560 | 0.512 |
| 511 | 0.5 | 0 | 1.47 | 1.86 | 185.9 | 148.7 | 0.372 | [0.835] | 0.608 |
| 512 | 0.5 | 0 | 1.56 | 1.66 | 162.8 | 129.6 | 0.332 | [0.985] | 0.557 |

## References


1. Yamamoto, C.; McIlwain, H. Electrical activities in thin sections from the mammalian brain maintained in chemically-defined media in vitro. *J. Neurochem.* **1966**, *13*, 1333–1343. [CrossRef] [PubMed]
2. Fountain, S.B.; Hennes, S.K.; Teyler, T.J. Aspartame exposure and in vitro hippocampal slice excitability and plasticity. *Fundam. Appl. Toxicol.* **1988**, *11*, 221–228. [CrossRef] [PubMed]

3. Lipton, P.; Aitken, P.G.; Dudek, F.E.; Eskessen, K.; Espanol, M.T.; Ferchmin, P.A.; Kelly, J.B.; Kreisman, N.R.; Landfield, P.W.; Larkman, P.M. Making the best of brain slices: Comparing preparative methods. *J. Neurosci. Methods* **1995**, *59*, 151–156. [CrossRef]
4. Voss, L.J.; van Kan, C.; Envall, G.; Lamber, O. Impact of variation in tissue preparation methodology on the functional outcome of neocortical mouse brain slices. *Brain Res.* **2020**, *1747*, 147043. [CrossRef]
5. Ramirez, O.T.; Mutharasan, R. Cell cycle- and growth phase-dependent variations in size distribution, antibody productivity, and oxygen demand in hybridoma cultures. *Biotechnol. Bioeng.* **1990**, *36*, 839–848. [CrossRef] [PubMed]
6. Deshpande, R.R.; Wittmann, C.; Heinzle, E. Microplates with integrated oxygen sensing for medium optimization in animal cell culture. *Cytotechnology* **2004**, *46*, 1–8. [CrossRef]
7. Eyer, K.; Oeggerli, A.; Heinzle, E. On-line gas analysis in animal cell cultivation: II. Methods for oxygen uptake rate estimation and its application to controlled feeding of glutamine. *Biotechnol. Bioeng.* **1995**, *45*, 54–62. [CrossRef]
8. Ganfield, R.A.; Nair, P.; Whalen, W.J. Mass transfer, storage, and utilization of $O_2$ in cat cerebral cortex. *Am. J. Physiol.* **1970**, *219*, 814–821. [CrossRef]
9. Fujii, T.; Buerk, D.G.; Whalen, W.J. Activation energy in the mammalian brain slice as determined by oxygen micro-electrode measurements. *Jpn. J. Physiol.* **1981**, *31*, 279–283. [CrossRef]
10. Ivanova, R.; Simeonov, G. A formula for the oxygen uptake of thin tissue slice in terms of its surface oxygen tension. *Comput. Math. Appl.* **2012**, *64*, 322–336.
11. Chen, P.Y.; Van Liew, H.D. Krogh constants for diffusion of nitrogen and carbon monoxide in bladder tissue. *Respir. Physiol.* **1975**, *24*, 43–49. [CrossRef]
12. Kawashiro, T.; Scheid, P. Measurement of Krogh's diffusion constant of $CO_2$ in respiring muscle at various $CO_2$ levels: Evidence for facilitated diffusion. *Pflugers Arch.* **1976**, *362*, 127–133. [CrossRef] [PubMed]
13. van der Laarse, W.J.; des Tombe, A.L.; van Beek-Harmsen, B.J.; Lee-de Groot, M.B.E.; Jaspers, R.T. Krogh's diffusion coefficient for oxygen in isolated Xenopus skeletal muscle fibers and rat myocardial trabeculae at maximum rates of oxygen consumption. *J. Appl. Physiol.* **2005**, *99*, 2173–2180. [CrossRef]
14. Sasaki, N.; Horinouchi, H.; Ushiyama, A.; Minamitani, H. A new method for measuring the oxygen diffusion constant and oxygen consumption rate of arteriolar walls. *Keio J. Med.* **2012**, *61*, 57–65. [CrossRef] [PubMed]
15. Poole, D.C.; Pittman, R.N.; Musch, T.I.; Østergaard, L. August Krogh's theory of muscle microvascular control and oxygen delivery: A paradigm shift based on new data. *J. Physiol.* **2020**, *598*, 4473–4507. [CrossRef] [PubMed]
16. Krogh, A. The rate of diffusion of gases through animal tissues, with some remarks on the coefficient of invasion. *J. Physiol.* **1919**, *52*, 391–408. [CrossRef] [PubMed]
17. Forstner, H.; Gnaiger, E. Calculation of Equilibrium Oxygen Concentration. In *Polarographic Oxygen Sensors*; Gnaiger, E., Forstner, H., Eds.; Springer: Berlin/Heidelberg, Germany, 1983; pp. 321–333. [CrossRef]
18. Han, P.; Bartels, D.M. Temperature dependence of oxygen diffusion in $H_2O$ and $D_2O$. *J. Phys. Chem.* **1996**, *100*, 5597–5602. [CrossRef]
19. Green, E.J.; Carritt, D.E. New tables for oxygen saturation of seawater. *J. Mar. Res.* **1967**, *25*, 140–147.
20. van Stroe, A.; Janssen, L.J.J. Determination of the diffusion coefficient of oxygen in sodium chloride solutions with a transient pulse technique. *Anal. Chim. Acta* **1993**, *279*, 213–219. [CrossRef]
21. Burkholder, J.B.; Sander, S.P.; Abbatt, J.; Barker, J.R.; Cappa, C.; Crounse, J.D.; Dibble, T.S.; Huie, R.E.; Kolb, C.E.; Kurylo, M.J.; et al. *Chemical Kinetics and Photochemical Data for Use in Atmospheric Studies*; Evaluation No. 19; JPL Publication 19-5; Jet Propulsion Laboratory: Pasadena, CA, USA, 2019.
22. Sander, R. Compilation of Henry's law constants (version 4.0) for water as solvent. *Atmos. Chem. Phys.* **2015**, *15*, 4399–4981. [CrossRef]
23. van Stroe-Biezen, S.A.M.; Janssen, A.A.; Janssen, L.J.J. Solubility of oxygen in glucose solutions. *Anal. Chim. Acta* **1993**, *280*, 217–222. .%2893%2985124-3. [CrossRef]
24. Thomas, M.G.; Covington, J.A.; Wall, M.J. A chamber for the perfusion of in vitro tissue with multiple solutions. *J. Neurophysiol.* **2013**, *110*, 269–277. [CrossRef] [PubMed]
25. Aram, J.A.; Lodge, D. Validation of a neocortical slice preparation for the study of epileptiform activity. *J. Neurosci. Methods* **1988**, *23*, 211–224. [CrossRef]
26. Anderson, W.W.; Lewis, D.V.; Swartzwelder, H.S.; Wilson, W.A. Magnesium-free medium activates seizure-like events in the rat hippocampal slice. *Brain Res.* **1986**, *398*, 215–219. [CrossRef] [PubMed]
27. Buerk, D.G.; Saidel, G.M. Local kinetics of oxygen metabolism in brain liver tissues. *Microvasc. Res.* **1978**, *16*, 391–405. [CrossRef]

International Journal of
*Molecular Sciences*

MDPI

*Article*

# *Dioscorea nipponica* Makino Rhizome Extract and Its Active Compound Dioscin Protect against Neuroinflammation and Scopolamine-Induced Memory Deficits

**Shofiul Azam [1,†], Yon-Suk Kim [2,†], Md. Jakaria [1,3], Ye-Ji Yu [1], Jae-Yong Ahn [1], In-Su Kim [1,4] and Dong-Kug Choi [1,4,*]**

1 BK21 Program, Department of Applied Life Sciences, Graduate School, Konkuk University, Chungju 27478, Korea
2 BKplus GLOCAL Education Program of Nutraceuticals Development, Konkuk University, Chungju 27478, Korea
3 Melbourne Dementia Research Centre, The Florey Institute of Neuroscience and Mental Health, The University of Melbourne, Parkville, VIC 3052, Australia
4 Department of Biotechnology, College of Biomedical and Health Science, Research Institute of Inflammatory Disease (RID), Konkuk University, Chungju 27478, Korea
* Correspondence: choidk@kku.ac.kr; Tel.: +82-43-840-3610; Fax: +82-43-840-3872
† These authors contributed equally to this work.

**Abstract:** Activation of microglial cells by intrinsic or extrinsic insult causes neuroinflammation, a common phenomenon in neurodegenerative diseases. Prevention of neuroinflammation may ameliorate many neurodegenerative disease progressions. *Dioscorea nipponica* Makino (DN) extract can alleviate muscular atrophy and inflammatory diseases; however, the efficacy and mechanism of action in microglial cells remain unknown. The current study investigates the possible anti-inflammatory effects and mechanisms of *Dioscorea nipponica* Makino ethanol extract and its steroidal saponin dioscin. Our in vitro study shows that *Dioscorea nipponica* rhizome ethanol extract (DNRE) and dioscin protect against lipopolysaccharide (LPS)-activated inflammatory responses in BV-2 microglial cells by inhibiting phosphorylation and the nuclear translocation of nuclear factor kappa-light-chain-enhancer of activated B cells (NF-κB), resulting in the downregulation of pro-inflammatory cytokines and enzymes. Consistent with our previous report of dioscin-mediated enhancement of neurotrophic factors in dopaminergic cells, here we found that dioscin upregulates brain-derived neurotrophic factor (BDNF) and cAMP-response element binding protein (CREB) phosphorylation (pCREB) in the cerebral cortex and hippocampus regions of the mouse brain. Scopolamine treatment increased pro-inflammatory enzyme levels and reduced the expression of BDNF and pCREB in the hippocampus and cortex regions, which led to impaired learning and referencing memory in mice. Pre-treatment of dioscin for 7 days substantially enhanced mice performances in maze studies, indicating amelioration in cognitive deficits. In conclusion, DNRE and its active compound dioscin protect against neurotoxicity most likely by suppressing NF-κB phosphorylation and upregulating neurotrophic factor BDNF.

**Keywords:** *Dioscorea nipponica* Makino; dioscin; anti-inflammatory; LPS; neuroinflammation; cognitive deficits

**Citation:** Azam, S.; Kim, Y.-S.; Jakaria, M.; Yu, Y.-J.; Ahn, J.-Y.; Kim, I.-S.; Choi, D.-K. *Dioscorea nipponica* Makino Rhizome Extract and Its Active Compound Dioscin Protect against Neuroinflammation and Scopolamine-Induced Memory Deficits. *Int. J. Mol. Sci.* **2022**, 23, 9923. https://doi.org/10.3390/ijms23179923

Academic Editors: Jacek Z. Kubiak, Patrick C. Baer and Ralf Schubert

Received: 3 July 2022
Accepted: 27 July 2022
Published: 1 September 2022

**Publisher's Note:** MDPI stays neutral with regard to jurisdictional claims in published maps and institutional affiliations.



## 1. Introduction

Neuroinflammation is a vital pathological feature of neurodegenerative disease progression, including Alzheimer's disease [1,2]. Brain-residing macrophages, the microglia, can sense exogenous neurotoxins such as lipopolysaccharide (LPS)—the outer membrane polysaccharide component of Gram-negative bacteria). LPS binds to toll-like receptor 4 (TLR4) in the microglia, which stimulates an innate immune response [2–4]. Following TLR4 signalling activation, a downstream signalling cascade activates nuclear factor kappa-B

(NF-κB), and this triggers the release of pro-inflammatory cytokines, including tumour necrosis factor-α (TNF-α), interleukin-1β (IL-1β), and interleukin-6 (IL-6) [5,6]. The inducible pro-inflammatory mediator isoforms of nitric oxide synthase (iNOS) and cyclooxygenase-2 (COX-2) are also upregulated in this process, and increase the production of nitric oxide (NO) and prostaglandin E2 (PGE2), which further exaggerate neuroinflammation [7].

The forebrain cholinergic neurons play an important role in controlling the central-nervous-system-regulated cognitive functions [8]. It has been seen that cholinergic decline impacts cognitive functions and disrupts anti-inflammatory pathways. Cholinergic signalling has been seen to regulate peripheral cytokine productions via anti-inflammatory mechanisms [9]. Scopolamine is an anticholinergic drug that impairs cognitive functions and learning memory, which resembles molecular and behavioural features of Alzheimer's disease by impairing cholinergic neurotransmission in the hippocampus and prefrontal cortex [10,11]. In addition, these neuropsychiatric changes also involve the upregulation of pro-inflammatory cytokines and inflammasome components in different brain regions [12].

*Dioscorea nipponica* (DN), a species of the Dioscorea genus belonging to the family Dioscoreaceae, is an important herb in Chinese traditional medicine [13]. It has commonly been prescribed to treat respiratory illnesses such as dry cough and bronchitis, rheumatism and analgesia, stimulate blood circulation, and enhance digestion and diuresis [13,14]. Major compounds found in *Dioscorea nipponica* are steroidal saponins, such as dioscin [13], and stilbene derivatives, such as diosniponol C and diosniposide A [15]. Recent findings have shown that bioactive compounds of *Dioscorea nipponica*, and steroidal saponin, attenuate cardiovascular diseases, improve immune functions, and prevent cancer progression [13]. *Dioscorea nipponica* has also been shown to promote muscle differentiation, restore muscle atrophy, and recover from injury by inhibiting the NF-κB pathway [16]. The total saponin fraction and phenolic compounds from *Dioscorea nipponica* extract have been shown as potential modulators of the MAPK signalling pathway in an inflammatory disease model [15,17]. However, the biological impact of *Dioscorea nipponica* in neuroinflammation progression or prevention is largely unknown. Therefore, this study investigates the possible anti-inflammatory effect of *Dioscorea nipponica* rhizome ethanol extract (DNRE) on microglial cell activation and pro-inflammatory cytokine production.

## 2. Results

### 2.1. DNRE Protects Microglial Cells from LPS-Induced Stress

To test the hypothesis that DNRE treatment could ameliorate inflammatory cell death and reduce NO release from microglia, we tested different doses of DNRE (10–400 μg/mL). The treatment with DNRE was shown to have a dose-dependent effect in reducing NO release, but high doses of DNRE (200–400 μg/mL) showed toxicity (Figure 1A). Thus, considering the non-toxic dose and $IC_{50}$ of MTT assay (152.2997 μg/mL), the better effect was attributed to the dose of 50–100 μg/mL (Figure 1A,B). Therefore, we chose 10–100 μg/mL for the rest of the study. This effect might be related to the antioxidant effect of DNRE (0.01–1.0 mg/mL), which was supported by the DPPH scavenging assay and LPS-induced ROS release by DCFDA assay (Figure 1C,D).

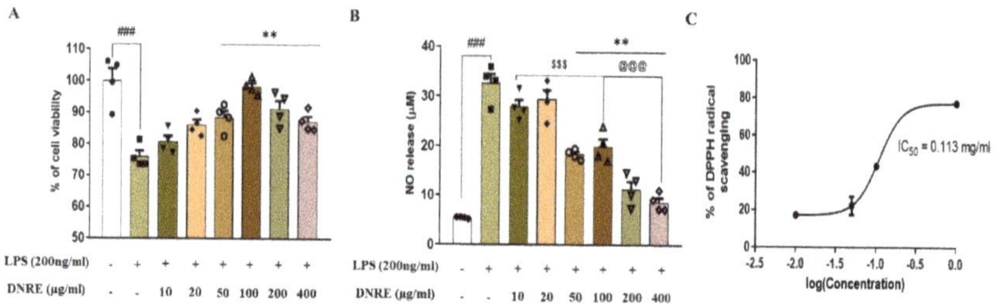


**Figure 1.** *Cont.*

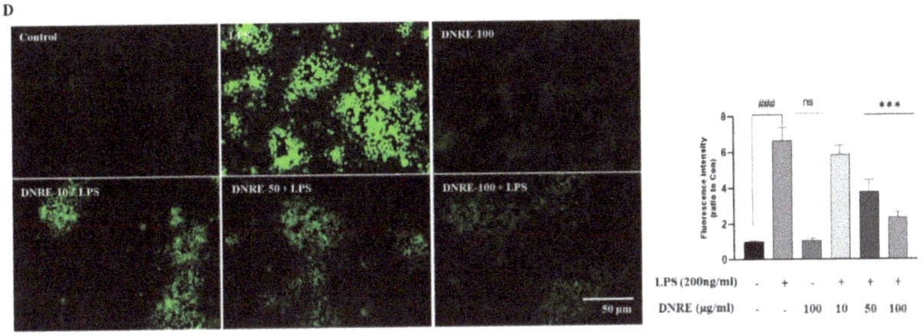


**Figure 1.** DNRE treatment protects microglial cells. (**A**) MTT assay ($IC_{50}$ = 152.2997 μg/mL) and (**B**) nitric oxide release assay for LPS-treated BV-2 cells with or without different doses of DNRE. (**C**) Sigmoid plot for determination of $IC_{50}$ value of DNRE of DPPH free radical scavenging. The half maximal inhibitory concentration ($IC_{50}$) was calculated from the equation (y = mx-c) obtained from the sigmoid curve. The value of 'y' was changed to 50, and we were subjected to determine value of 'x', which is the $IC_{50}$ value. (**D**) Fluorescence microscopy of DCFDA assay for LPS-treated BV-2 cells with or without DNRE, and measurement of fluorescence intensity using ImageJ software. ns—not significant compared with non-treated; each shape (circles, triangles, squares, etc.) is representative of the number of repetitions of a particular group; ### $p < 0.001$ compared with non-treated; ** $p < 0.01$, *** $p < 0.001$ comparing LPS vs. LPS + DNRE; $^{\$\$\$}$ $p < 0.001$ comparing low-dose DNRE (10–20 μg/mL) vs. mid-to-high-dose DNRE (50–400 μg/mL), $^{@@@}$ $p < 0.001$ comparing mid-dose DNRE (50–100 μg/mL) vs. high-dose DNRE (200–400 μg/mL).

### 2.2. DNRE Pre-Treatment Suppresses Inflammatory Mediators following LPS Toxication

In response to the microglial activation, pro-inflammatory mediator iNOS and COX-2 levels also increase via a mechanism involving increased NO and PGE2 level [7]. Several studies reported that systemic injection of LPS triggers neuroinflammation, leading to neurodegeneration and cognitive dysfunction [18–20]. As we have seen, DNRE treatment significantly reduced NO release and scavenged DPPH radicals; therefore, we investigated DNRE's impact on post-LPS-treated inflammatory biomarkers. Pre-treatment of DNRE showed a dose-dependent effect in reducing inflammatory enzymes COX-2 and iNOS mRNA and protein levels (Figure 2A,B). DNRE pre-treatment was shown to reduce, by at least twofold, iNOS mRNA or protein level (Figure 2A,B), signifying DNRE-treated decrease in NO release. This result was further justified by the reduction in mRNA level of IL-1β and IL-6 at least twofold in LPS-induced BV-2 microglial cells (Figure 2C). DNRE also significantly downregulated LPS-induced increase inflammatory cytokine TNF α production in BV-2 microglial cells (Figure 2C).

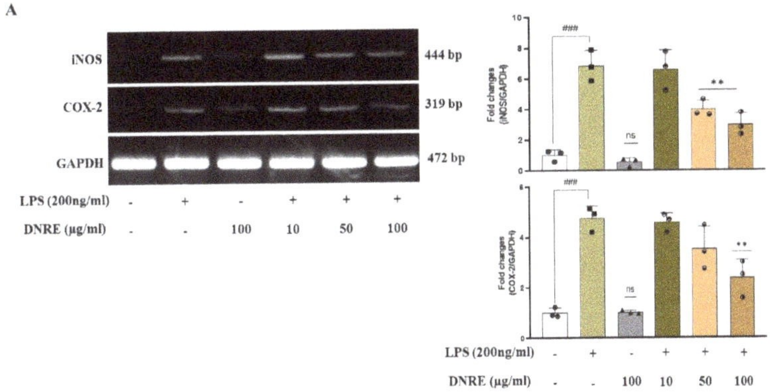


**Figure 2.** *Cont.*

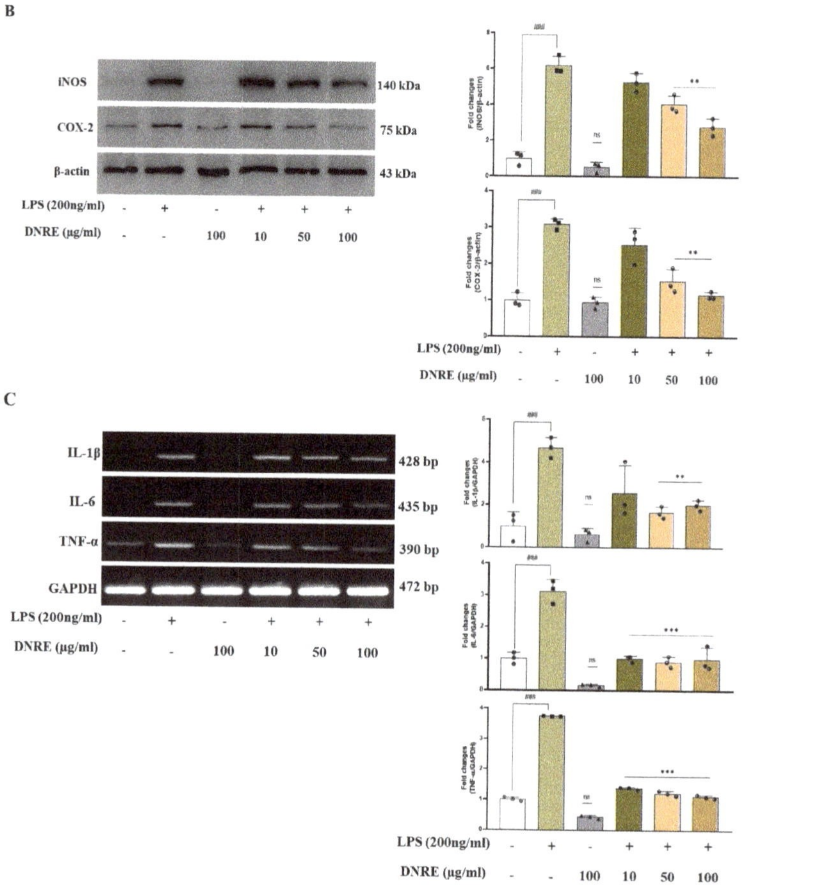


**Figure 2.** DNRE activities on inflammatory mediators. (**A**) mRNA and (**B**) protein expression of inflammatory enzymes in LPS (200 ng/mL)-treated BV-2 cells with or without different doses of DNRE. (**C**) mRNA expression of pro-inflammatory cytokines in LPS-treated BV-2 cells with or without DNRE. ns—not significant compared with non-treated; each shape (circles, triangles, squares, etc.) is representative of the number of repetitions of a particular group; ### $p < 0.001$ compared with non-treated; ** $p < 0.01$, *** $p < 0.001$ comparing LPS vs. LPS + DNRE.

### *2.3. DNRE Inhibits Nuclear Translocation of NF-κB*

LPS-induced microglial activation hyperactivates the MAPK pathway and exaggerates inflammation by promoting transcription factors such as NF-κB [4]. DNRE treatment dose-dependently decreased phosphorylation of MAPK signalling by reducing p-ERK, p-JNK, and p-p38 (Supplementary Figure S1) in LPS-treated BV-2 cells.

Phosphorylation of the p65 subunit of NF-κB translocates into the nucleus to translate inflammatory cytokines such as IL-1β, and causes inflammation. As we found that DNRE reduced phosphorylation of several biomarkers of MAPK signalling, we investigated whether DNRE treatment inhibited nuclear translocation of transcription factor NF-κB. LPS induction activates NF-κB by phosphorylating IκB at the cytosol, promoting the p65 subunit of NF-κB translocation into the nucleus [4]. DNRE treatment (100 μg/mL) was shown to inhibit IκB phosphorylation (Figure 3A) and reduce p65 permeation into the nucleus (Figure 3B). To confirm this finding, we used fluorescence microscopy to visualise DNRE impact on phosphorylated p65 subunit, and found a similar decrease to that observed with the WB probe (Supplementary Figure S2).

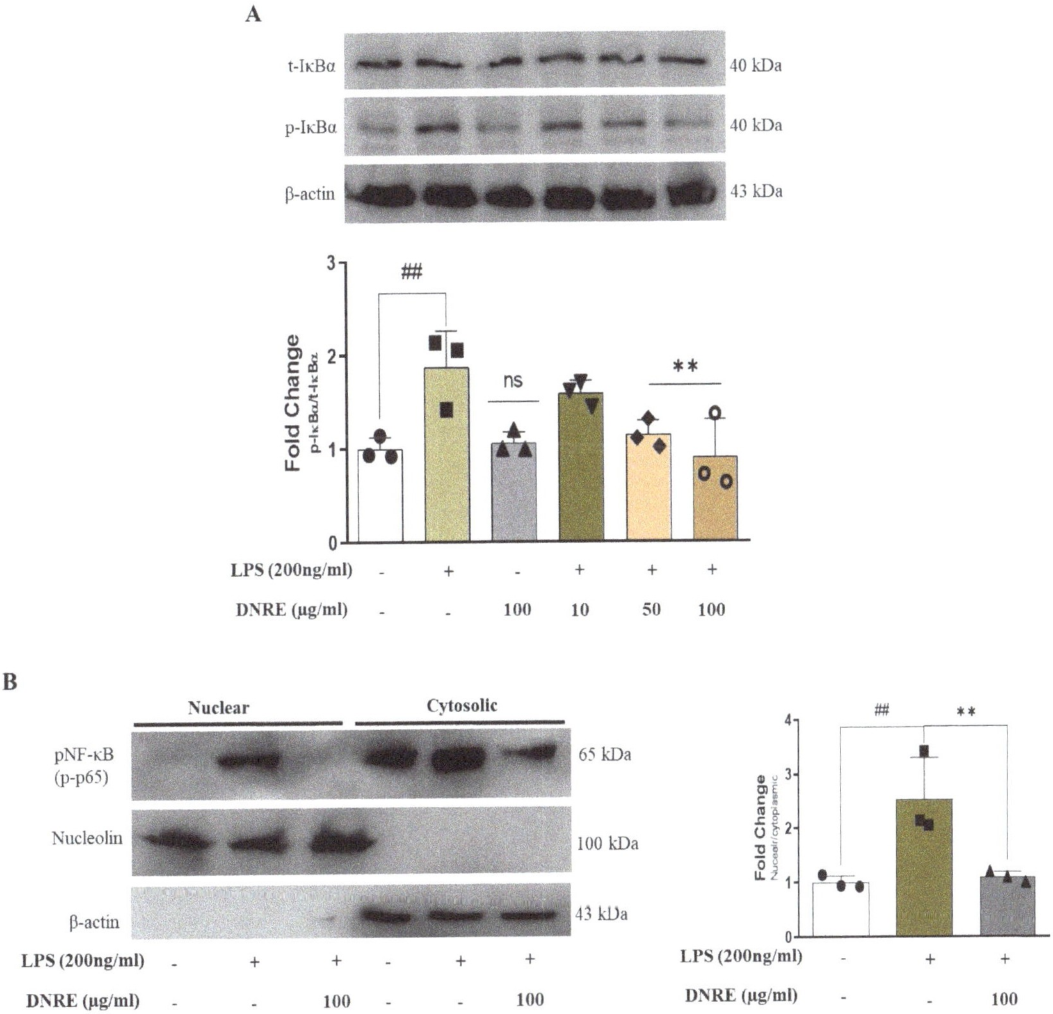


**Figure 3.** DNRE effects on nuclear factor. (**A**) Protein expression of IκB and densitometric calculation. (**B**) Protein expression of phosphorylated NF-κB (p-p65) of nuclear and cytoplasmic fractions of BV-2 cells treated with or without LPS (200 ng/mL). ns—not significant compared with non-treated; each shape (circles, triangles, squares, etc.) is representative of the number of repetitions of a particular group; ## $p < 0.01$ compared with non-treated; ** $p < 0.01$ comparing LPS vs. LPS + DNRE.

### 2.4. *Anti-Neuroinflammatory Activities of Dioscin*

Multiple studies have indicated that DNRE consists of steroidal saponin derivatives; dioscin is the most abundant among them [21]. In addition, our recent study found that dioscin ameliorates $MPP^+$-induced neurotoxicity via autophagy [22]. We observed that dioscin substantially downregulates inflammatory enzymes iNOS and COX-2 in LPS-activated BV-2 cells. Thus, we used HPLC fingerprinting to identify and quantify dioscin in DNRE (Figure 4A). Comparing the retention times with standard dioscin (Figure 4A top), we confirmed the identity of dioscin in DNRE (Figure 4A bottom). Subsequently, we calculated the concentrations of dioscin using peak area, and found that 7.07 ± 2.95 µg of dioscin is present per milligram of our ethanolic crude extract of DN. Interesting unknown peaks were observed between retention time 5–10 min, before dioscin peaks, in the DNRE chromatogram. We speculate that these could be other saponin constituents of DNRE; further analysis should be performed to identify these constituents.

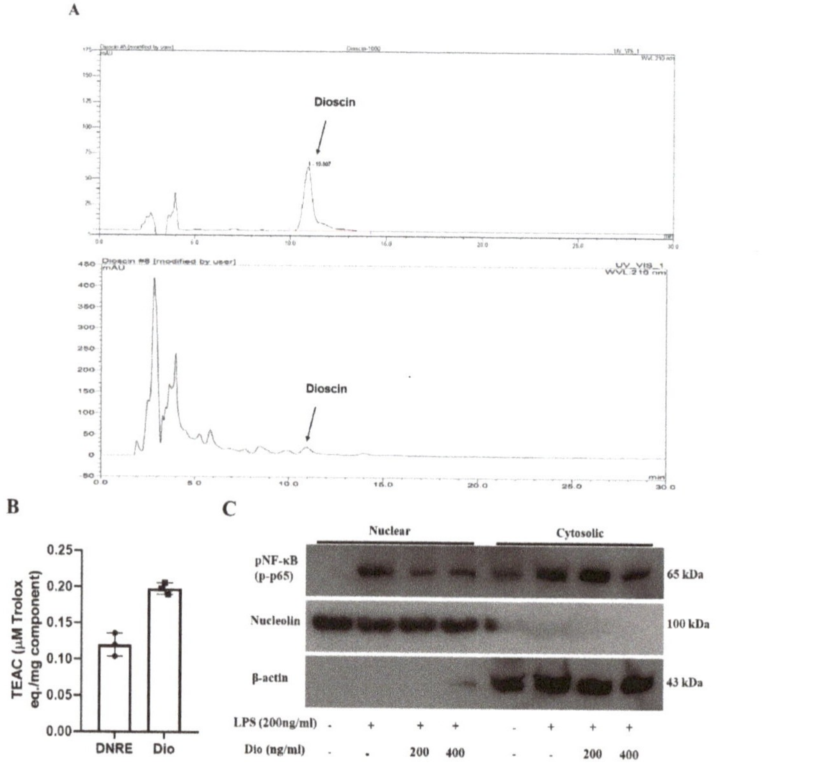


**Figure 4.** HPLC screening was conducted to identify the presence of and quantify dioscin in DNRE. (**A**) Representative chromatogram showing retention time and area of a standard steroidal saponin dioscin (**top**) and dioscin in DNRE (**bottom**). (**B**) The antioxidative capacity of dioscin (Dio) was measured using ABTS assay; an equimolar amount of DNRE and Dio was used, and the spectrophotometric measurements are expressed as µM of Trolox per mg of DNRE or Dio. (**C**) Nuclear and cytosolic fractions were isolated from LPS-activated BV-2 microglial cells pre-treated with or without different doses of dioscin and probed with anti-phosphorylated NF-κB (p-p65).

Since DNRE-mediated neuroprotection partly relies on its potential antioxidant effects, we tested whether dioscin possesses an antioxidant effect. In the ABTS analysis, dioscin showed more efficiency in free radicals scavenging than DNRE (Figure 4B). Previously, we showed that dioscin is a potential neuroprotector in an $MPP^+$-induced Parkinson's disease model [22], where dioscin showed a substantial reduction of apoptosis and upregulated neurotrophic factors such as BDNF and pCREB. Additionally, dioscin ameliorated inflammatory enzyme upregulation in LPS-stimulated in vitro model. Thus, we tested the speculation that dioscin could ameliorate phosphorylation and nuclear translocation of NF-κB p65 subunits (Figure 4C). We used two doses of dioscin and found that both doses are highly capable of downregulating p65's redistribution. Since NF-κB p65 phosphorylation and nuclear localisation could transcribe multiple inflammatory cytokines, one of the therapeutic strategies is to prevent p65 nuclear translocation [23]. Our data suggest that the dioscin of DNRE could be an inhibitor of p65 phosphorylation and translocation, which could prevent neuroinflammation.

### 2.5. Dioscin Ameliorates Scopolamine-Induced Learning Deficits

Based on our previous finding of dioscin-induced improvement on neurotrophic factors in neuronal cells [22], we tested dioscin's impact on the scopolamine-induced amnesic mice model. Scopolamine is an anticholinergic drug that antagonises cholinergic receptors and impairs hippocampal and prefrontal cortex transmission that deficits learning acquisition [10]. Intraperitoneal injection of scopolamine also stimulates systemic and consequent neuroinflammation in C57BL/6 mice [12]. Dioscin (60 mg/kg) treatment for 7 days ameliorated scopolamine-induced deficits in spatial learning memory in mice (Figure 5A–C).

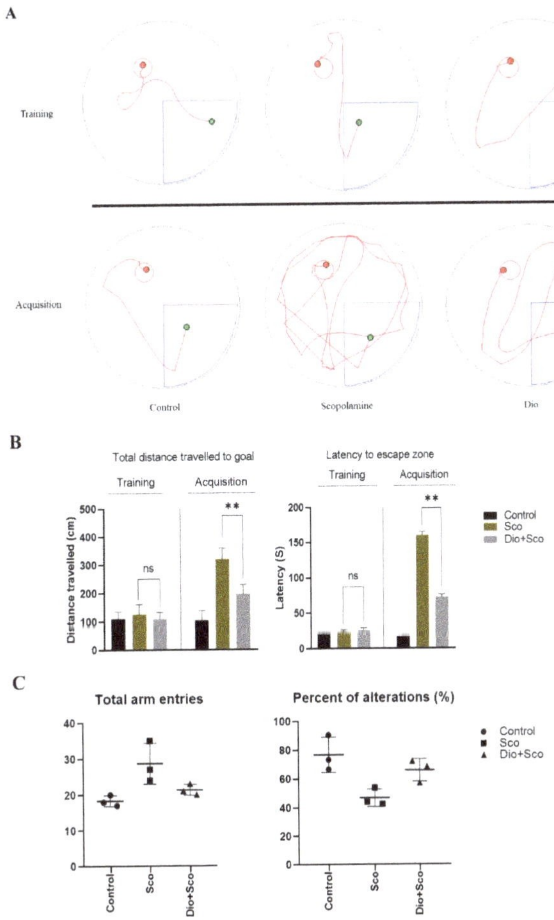


**Figure 5.** Dioscin effect on scopolamine-induced amnesic mice model. (**A**) Trajectory map used by animals during training and data acquisition in MWM pool; B = starting point and E = hidden platform. (**B**) Distance travelled by and latency to find a hidden platform in MWM; ns—not significant compared scopolamine vs. dioscin + scopolamine, ** $p < 0.01$ comparing scopolamine vs. dioscin + scopolamine. (**C**) Mice intensity of arms entry and alterations in Y-maze assay followed by treatment vs. non-treated.

*2.6. Dioscin Ameliorates Scopolamine-Induced Neurotrophic Factor Deficits*

Scopolamine treatment significantly upregulated pro-inflammatory enzyme iNOS and COX-2 expression (Figure 6A) in the cerebral cortex and hippocampus, which was significantly reduced by dioscin treatment. In addition, dioscin treatment also significantly improved neurotrophic factors BDNF and pCREB expressions (Figure 6B). These results indicate that dioscin ameliorates pro-inflammatory enzymes and neurotrophic factors induced by scopolamine, and improves spatial learning memory. Our findings also suggest that the dioscin might have a better effect on the cerebral cortex region than on the hippocampus; however, further investigation needs to be conducted to understand region-specific effects.

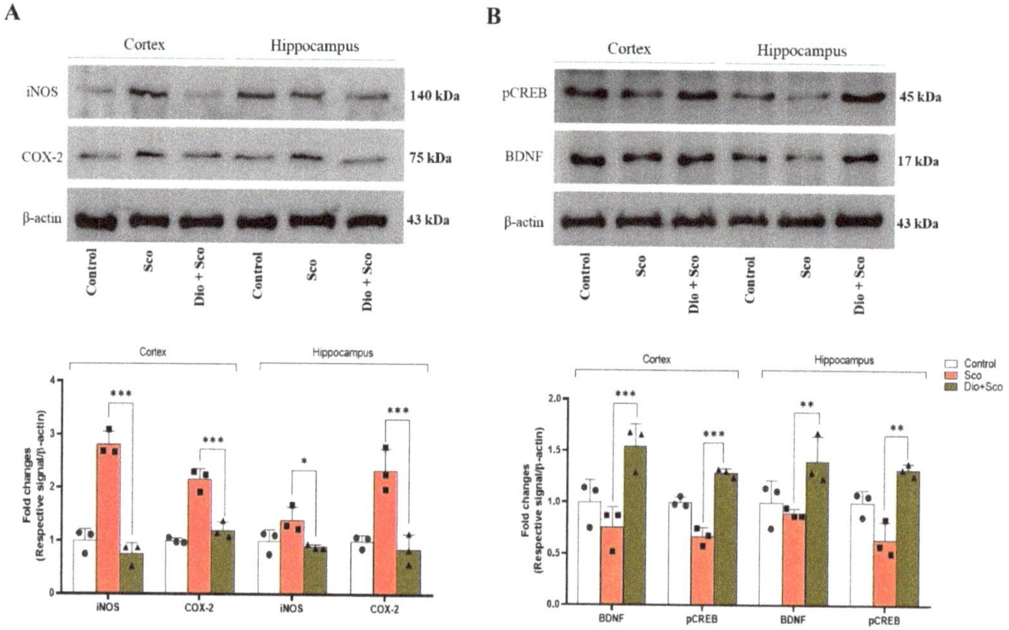


**Figure 6.** Dioscin effect on scopolamine-stimulated inflammatory response. (**A**) Dioscin treatment and pro-inflammatory response in cerebral cortex and hippocampus, and (**B**) neurotrophic factors. Sco, Scopolamine; Dio, Dioscin; each shape (circles, triangles, squares, etc.) is representative of the number of repetitions of a particular group; * $p < 0.05$, ** $p < 0.01$, and *** $p < 0.001$ comparing scopolamine vs. dioscin + scopolamine.

## 3. Discussion

Microglia are tissues residing macrophages that maintain the neuronal microenvironment and respond to endogenous and/or exogenous injury. Unlike peripheral macrophages, microglia possess phenotype-based polarisation features, which allow them to function as innate immune system modulators [24]. LPS can stimulate the microglial M1 phenotype and induce an inflammatory response via increasing cytokine release. This study hypothesised that DNRE could ameliorate LPS-induced microglial shifting and downregulate inflammatory response. DNRE has substantially decreased NO release and phosphorylation of different downstream effectors of MAPK signalling in LPS-activated BV-2 microglia.

LPS activates microglia and releases pro-inflammatory enzymes and cytokines, such as iNOS, IL-1β, IL-6, and TNF-α [25]. These are the key mediators of acute and chronic inflammation, leading to neurodegenerative diseases. In this study, LPS treatment substantially increased NO release and, subsequently, mRNA level of iNOS, IL-1β, IL-6, and TNF-α, and protein level of iNOS in BV-2 microglial cells. This indicates that DNRE pre-treatment might have reversed the M1 polarisation and reduced pro-inflammatory cytokines mRNA and protein expression. COX-2 is another pro-inflammatory enzyme synthesising pro-inflammatory prostaglandin E2 (PGE2) when exposed to LPS [26]. Consistently, LPS treatment significantly increased both COX-2 mRNA and protein levels that were alleviated by DNRE pre-treatment. Thus, we assume that DNRE pre-treatment could inactivate the NF-κB phosphorylation and MAPK signal transduction pathways (Supplementary Figure S1) responsible for pro-inflammatory cytokine production [17,27,28].

Several previous studies have published that DNRE contains bioactive compounds, including phenols and saponins [15,29], and these compounds might have been responsible for anti-inflammatory activity in this study. Among the several compounds, dioscin has been found to act as a primary bioactive compound in DNRE [29], which has shown anti-inflammatory activity in several independent studies [30,31]. In this study, HPLC fingerprinting assay found substantial levels of dioscin in DNRE. Dioscin also showed significant inhibition of NF-κB (p65) nuclear translocation, suggesting that DNRE-mediated inhibition of neuroinflammation could be associated with the presence of dioscin.

Our previous study [22] showed that dioscin is a potential bioactive compound that enhances neurotrophic factors in neuronal cells. In corroboration to prior findings, we found that dioscin treatment improves spatial and reference memory in scopolamine-treated animal models. MWM is not a maze in the usual sense–a labyrinth; instead, it uses hippocampal synaptic plasticity and NMDA receptor function [32]. We found that dioscin treatment improved mice performance in the search-to-platform area and avoided latency, which indicated improvement in synaptic plasticity. BDNF and other neurotrophic factors regulate the neural plasticity network [33]; dioscin treatment significantly upregulated neurotrophic factors in the hippocampus and cortex region. Our data validate the mice performances in MWM and Y-maze analyses.

Scopolamine inhibits cholinergic transmission in the hippocampus and prefrontal cortex region, affecting spatial and related learning memory. In this process of neurocognitive impairments, scopolamine increases inflammatory cytokines TNF-α, IL-6 and IL-1β [12], and enzymes iNOS and COX-2 [34]. In the animal model of scopolamine, dioscin treatment for 7 days reduced pro-inflammatory mediators, which follows the inhibition of NF-κB phosphorylation. In agreement with our previous reports, this study found that dioscin provides neuroprotection possibly via the upregulation of neurotrophic factors in different brain regions. This study provides a scientific basis for ethnobotanical use of DNRE and suggests further investigation needs to be conducted with its potential active compound, dioscin, emphasising the regulation of neurotrophic factors.

## 4. Materials and Methods

### *4.1. Chemicals*

DNRE was purchased from Korea Plant Extract Bank (ID: PBC-464AS), lipopolysaccharide (LPS; *Escherichia coli*; 055:B5), dimethyl sulfoxide (DMSO), 3-(3,4-dimehylthiazol-2-yl)-2,5-diphenyltetrazolium bromide (MTT), and scopolamine hydrochloride were obtained from Sigma-Aldrich (St. Louis, MO, USA). Foetal bovine serum (FBS) (#16000-442; Gibco, NY14072, USA), phosphate-buffered saline (PBS), and Dulbecco's modified Eagle's medium (DMEM) were purchased from Gibco-BRL Technologies (Gaithersburg, MD, USA). Trizol was purchased from Invitrogen Life Technologies (Carlsbad, CA, USA). RIPA buffer (10x) was purchased from Millipore (Milford, MA, USA), and protease and phosphatase inhibitors were obtained from Roche (Indianapolis, IN, USA).

### *4.2. Cell Culture and Treatment*

The BV-2 microglial cells were a generous gift from Dr K. Suk (Kyung-Pook National University, South Korea) [35]. Cells were grown in DMEM supplemented with 5% ($v/v$) FBS and 1% ($v/v$) of penicillin/streptomycin in a maintained incubator (37 °C and 5% $CO_2$).

Upon reaching confluency (80–90%), cells were trypsinised (0.05% trypsin-EDTA) for sub-culture and/or for seeding before respective treatment. For each experiment, cells from at least three consecutive passages were used. The BV-2 cells were exposed to different doses of DNRE (10–400 μg/mL) and dioscin (200 and 400 ng/mL) for 2 h before being exposed to LPS (200 ng/mL) for 24 h.

### 4.3. Cell Viability and Nitric Oxide Assay

For assessing cell viability and nitric oxide (NO) release, the BV-2 cells were seeded at $5 \times 10^3$ cells/well in a 96-well plate. Upon reaching approx. 80% confluency, cells were treated with various doses of DNRE (10–400 μg/mL) with or without LPS (200 ng/mL), and incubated for 24 h. Subsequently, the culture medium was collected for colourimetric NO release assay using Griess reagent; measurements were detected at 540 nm on the UV spectrum using a microplate reader (SunriseTM, Tecan Trading AG, Switzerland).

For cell viability, 20 μL of MTT (5 mg/mL) was added and incubated for 45 min. Following this, culture media was removed, and 200 μL of DMSO was added to dissolve the formazan crystals. Absorption was measured at 552 nm using a microplate reader.

### 4.4. DPPH Free Radical Scavenging Assay

2,2-Diphenyl-1-picrylhydrazyl (DPPH; 95%) was purchased from Thermo Fisher Scientific (Ward Hill, MA 01835, USA). A working stock of DPPH (200 μM/L) was prepared in ethanol (99%), and was added to different concentrations of DNRE to make up a final volume of 300 μL/well in a 96-well plate. The first lane contained DPPH + ethanol as the control, and an additional lane was filled with ethanol alone as a blank (negative control). Absorbance was measured in a microplate reader at 520 nm. The calculation for DPPH radical inhibition was as follows:

$$\%\,of\ DPPH\ radical\ inhibition = \left[\frac{Abs_{control} - (Abs_{sample} - Abs_{blank})}{Abs_{control}}\right] \times 100 \quad (1)$$

### 4.5. DCFDA Assay

To detect the intracellular ROS release after LPS induction, we used ROS-sensitive fluorescent dye, 2′,7′-dichlorofluorescein diacetate (DCFDA; Sigma-Aldrich). BV-2 cells ($2.5 \times 10^5$ cells/mL) were cultured in 6-well plates in regular media. Upon reaching the confluency (60–70%), LPS (200 ng/mL) followed by DNRE (10–100 μg/mL) was treated for 2 h in a serum-free media. Later, treated or non-treated cells were incubated for an additional 30 min at 37 °C with DCFDA (10 μM final concentration) and then washed twice with ice-cold PBS. Images were immediately captured using a fluorescence microscopy system (Nikon Eclipse Ts2R). The fluorescence intensity was calculated using ImageJ (Java 1.8.0_172, NIH).

### 4.6. ABTS Radical Scavenging Activity

The total antioxidant activity of the DNRE and Dio was measured using the ABTS + radical cation decolourisation assay [36]. A 7.4 mM of ABTS was mixed in equal volume with 2.6 mM potassium persulfate. The mixture was then stored in the dark at room temperature (RT) for 12–14 h before use. After radical generation, the ABTS radical solution was diluted with deionised water until its absorbance was 0.70 ± 0.02 at 734 nm. Then 0.9 mL of ABTS radical solution was mixed with 0.1 mL of samples, and the absorbance was measured at 734 nm. The antioxidant activity of the DNRE and Dio were expressed as Trolox equivalents antioxidant capacity (TEAC) or mM of Trolox equivalent per mg of extract (mM Trolox eq./mg extract).

### 4.7. Nucleic–Cytosolic Fraction Preparation

BV-2 cells were seeded into a 60 mm cell culture Petri dish at $2.5 \times 10^5$ cells/mL and ~80% confluency; cells were treated with LPS or LPS + DNRE, or were not treated, for

24 h. Washed twice with ice-cold PBS, nucleic acid lysis buffer and 10% NP-40 (detergent) were used to collect cytosolic fractions at 4000 rpm for 10 min. Pellets were dissolved in extraction buffer (containing 20 mM HEPES, 20% glycerol and 0.2 mM EDTA) and incubated on ice for 30 min with occasional tapping to liberate the mixture to collect nucleic acid fractions. The mixture was centrifuged at 13,000 rpm for 15 min to collect nucleic acid fraction as supernatant, and both fractions were quantified and prepared for Western blot analysis.

### 4.8. Immunofluorescence

As described by Runwal et al. (2019) [37], treated and non-treated cells were washed with cold PBS once, followed by fixation with 4% cold PFA. Next, cells were permeabilised with 0.1% (*v*/*v*) Triton X-100 for 10 min at room temperature and washed (2x). Subsequently, primary antibody (anti-pNF-κB) at 2 μg/mL was added and incubated overnight at 4 °C. The next day, cells were counterstained with chicken anti-rabbit secondary antibodies (CAR-594; A21201 Invitrogen) at room temperature for 1 h. Finally, cells were stained with DAPI (2 μg/mL), and images were captured using a wide-angle fluorescence microscope, and were processed by NIS-Elements software (BR-2.01.00, NY 11747-3064, New York, USA).

### 4.9. Determination of a Possible Active Compound in the Extract

We performed high-performance liquid chromatography (HPLC) to determine and quantify a marker compound in DNRE [21,38,39]. A Thermo Scientific (Ultimate 3000) HPLC system (Thermo Scientific, Korea) equipped with a UV detector and 20 μL injection loop was used. We used a gradient system consisting of (A) 0.1% acetic acid in distilled water, and (B) 100% acetonitrile, and separated using Diamonsil (C18; 250 × 4.6 mm; 5 μm) column. The gradient ratio was set to 45:55 (A:B) for 30 min at the flow rate of 1.0 mL/min, and the UV detection was set to 210 nm. Using these conditions, we detected and quantified dioscin in the extract.

### 4.10. Animals Handling and Treatment

Male 8-week-old C57BL/6 (20–25 g) mice were purchased from Daehan Bio-Link, South Korea. All animals were housed in a controlled environment, as described by Jo and colleagues [40]. Five animals per group (control, scopolamine, and dioscin + scopolamine) were acclimatised for 1 week in different cages before experiments (Figure 7). Scopolamine (2 mg/kg) was administered intraperitoneally, and dioscin (60 mg/kg) and saline (control) were administered orally, for 7 consecutive days. All experiments were approved by the Institutional Animal Care and Use Committee (IACUC), Konkuk University (IACUC no- KUB201101).

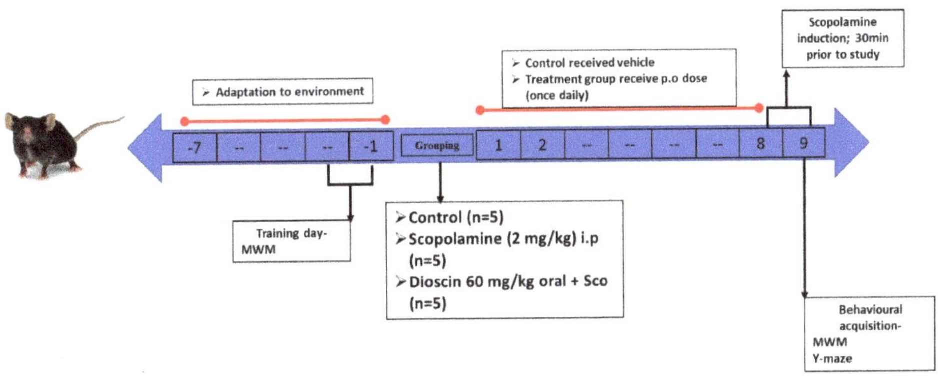


**Figure 7.** Schematic presentation of scheduling for animal grouping, treatment, behaviour, and sacrifice.

### 4.11. Y-Maze Test

We used a Y-maze assay for spatial and reference memory assessment [41]. A 'Y' shaped maze with three identical arms, each arm separated at a 120° angle, was used for the test. Mice were placed at the centre of the maze, and latency to move and altered entry into different arms were noted for 5 min. Mice with intact memory should show less interest to re-visit recently entered arm, while those with impaired memory have high interest to re-visit.

### 4.12. Morris Water Maze Test

A Morris water maze (MWM) [32] evaluates the spatial memory of rodents, which relies on rodents' navigation skills to find the submerged escape platform. This test used a circular pool (122 cm) with a hidden platform. The aim of the animal was to find a short and direct path to the hidden platform from the start position. We used two back-to-back training sessions with 2 h intervals for 2 days before treatment. Post-treatment, for the respective treatment plan, the spatial acquisition was measured and data analysed using SMART 3.0 software (Ver. 3.0, Harvard Apparatus, Holliston, MA 01746, USA).

### 4.13. Reverse Transcription–Polymerase Chain Reaction (RT–PCR)

BV-2 cells were washed twice with ice-cold PBS, and 1 mL of trizol reagent was added to each well and incubated at room temperature for 30 s. Cells were collected carefully to an Eppendorf tube followed by centrifugation. We used the ReverTra Ace-α kit (Toyobo, Osaka, Japan) to isolate first-strand cDNA, according to the manufacturer's instructions. In addition, 1 μL of RT-mixture templet was used for further amplification in the presence of specific primers (Table 1). PCR products were electrophoresed in 1.5% agarose gel containing GelRed® Nucleic Acid Gel Stain (Biotium Inc., Fremont, CA 94538, USA).

**Table 1.** Specific primer sequences used in this study.

| Gene | Primer Sequence | Size (bp) |
|---|---|---|
| iNOS | F 5′-GAG GTA CTC AGC GTC CTC CA-3′<br>R 5′-AGG GAG GAA AGG GAG AGA GG-3′ | 444 |
| COX-2 | F 5′-TGA GTG GTA GCC AGC AAA GC-3′<br>R 5′-CTG CAG TCC AGG TTC AAT GG-3′ | 319 |
| TNF-α | F 5′-TTC GAG TGA CAA GCC TGT AGC-3′<br>R 5′-AGA TTG ACC TCA GCG CTG AGT-3′ | 390 |
| IL-1β | F 5′-CAA GGA GAA CCA AGC AAC GA-3′<br>R 5′-TTG GCC GAG GAC TAA GGA GT-3′ | 428 |
| IL-6 | F 5′-GGA GGC TTA ATT ACA CAT GTT-3′<br>R 5′-TGA TTT CAA GAT GAA TTG GAT-3′ | 435 |
| GAPDH | F 5′- ACC ACA GTC CAT GCC ATC AC-3′<br>R 5′- CCA CCA CCC TGT TGC TGT AG-3′ | 472 |

### *4.14. Western Blot Analysis*

The cells were washed twice with ice-cold PBS and lysed using lysis buffer (1x RIPA lysis buffer containing a protease and phosphatase inhibitor (1:1) cocktail). Equal protein (20 μg/10 μL) were loaded to each lane for electrophoresis in ~8–15% sodium dodecyl sulphate-polyacrylamide electrophoresis (SDS-PAGE) gel. Electrophoresed proteins were then transferred onto polyvinylidene-difluoride (PVDF) membranes (Millipore, Bedford, MA, USA). The membranes were incubated at room temperature with 5% skim milk to prevent nonspecific binding. Then the blots were incubated overnight at 4 °C on a rocking platform with specific primary antibodies, including anti-iNOS (#A0312, ABclonal), anti-COX-2 (M-19 #sc1747, Santa Cruz biotechnology), anti-p38 (#9212S, cell signaling), anti-pp38 (#9201S, cell signaling), anti-JNK (#9253S, cell signaling), anti-pJNK (#9251S, cell signaling), anti-ERK (#91012S cell signaling), anti-pERK (#9101S, cell signaling), anti-NF-κB/p65 (F-6 #sc-8008, Santa Cruz biotechnology), anti-pNF-κB/pp65 (#3033S, cell signaling), anti-IκB (#4812S, cell signaling), anti-pIκB (#2859S, cell signaling), anti-nucleolin (#14574; cell signaling) at concentration of 1:1000, and anti-β-actin (#C4; Santa Cruz biotechnology) at concentration of 1:5000. The next day, each blot was incubated at room temperature with either an anti-mouse or anti-rabbit (1:10,000) secondary antibodies. The blots were visualised with an enhanced chemiluminescence detection system (LAS 500; GE Healthcare Bio-Sciences AB, 751 25, Uppsala, Sweden) per the recommended protocol.

### *4.15. Statistical Analysis*

All statistical analyses were performed using GraphPad Prism (by Dotmatics, version-8.0.1; La Jolla, CA, USA) software. Data represent the mean ± SEM (standard error mean) of three independent experiments, at least. We used one-way ANOVAs followed by Sidak's multiple comparisons to determine the statistical significance. The *p*-values were considered significant at $<0.05$.

## 5. Conclusions

DNRE pre-treatment significantly reversed LPS-mediated microglial activation, resulting in the downregulation of NF-κB (p65) phosphorylation and subsequent neuroinflammation. We found that dioscin, an active compound of DNRE, reduces p65 phosphorylation in vitro and upregulates neurotrophic factors in the hippocampus and cortex region of scopolamine-induced mice. Our data indicate that the DNRE-mediated amelioration of neuroinflammatory responses and memory deficits is partly attributed to the dioscin effect.

**Supplementary Materials:** The following supporting information can be downloaded at: https://www.mdpi.com/article/10.3390/ijms23179923/s1.

**Author Contributions:** Conceptualization, S.A.; methodology, S.A., M.J., and Y.-S.K.,.; software, S.A.; validation, M.J. and Y.-S.K.; formal analysis, S.A. and Y.-S.K.; investigation, S.A.; resources, Y.-J.Y. and J.-Y.A.; data curation, S.A.; writing—original draft preparation, S.A.; writing—review and editing, M.J.; visualization, I.-S.K.; supervision, D.-K.C.; project administration, I.-S.K. and D.-K.C.; funding acquisition, D.-K.C. All authors have read and agreed to the published version of the manuscript.

**Funding:** This research received no external funding.

**Institutional Review Board Statement:** Institutional review board has approved animal experiments for this study (IACUC no- KUB201101).

**Informed Consent Statement:** Not applicable.

**Data Availability Statement:** The datasets generated and/or analysed during the current study are publicly available upon acceptance of this manuscript.

**Acknowledgments:** This paper was supported by Konkuk University in 2022.

**Conflicts of Interest:** The authors declare no conflict of interest.

## References


1. Glass, C.K.; Saijo, K.; Winner, B.; Marchetto, M.C.; Gage, F.H. Mechanisms underlying inflammation in neurodegeneration. *Cell* **2010**, *140*, 918–934. [CrossRef] [PubMed]
2. Yang, L.; Zhou, R.; Tong, Y.; Chen, P.; Shen, Y.; Miao, S.; Liu, X. Neuroprotection by dihydrotestosterone in lps-induced neuroinflammation. *Neurobiol. Dis.* **2020**, *140*, 104814. [CrossRef] [PubMed]
3. Lehnardt, S.; Massillon, L.; Follett, P.; Jensen, F.E.; Ratan, R.; Rosenberg, P.A.; Volpe, J.J.; Vartanian, T. Activation of innate immunity in the cns triggers neurodegeneration through a toll-like receptor 4-dependent pathway. *Proc. Natl. Acad. Sci. USA* **2003**, *100*, 8514–8519. [CrossRef]
4. Azam, S.; Jakaria, M.; Kim, I.S.; Kim, J.; Haque, M.E.; Choi, D.K. Regulation of toll-like receptor (tlr) signaling pathway by polyphenols in the treatment of age-linked neurodegenerative diseases: Focus on tlr4 signaling. *Front. Immunol.* **2019**, *10*, 1000. [CrossRef] [PubMed]
5. He, P.; Yan, S.; Zheng, J.; Gao, Y.; Zhang, S.; Liu, Z.; Liu, X.; Xiao, C. Eriodictyol attenuates lps-induced neuroinflammation, amyloidogenesis, and cognitive impairments via the inhibition of nf-κb in male c57bl/6j mice and bv2 microglial cells. *J. Agric. Food Chem.* **2018**, *66*, 10205–10214. [CrossRef] [PubMed]
6. Lu, Y.C.; Yeh, W.C.; Ohashi, P.S. Lps/tlr4 signal transduction pathway. *Cytokine* **2008**, *42*, 145–151. [CrossRef]
7. Zhou, Y.L.; Yan, Y.M.; Li, S.Y.; He, D.H.; Xiong, S.; Wei, S.F.; Liu, W.; Hu, L.; Wang, Q.; Pan, H.F.; et al. 6-o-angeloylplenolin exerts neuroprotection against lipopolysaccharide-induced neuroinflammation in vitro and in vivo. *Acta Pharmacol. Sin.* **2020**, *41*, 10–21. [CrossRef] [PubMed]
8. Kilimann, I.; Hausner, L.; Fellgiebel, A.; Filippi, M.; Würdemann, T.J.; Heinsen, H.; Teipel, S.J. Parallel atrophy of cortex and basal forebrain cholinergic system in mild cognitive impairment. *Cereb. Cortex* **2017**, *27*, 1841–1848. [CrossRef] [PubMed]
9. Hoover, D.B. Cholinergic modulation of the immune system presents new approaches for treating inflammation. *Pharmacol. Ther.* **2017**, *179*, 1–16. [CrossRef]
10. Rahimzadegan, M.; Soodi, M. Comparison of memory impairment and oxidative stress following single or repeated doses administration of scopolamine in rat hippocampus. *Basic Clin. Neurosci.* **2018**, *9*, 5–14. [CrossRef]
11. Wong-Guerra, M.; Jiménez-Martin, J.; Pardo-Andreu, G.L.; Fonseca-Fonseca, L.A.; Souza, D.O.; de Assis, A.M.; Ramirez-Sanchez, J.; Del Valle, R.M.; Nuñez-Figueredo, Y. Mitochondrial involvement in memory impairment induced by scopolamine in rats. *Neurol. Res.* **2017**, *39*, 649–659. [CrossRef]
12. Cheon, S.Y.; Koo, B.N.; Kim, S.Y.; Kam, E.H.; Nam, J.; Kim, E.J. Scopolamine promotes neuroinflammation and delirium-like neuropsychiatric disorder in mice. *Sci. Rep.* **2021**, *11*, 8376. [CrossRef] [PubMed]
13. Ou-Yang, S.H.; Jiang, T.; Zhu, L.; Yi, T. Dioscorea nipponica makino: A systematic review on its ethnobotany, phytochemical and pharmacological profiles. *Chem. Cent. J.* **2018**, *12*, 57. [CrossRef]
14. Feng, J.F.; Tang, Y.N.; Ji, H.; Xiao, Z.G.; Zhu, L.; Yi, T. Biotransformation of dioscorea nipponica by rat intestinal microflora and cardioprotective effects of diosgenin. *Oxidative Med. Cell. Longev.* **2017**, *2017*, 4176518. [CrossRef] [PubMed]
15. Woo, K.W.; Kwon, O.W.; Kim, S.Y.; Choi, S.Z.; Son, M.W.; Kim, K.H.; Lee, K.R. Phenolic derivatives from the rhizomes of dioscorea nipponica and their anti-neuroinflammatory and neuroprotective activities. *J. Ethnopharmacol.* **2014**, *155*, 1164–1170. [CrossRef] [PubMed]
16. Park, K.-S.; Kim, H.J.; Hwang, J.T.; Ko, B.S. Dioscorea nipponica extracts enhance recovery from skeletal muscle atrophy by suppressing nf-κb expression. *J. Funct. Foods* **2020**, *73*, 104109. [CrossRef]
17. Zhou, Q.; Sun, H.J.; Liu, S.M.; Jiang, X.H.; Wang, Q.Y.; Zhang, S.; Yu, D.H. Anti-inflammation effects of the total saponin fraction from dioscorea nipponica makino on rats with gouty arthritis by influencing mapk signalling pathway. *BMC Complement. Med. Ther.* **2020**, *20*, 261. [CrossRef]
18. Huffman, W.J.; Subramaniyan, S.; Rodriguiz, R.M.; Wetsel, W.C.; Grill, W.M.; Terrando, N. Modulation of neuroinflammation and memory dysfunction using percutaneous vagus nerve stimulation in mice. *Brain Stimul.* **2019**, *12*, 19–29. [CrossRef] [PubMed]
19. Qin, L.; Wu, X.; Block, M.L.; Liu, Y.; Breese, G.R.; Hong, J.S.; Knapp, D.J.; Crews, F.T. Systemic lps causes chronic neuroinflammation and progressive neurodegeneration. *Glia* **2007**, *55*, 453–462. [CrossRef] [PubMed]
20. Vasconcelos, A.R.; Yshii, L.M.; Viel, T.A.; Buck, H.S.; Mattson, M.P.; Scavone, C.; Kawamoto, E.M. Intermittent fasting attenuates lipopolysaccharide-induced neuroinflammation and memory impairment. *J. Neuroinflamm.* **2014**, *11*, 85. [CrossRef]
21. Lin, S.; Wang, D.; Yang, D.; Yao, J.; Tong, Y.; Chen, J. Characterization of steroidal saponins in crude extract from dioscorea nipponica makino by liquid chromatography tandem multi-stage mass spectrometry. *Anal. Chim. Acta* **2007**, *599*, 98–106. [CrossRef] [PubMed]
22. Azam, S.; Haque, M.E.; Cho, D.Y.; Kim, J.S.; Jakaria, M.; Kim, I.S.; Choi, D.K. Dioscin-mediated autophagy alleviates mpp$^+$-induced neuronal degeneration: An in vitro parkinson's disease model. *Molecules* **2022**, *27*, 2827. [CrossRef] [PubMed]
23. Giridharan, S.; Srinivasan, M. Mechanisms of nf-κb p65 and strategies for therapeutic manipulation. *J. Inflamm. Res.* **2018**, *11*, 407–419. [CrossRef] [PubMed]
24. Orihuela, R.; McPherson, C.A.; Harry, G.J. Microglial m1/m2 polarization and metabolic states. *Br. J. Pharmacol.* **2016**, *173*, 649–665. [CrossRef]
25. Wang, W.Y.; Tan, M.S.; Yu, J.T.; Tan, L. Role of pro-inflammatory cytokines released from microglia in alzheimer's disease. *Ann. Transl. Med.* **2015**, *3*, 136. [PubMed]

26. Ricciotti, E.; FitzGerald, G.A. Prostaglandins and inflammation. *Arterioscler. Thromb. Vasc. Biol.* **2011**, *31*, 986–1000. [CrossRef] [PubMed]
27. Zhang, P.; Martin, M.; Michalek, S.M.; Katz, J. Role of mitogen-activated protein kinases and nf-kappab in the regulation of proinflammatory and anti-inflammatory cytokines by porphyromonas gingivalis hemagglutinin b. *Infect. Immun.* **2005**, *73*, 3990–3998. [CrossRef] [PubMed]
28. Liu, T.; Zhang, L.; Joo, D.; Sun, S.C. Nf-κb signaling in inflammation. *Signal Transduct. Target. Ther.* **2017**, *2*, 17023. [CrossRef] [PubMed]
29. Park, K.J.; Suh, W.S.; Cha, J.M.; Park, J.E.; Woo, K.W.; Lee, K.R.; Kim, S.Y. Steroidal saponins from dioscorea nipponica rhizomes and their biological activity. *Korean J. Pharmacogn.* **2017**, *48*, 261–267.
30. Yao, H.; Hu, C.; Yin, L.; Tao, X.; Xu, L.; Qi, Y.; Han, X.; Xu, Y.; Zhao, Y.; Wang, C.; et al. Dioscin reduces lipopolysaccharide-induced inflammatory liver injury via regulating tlr4/myd88 signal pathway. *Int. Immunopharmacol.* **2016**, *36*, 132–141. [CrossRef]
31. Yao, H.; Sun, Y.; Song, S.; Qi, Y.; Tao, X.; Xu, L.; Yin, L.; Han, X.; Xu, Y.; Li, H.; et al. Protective effects of dioscin against lipopolysaccharide-induced acute lung injury through inhibition of oxidative stress and inflammation. *Front. Pharmacol.* **2017**, *8*, 120. [CrossRef] [PubMed]
32. Vorhees, C.V.; Williams, M.T. Morris water maze: Procedures for assessing spatial and related forms of learning and memory. *Nat. Protoc.* **2006**, *1*, 848–858. [CrossRef]
33. Castrén, E.; Antila, H. Neuronal plasticity and neurotrophic factors in drug responses. *Mol. Psychiatry* **2017**, *22*, 1085–1095. [CrossRef] [PubMed]
34. Karthivashan, G.; Park, S.-Y.; Kweon, M.-H.; Kim, J.; Haque, M.E.; Cho, D.-Y.; Kim, I.-S.; Cho, E.-A.; Ganesan, P.; Choi, D.-K. Ameliorative potential of desalted salicornia europaea l. Extract in multifaceted alzheimer's-like scopolamine-induced amnesic mice model. *Sci. Rep.* **2018**, *8*, 7174. [CrossRef] [PubMed]
35. Jakaria, M.; Azam, S.; Cho, D.Y.; Haque, M.E.; Kim, I.S.; Choi, D.K. The methanol extract of allium cepa l. Protects inflammatory markers in lps-induced bv-2 microglial cells and upregulates the antiapoptotic gene and antioxidant enzymes in n27-a cells. *Antioxidants* **2019**, *8*, 348. [CrossRef]
36. Tachakittirungrod, S.; Okonogi, S.; Chowwanapoonpohn, S. Study on antioxidant activity of certain plants in thailand: Mechanism of antioxidant action of guava leaf extract. *Food Chem.* **2007**, *103*, 381–388. [CrossRef]
37. Runwal, G.; Stamatakou, E.; Siddiqi, F.H.; Puri, C.; Zhu, Y.; Rubinsztein, D.C. Lc3-positive structures are prominent in autophagy-deficient cells. *Sci. Rep.* **2019**, *9*, 10147. [CrossRef] [PubMed]
38. Liu, C.Z.; Zhou, H.Y.; Yan, Q. Fingerprint analysis of dioscorea nipponica by high-performance liquid chromatography with evaporative light scattering detection. *Anal. Chim. Acta* **2007**, *582*, 61–68. [CrossRef] [PubMed]
39. Yu, H.; Zheng, L.; Yin, L.; Xu, L.; Qi, Y.; Han, X.; Xu, Y.; Liu, K.; Peng, J. Protective effects of the total saponins from dioscorea nipponica makino against carbon tetrachloride-induced liver injury in mice through suppression of apoptosis and inflammation. *Int. Immunopharmacol.* **2014**, *19*, 233–244. [CrossRef] [PubMed]
40. Jo, S.H.; Kang, T.B.; Koppula, S.; Cho, D.Y.; Kim, J.S.; Kim, I.S.; Choi, D.K. Mitigating effect of lindera obtusiloba blume extract on neuroinflammation in microglial cells and scopolamine-induced amnesia in mice. *Molecules* **2021**, *26*, 2870. [CrossRef] [PubMed]
41. Kraeuter, A.K.; Guest, P.C.; Sarnyai, Z. The y-maze for assessment of spatial working and reference memory in mice. *Methods Mol. Biol.* **2019**, *1916*, 105–111. [PubMed]

International Journal of
*Molecular Sciences*

MDPI

*Article*

# Histomorphometry Changes and Decreased Reactivity to Angiotensin II in the Ileum and Colon of Streptozotocin-Induced Diabetic Rats

**Marisa Esteves-Monteiro [1,2,3], Daniela Menezes-Pinto [3], Mariana Ferreira-Duarte [1,3], Patrícia Dias-Pereira [4], Manuela Morato [1,3,*] and Margarida Duarte-Araújo [1,2,*]**

1 LAQV-REQUIMTE, Faculty of Pharmacy, University of Porto, 4050-313 Porto, Portugal
2 Department of Immuno-Physiology and Pharmacology, Institute of Biomedical Sciences Abel Salazar, University of Porto (ICBAS-UP), 4050-313 Porto, Portugal
3 Laboratory of Pharmacology, Department of Drug Sciences, Faculty of Pharmacy, University of Porto (FFUP), 4050-313 Porto, Portugal
4 Department of Pathology and Molecular Immunology, Institute of Biomedical Sciences Abel Salazar, University of Porto (ICBAS-UP), 4050-313 Porto, Portugal
* Correspondence: mmorato@ff.up.pt (M.M.); mdcma@icbas.up.pt (M.D.-A.); Tel.: +351-220428605 (M.M.); +351-220428227 (M.D.-A.)

**Abstract:** Diabetes mellitus (DM) is a chronic progressive metabolic disorder associated with several gastrointestinal complications, affecting up to 75% of patients. Knowing that Angiotensin II (AngII) also regulates intestinal contraction, we decided to evaluate changes in ileum and colon histomorphometry and AngII reactivity in a rat model of DM. Streptozotocin (STZ, 55 mg/kg) was administered to induce DM to 24 adult male Wistar rats. Diabetic rats displayed all the characteristic signs of type 1 DM (T1DM) and fecal excretion increased about 4-fold over 14 days, while the excretion of controls remained unaltered. Compared to controls, diabetic ileum and colon presented an increase in both macroscopic (length, perimeter and weight) and microscopic (muscular wall thickness) parameters. Functionally, AngII-induced smooth muscle contraction was lower in diabetic rats, except in the distal colon. These differences in the contractile response to AngII may result from an imbalance between AngII type 1 (antagonized by candesartan, 10 nM) and type 2 receptors activation (antagonized by PD123319, 100 nM). Taken together, these results indicate that an early and refined STZ-induced T1DM rat model already shows structural remodelling of the gut wall and decreased contractile response to AngII, findings that may help to explain diabetic dysmotility.

**Keywords:** diabetes mellitus; STZ; ileum histomorphometry; colon histomorphometry; smooth muscle contraction; Angiotensin II receptors

**Citation:** Esteves-Monteiro, M.; Menezes-Pinto, D.; Ferreira-Duarte, M.; Dias-Pereira, P.; Morato, M.; Duarte-Araújo, M. Histomorphometry Changes and Decreased Reactivity to Angiotensin II in the Ileum and Colon of Streptozotocin-Induced Diabetic Rats. *Int. J. Mol. Sci.* **2022**, *23*, 13233. https://doi.org/10.3390/ijms232113233

Academic Editors: Patrick C. Baer and Ralf Schubert

Received: 30 August 2022
Accepted: 27 October 2022
Published: 31 October 2022



## 1. Introduction

Diabetes mellitus (DM) is a complex chronic progressive metabolic disorder, medically incurable, that can affect almost every organ system [1]. There are different animal models of DM, but streptozotocin (STZ) has been the agent of choice to chemically induce diabetes in rats and mice, causing the selective destruction of pancreatic β-cells. High doses of STZ are associated with type 1 DM (T1DM) induction, while multiple low doses are usually associated with a high fat diet to cause insulin resistance, characteristic of type 2 DM (T2DM) [2–4]. In this animal model of T1DM structural, functional and biochemical alterations resemble those observed in human diabetic patients [5]. Over time, several investigators have used this model with different induction times (raising questions about animal welfare for longer protocols) in different portions of the intestine, making it harder to compare results [6–8]. For that reason, we decided to assess whether two weeks is sufficient to induce ileum and colon alterations that resemble those observed in long-lasting STZ models [9,10].

Gastrointestinal (GI) complications of DM are very important as they can be associated with significant morbidity, affecting up to 75% of patients [11]. The most common GI complications include esophageal dysmotility, gastroparesis, enteropathy and colonic disorders, such as chronic constipation and diarrhea [7,12]. Since these symptoms are not considered important causes of mortality in patients with DM they are often neglected [13]. However, it is important to recognize that they negatively influence health status and quality of life [13,14].

The pathogenesis of diabetic intestinal dysfunction seems to be multifactorial, related to the accumulation of advanced glycation end-products (AGE), injury of the enteric nervous system (ENS) or interstitial cells of Cajal, and muscular layers fibrosis [8]. Several studies also indicate that diabetic autonomic neuropathy causes damage to the ENS and changes the number and size of myenteric neurons throughout the entire GI tract in rats [15–20]. It has also been described as a deficit in the intestine's cholinergic neurotransmission, since the response to exogenous acetylcholine (ACh) seems to be impaired in the ileum (30 days after STZ-induction) and colon of long-term diabetic rats (60 weeks) [21,22]. Mechanical factors can also contribute to intestinal disorders, since DM seems to cause structural remodeling that can affect histomorphometry and biomechanical properties, increasing stiffness, and decreasing the resting compliance and relaxation capacity of the intestinal wall [9,10,23].

The renin–angiotensin system (RAS) is mostly known for its effects in the cardiovascular and renal systems but it also has an influence in other systems, such as the GI tract, which expresses all of the RAS components [24,25]. Angiotensin II (Ang II) is the major effector peptide of this system, and most of its functions are mediated by the Ang II type 1 receptor ($AT_1R$), while activation of the Ang II type 2 receptor ($AT_2R$) usually counteracts them [26,27]. In the colon, Ang II contracts circular and longitudinal smooth muscle in response to direct activation of post-junctional $AT_1R$ and indirect activation of pre-junctional $AT_1R$ in myenteric and submucosal neurons [26–29]. Curiously, the human colonic smooth muscle is more sensitive to Ang II than to acetylcholine (ACh), but the physiological importance of Ang II in the GI tract is still not completely understood [25,30,31]. Interestingly, there is little information on RAS alterations in the intestine of diabetic individuals, but recently one study concluded that ACE gene polymorphism in patients with T2DM influences intestinal motility, since those patients presented a prevalent genotype that was associated to constipation [32].

Considering the above, the aim of this study was to evaluate the structural (macro and microscopic histomorphometry) and functional (smooth muscle reactivity to Ang II) impact of T1DM in the ileum and colon of a refined rat model, just two weeks after induction.

## 2. Results

### 2.1. Animal Welfare and Monitorization

STZ-induced rats had an initial glycemia of 99.30 ± 3.29 mg/dL that increased to 395.09 ± 13.80 mg/dL within 48 h ($p < 0.0001$, $n = 23$), while control rats had an initial glycaemia of 105.63 ± 6.31 mg/dL that was roughly the same within 48 h (111.14 ± 5.41 mg/dL; $p > 0.05$, $n = 8$). On d7 and d14, almost all STZ rats had glycemia above 500 mg/dL, while control animals presented glycemic values of 105.57 ± 4.76 mg/dL ($n = 8$) on the 14th day.

The parameters documented during the daily monitorization (body weight, water/food intake and fecal excretion) are shown in Figure 1. In the control group ($n = 8$), rats progressively gained weight, their weight being 7.8% ± 0.73% higher by d14 than on d0 (before fasting). Diabetic rats ($n = 21$) had a consistent weight loss that was more pronounced on d2 (5% less compared to the previous day) and then maintained that weight for the remainder of the protocol (7.66 ± 1.04% lower at d14 when compared to the initial weight before fasting) (Figure 1a). Water intake was significantly higher in diabetic rats comparing to controls that maintained a constant water intake through all the experimental protocol: 37.54 ± 0.53 mL/day ($n = 8$). The STZ group drank more water since d1 (48.38 ± 1.16 mL), but their water intake increased progressively throughout the protocol, reaching values 7 times higher than those of control animals at d14: 264.08 ± 12.18 mL ($n$ = 16) (Figure 1b). Despite the weight loss, STZ rats' food intake was significantly higher than controls after the 3rd day. Diabetic rats started the experimental protocol eating 13.25 ± 1.86 g in the first day, and progressively increased food consumption until the last day, when the intake was 49.08 ± 2.64 g/rat ($n$ = 16). The control group maintained a constant food intake during the experimental time, with a mean consumption of 22.44 ± 0.38 g/day ($n$ = 8) (Figure 1c).

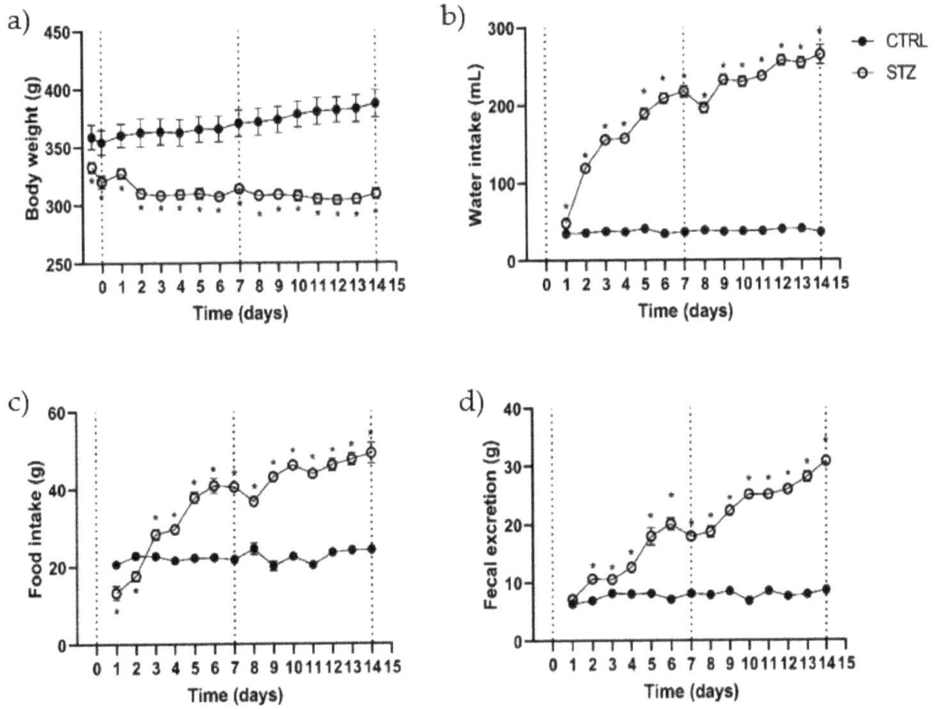


**Figure 1.** Evaluation during the experimental protocol (14 days) in control (CTRL, $n$ = 8) and streptozotocin-induced diabetic rats (STZ, $n$ = 16–21) of: (**a**) body weight; (**b**) water intake; (**c**), food intake and (**d**) fecal excretion. Values are mean ± SEM and unpaired student's $t$ test was used to compare the two experimental groups (CTRL and STZ). * Statistical difference, $p < 0.05$.

To our knowledge, this is the first study to quantify fecal excretion in STZ-induced diabetic animals. Non-diabetic animals maintained a relatively stable fecal excretion during the entire experimental period (7.75 ± 0.18 g/day/rat, $n$ = 8), whereas diabetic rats gradually increased their fecal excretion, reaching values 4 times higher than those obtained in the first day (d1: 7.11 ± 0.34 g/rat; d14: 30.79 ± 0.73 g/rat; $p < 0.0001$, $n$ = 16) (Figure 1d).

### *2.2. Ileum and Colon Macroscopic Evaluation*

Comparing to control animals, all segments of the intestines of STZ rats seemed enlarged. In addition, upon the opening of the abdomen of STZ-induced rats it was easy to perceive an extremely dilated cecum that produced a "mass effect", pushing the intestine to the side. The colon length was significantly higher in diabetic animals compared to the control group (Figure 2a,b: 25.75 ± 0.77 cm, $n$ = 14 vs. 19.63 ± 0.47 cm, $n$ = 12, $p < 0.05$). Since some animals were heavier than others, colon length *per* body weight was measured and the difference between the two groups was maintained (Figure 2b). The circumferential perimeter of the intestinal portions was also measured, being significantly higher in the STZ-induced rats ($n = 11$) compared to non-diabetic rats ($n = 8$) both in the colon (15.45 ± 0.58 mm vs. 11 ± 0.46 mm, $p < 0.0001$, respectively) and ileum (12.55 ± 0.31 mm vs. 9.38 ± 0.32 mm, $p < 0.0001$, respectively) (Figure 2c). The relative weight of the whole intestine segment studied (with fecal content) was higher in STZ-induced animals than in controls (2.69 ± 0.10 g/g of body weight, $n = 21$ vs. 1.80 ± 0.05 g/g of body weight, $n = 12$; $p < 0.0001$, respectively). This increase was also observed at the individual intestinal segments free of fecal content (Figure 2d). Furthermore, no differences were found between STZ-induced animals and controls in the wet-to-dry ratio of all the segments studied (ileum: 5.23 ± 0.37 vs. 5.61 ± 0.33; PC: 5.17 ± 0.24 vs. 4.52 ± 0.20; MC: 4.84 ± 0.30 vs. 5.16 ± 0.21; DC: 5.07 ± 0.20 vs. 4.86 ± 0.28, respectively, $p > 0.05$ for all). The 2-way ANOVA results showed an interaction between the experimental group (control or STZ) and the intestinal segments ($p < 0.0001$), in accordance with our visual observation of the marked dilatation of the intestine in STZ-induced animals. The relative fecal content weight was also higher in STZ-induced animals than in controls (7.10 ± 0.15 g/g of body weight, $n = 21$ vs. 2.66 ± 0.11 g/g of body weight, $n = 12$; $p < 0.0001$). To our knowledge, this is the first time that the weight of intestinal content is reported in STZ rats.

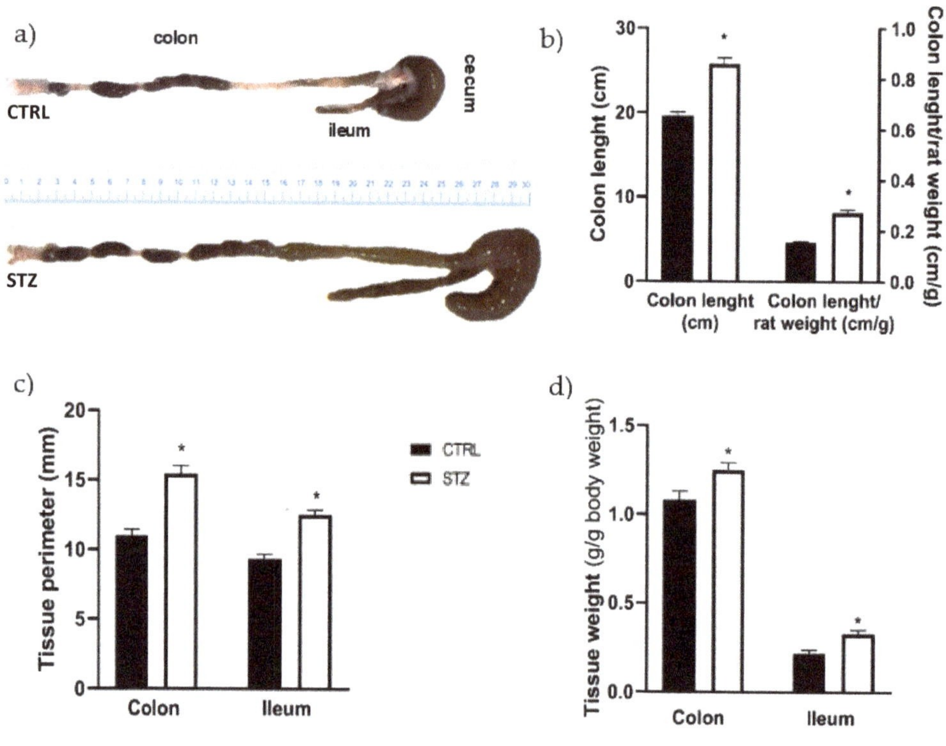


**Figure 2.** Macroscopic evaluation of the ileum and colon of control (CTRL, black bars, $n = 8$–12) and streptozotocin-induced diabetic rats (STZ, white bars, $n = 11$–14): (**a**) representative images of the colon length; (**b**) quantitative analysis of colon length (left $y$ axis) and colon length *per* rat weight (right $y$ axis); (**c**) tissue circumferential perimeter of the colon and ileum and (**d**) relative weight of intestinal segments (without fecal content) expressed as g of colon or ileum/g of body weight. Values are mean ± SEM and unpaired student's $t$ test was used to compare the two experimental groups (CTRL and STZ). * Statistical difference, $p < 0.05$.

### 2.3. Ileum and Colon Microscopic Evaluation

The results of the histomorphometric evaluation of the intestines of STZ-induced animals ($n$ = 8) were concordant with the macroscopic data, showing an increase in the thickness of the intestinal wall of the ileum, proximal colon (PC), middle colon (MC) and distal colon (DC) compared to controls ($n$ = 4), as can be observed in Figures 3 and 4a (ileum: 671.64 ± 74.34 μm vs. 404.97 ± 82.04 μm; PC: 666.66 ± 32.340 μm vs. 389.24 ± 39.03 μm; MC: 589.03 ± 17.88 μm vs. 376.06 ± 50.62 μm; DC: 570.93 ± 27.16μm vs. 430.42 ± 26.26μm, respectively, $p < 0.01$ for all). The intestinal wall thickness increase was similar for all the intestinal segments, as 2-way ANOVA showed a non-significant association ($p = 0.1681$) between experimental group and intestinal segment. Both ileum (longitudinal muscle: 81.02 ± 7.66 μm vs. 31.18 ± 5.44 μm, circular muscle: 116.12 ± 4.59 μm vs. 44.47 ± 10.40 μm, submucosa: 41.68 ± 1.68 μm vs. 17.47 ± 2.13 μm, mucosa: 432.82 ± 20.59 μm vs. 311.85 ± 24.51 μm, respectively, $p < 0.01$ for all) and middle colon (longitudinal muscle: 48.93 ± 2.93 μm vs. 29.66 ± 4.25 μm, circular muscle: 142.55 ± 8.37 μm vs. 74.31 ± 10.9 μm, submucosa: 56.39 ± 4.09 μm vs. 35.63 ± 6.47 μm, mucosa: 341.17 ± 13.79 μm vs. 236.46 ± 34.58 μm, respectively, $p < 0.05$ for all) presented increased thickness of all the intestinal layers assessed in STZ-induced rats compared to controls (Figure 4b). In the PC, the submucosa was the only layer that presented a similar thickness between STZ-induced animals and controls (50.47 ± 7.33 μm vs. 33.81 ± 6.00 μm, respectively, $p = 0.1104$), while all the other segments were thicker in diabetic animals compared to controls (longitudinal muscle: 57.02 ± 6.90 μm vs. 34.64 ± 4.29 μm, circular muscle: 205.2 ± 17.00 μm vs. 90.14 ± 11.33 μm, mucosa: 353.97 ± 14.27 μm vs. 230.64 ± 26.18 μm, respectively, $p < 0.05$ for all). DC only showed an increase in the muscle thickness (longitudinal muscle: 52.51 ± 2.72 μm vs. 28.51 ± 1.67 μm, circular muscle: 150.54 ± 14.58 μm vs. 87.21 ± 7.06 μm, $p < 0.01$ for both; submucosa: 66.11 ± 7.70 μm vs. 53.27 ± 7.54 μm and mucosa: 301.77 ± 10.00 μm vs. 261.42 ± 16.49 μm, $p > 0.05$ for both) (Figure 4b). The 2-way ANOVA showed an association between the experimental group (control vs. STZ) and the intestinal layers thickness (longitudinal muscle, circular muscle, submucosa and mucosa) for the ileum ($p = 0.0058$), PC ($p = 0.0002$), MC ($p = 0.0027$) but not for the DC ($p = 0.1109$).

### 2.4. Ileum and Colon Functional Evaluation

To assess whether intestinal muscle contraction is altered in diabetic animals, ileum and colon reactivity to exogenously applied KCl (Figure 5), ACh (Figure 6) and Ang II (Figure 7) was evaluated. For the concentration–response curves to ACh and Ang II the results were expressed using two recognized pharmacological concepts: the maximum contractile effect ($E_{max}$, expressed in mN/g) and the concentration of agonist capable of causing 50% of the maximal contraction ($EC_{50}$, expressed in uM). In all intestinal segments (ileum, PC, MC and DC) the contractile response to 125 mM KCl (and the ACh concentration-dependent contraction were similar in both control and STZ-induced animals (Figure 6), with comparable $E_{max}$ and $EC_{50}$ values, presented in Table 1.

Regarding reactivity to Ang II, this RAS effector peptide caused a concentration-dependent contraction in control and diabetic animals (Figure 7). The contractile response to Ang II normalized to the tissue weight was lower (but with the same $EC_{50}$) in the ileum, PC and MC of STZ-induced animals. Interestingly, the maximum response in the DC was similar between control and STZ-induced animals, but the $EC_{50}$ of that portion of diabetic colon was significantly lower than that of controls (Table 2).

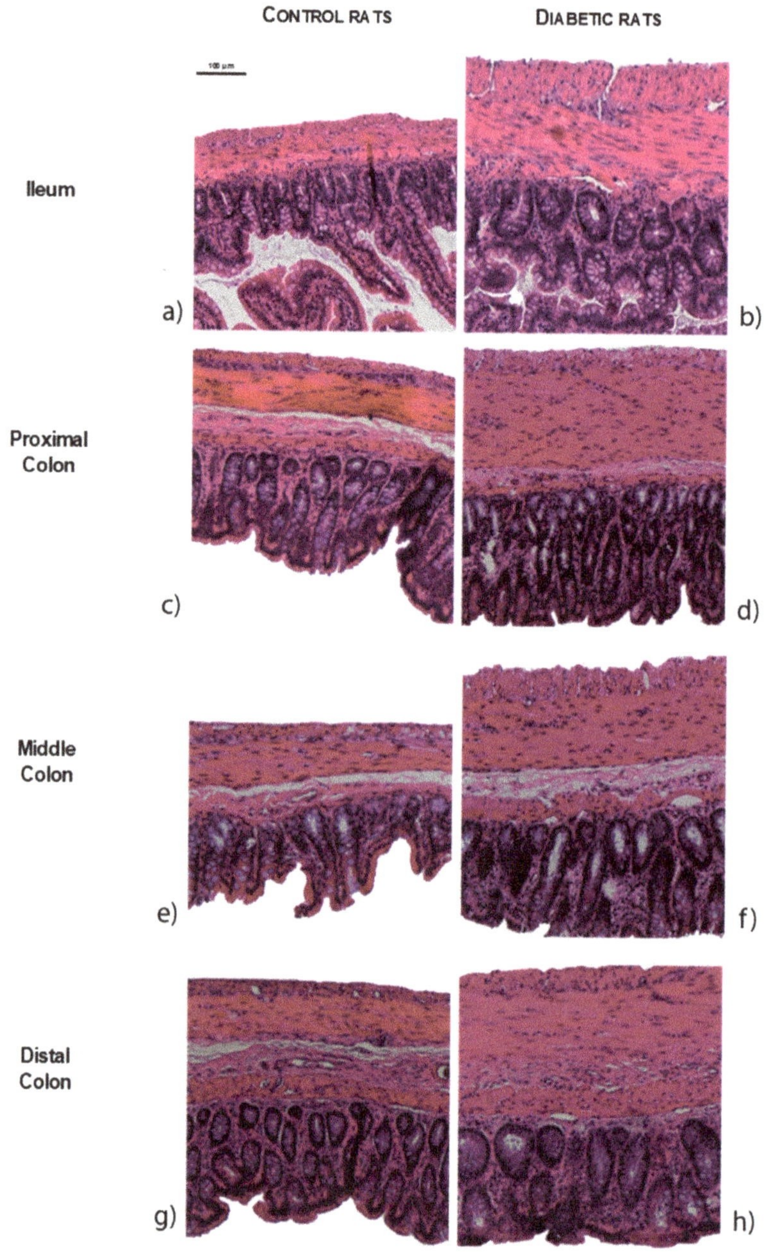


**Figure 3.** Representative microscopic photographs of intestinal segments of control (CTRL, **a**,**c**,**e**,**g**) and streptozotocin-induced diabetic rats (STZ, **b**,**d**,**f**,**h**), stained with hematoxylin and eosin: ileum (**a**,**b**); proximal colon (**c**,**d**); middle colon (**e**,**f**) and distal colon (**g**,**h**). The scale bar (100 µm) is valid for all images.

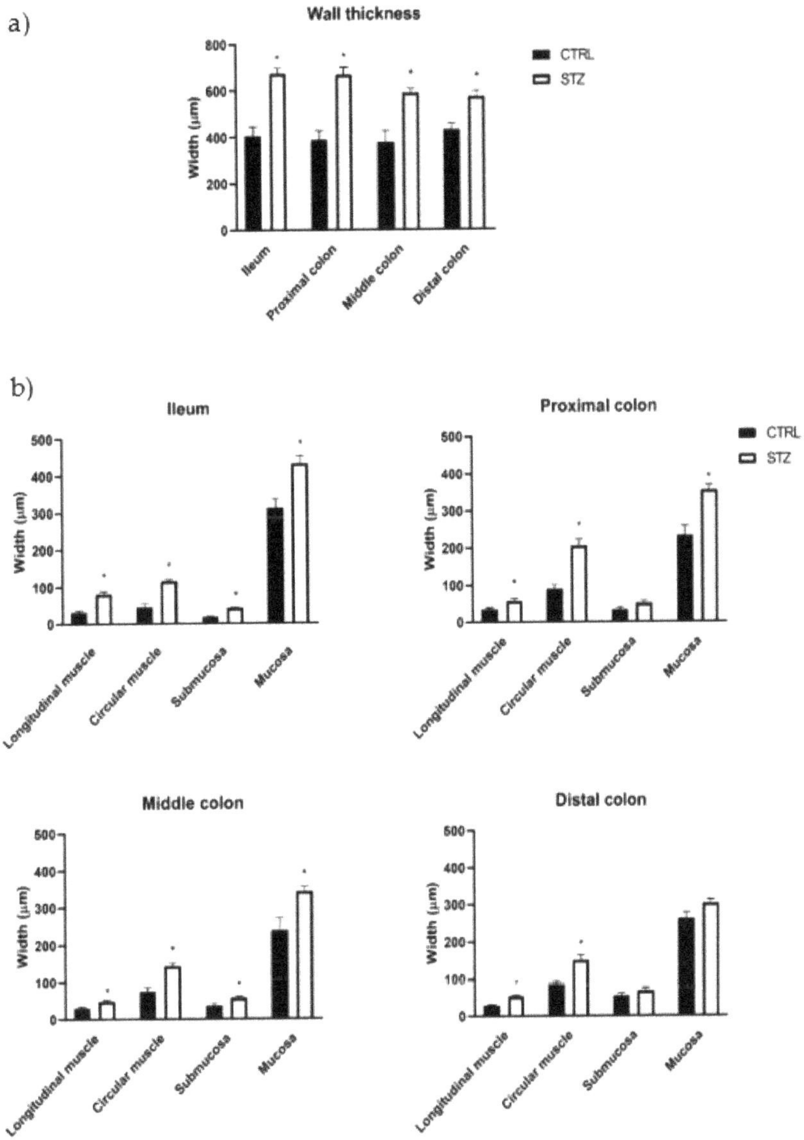


**Figure 4.** Morphometric evaluation of intestinal segments (ileum, proximal colon, middle colon and distal colon) of control (CTRL, $n = 4$) and streptozotocin-induced diabetic (STZ, $n = 8$) rats: (**a**) total wall thickness (µm) of each intestinal segment; (**b**) thickness (µm) of the intestinal layers (longitudinal muscle, circular muscle, submucosa and mucosa) of each intestinal segment. Values are mean ± SEM and a 2-way ANOVA followed by an unpaired t test with Welch's correction was used to compare the two experimental groups (CTRL and STZ). * Statistical difference $p < 0.05$ vs. correspondent control.

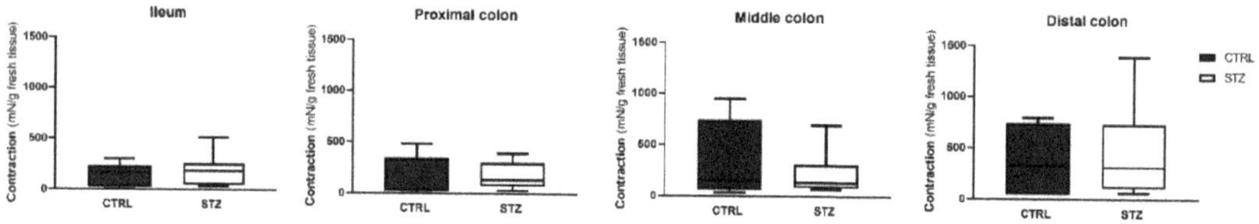


**Figure 5.** Contractile response to KCl (125 mM) in the ileum, proximal colon, middle colon and distal colon of control (CTRL, $n = 6$) and streptozotocin-induced diabetic rats (STZ, $n = 10$). Data are expressed as mN of force *per* g of fresh tissue (mN/g). Values represent the median (95% confidence limits) and a Mann–Whitney test was used to compare the two experimental groups (CTRL and STZ).

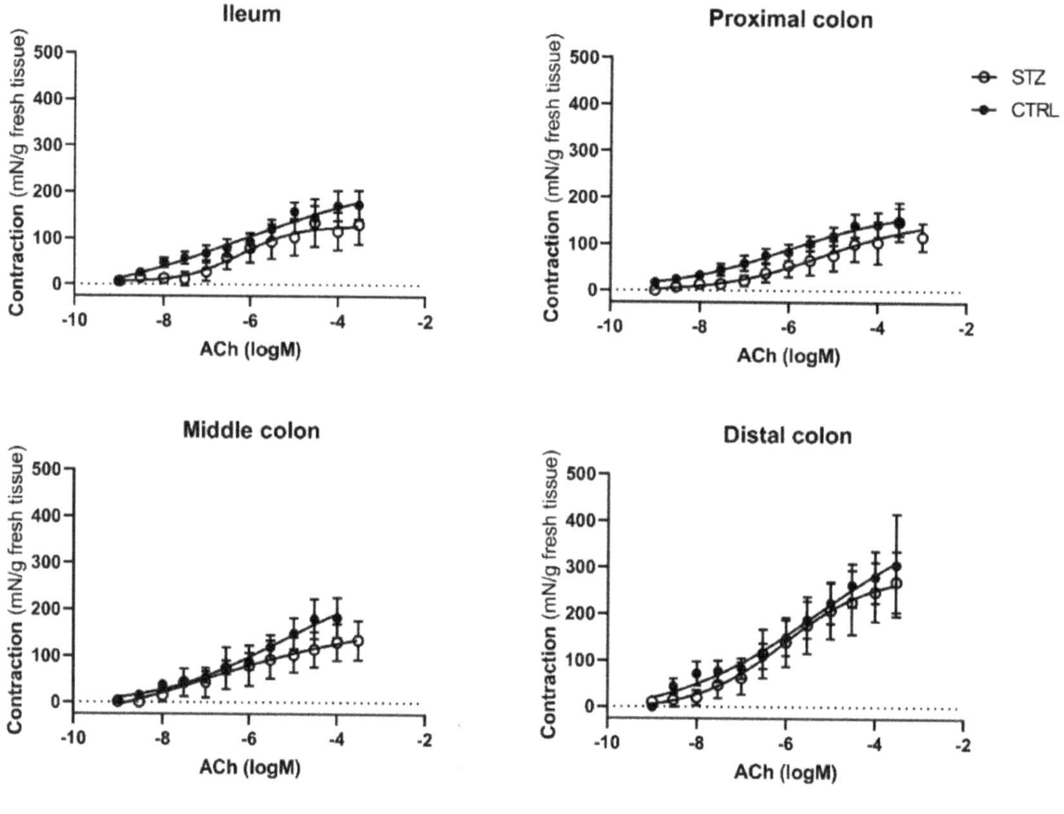


**Figure 6.** Concentration-response curves to ACh in the ileum, proximal colon, middle colon and distal colon of control (CTRL, $n = 6$–$7$) and streptozotocin-induced diabetic rats (STZ, $n = 10$). Data are expressed as mN of force per g of fresh tissue (mN/g). Values are mean ± SEM.

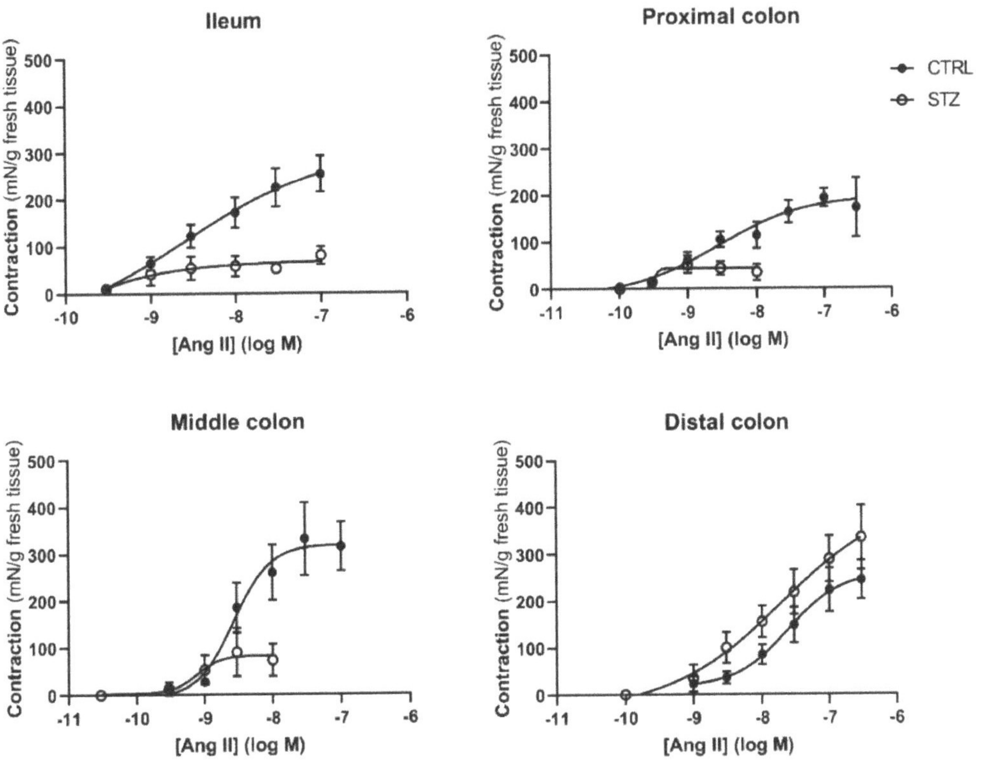


**Figure 7.** Concentration-response curves to Angiotensin II in the ileum, proximal colon, middle colon and distal colon of control (CTRL, $n$ = 5–8) and streptozotocin-induced diabetic rats (STZ, $n$ = 5). Data are expressed as mN of force *per* g of fresh tissue (mN/g). Values are mean ± SEM.

**Table 1.** $E_{max}$ (mN/g) and $EC_{50}$ (μM) values of smooth muscle contraction induced by ACh application in the ileum, proximal colon, middle colon and distal colon of control (CTRL, $n$ = 6–7) and streptozotocin-induced diabetic rats (STZ, $n$ = 10).

| | Ileum | Proximal Colon | Middle Colon | Distal Colon |
|---|---|---|---|---|
| | | Control | | |
| $E_{max}$ (mN/g) | 165.9 [116.4–216.0] | 141.0 [116.7–278.5] | 184.7 [68.95–378.8] | 313.4 [176.2–823.1] |
| $EC_{50}$ (μM) | 0.85 [0.32–3.53] | 1.15 [0.22–14.70] | 3.41 [1.1–4.8] | 2.74 [0.94–7.47] |
| | | STZ | | |
| $E_{max}$ (mN/g) | 79.06 [34.65–338.9] | 158.0 [75.0–569.5] | 143.6 [86.56–411.3] | 271.7 [163.6–370.9] |
| $EC_{50}$ (μM) | 0.82 [0.27–1.87] | 114.0 [8.31–3408] | 18.96 [0.87–75.7] | 2.94 [0.28–142.0] |

For comparison between the two experimental groups (CTRL and STZ) we used a Mann–Whitney test. Values are median (95% confidence limits).

**Table 2.** $E_{max}$ (mN/g) and $EC_{50}$ (μM) values of smooth muscle contraction induced by Angiotensin II application in the ileum, proximal colon, middle colon, and distal colon of control (CTRL, $n$ = 5–8) and streptozotocin-induced diabetic rats (STZ, $n$ = 5).

| | Ileum | Proximal Colon | Middle Colon | Distal Colon |
|---|---|---|---|---|
| | | Control | | |
| $E_{max}$ (mN/g) | 305.3 [138.6–620.5] | 181.5 [136.0–297.0] | 276.6 [246.4–451.1] | 344.4 [222.4–433.5] |
| $EC_{50}$ (μM) | 8.29 [1.24–24.68] | 1.10 [0.36–2.12] | 3.80 [1.95–4.76] | 40.50 [17.08–309.3] |
| | | STZ | | |
| $E_{max}$ (mN/g) | 71.20 [12.3–100.6] * | 50.46 [15.32–78.15] * | 100.6 [22.86–163.5] * | 263.5 [165.0–415.9] |
| $EC_{50}$ (μM) | 7.985 [0.31–8.89] | 0.59 [0.35–14.93] | 2.60 [0.89–7.81] | 4.17 [0.84–8.38] * |

For comparison between the two experimental groups (CTRL and STZ) we used a Mann–Whitney test. Values are median (95% confidence limits). * $p < 0.05$ vs. correspondent control.

Knowing that the differences observed in the contractile response to Ang II could result from an imbalance between $AT_1R$ and $AT_2R$ mediated effect, we decided to further characterize the response to Ang II. The contractile response to Ang II was antagonized by candesartan (10 nM), an $AT_1R$ antagonist, in all four intestinal segments of both control (in mN/g for all, ileum: 54.20 ± 4.50 vs. 2.35 ± 1,60; PC: 17.37 ± 3.14 vs. 1.07 ± 0.49; MC: 12.42 ± 2.23 vs. 0.28 ± 0.15; DC:15.85 ± 1.32 vs. 0.1 6± 0.08; $p < 0.05$ for all) and STZ-induced rats (ileum: 35,75 ± 11,06 vs. −0,87 ± 2,78; PC: 24.80 ± 9.45 vs. 0.7 8± 1.19; MC :95.86 ± 29.03 vs. 5.20 ± 6.39; DC: 288.48 ± 49.08 vs. 5.57 ± 5.54; $p < 0.05$ for all) (Figure 8a). Differently, PD123319 ($AT_2R$ antagonist, 100 nM) decreased the response to Ang II in the ileum (12.43 ± 1.03 mN/g vs. 11.02 ± 1.21 mN/g, $p < 0.05$) and increased the response in all colonic segments of control animals (in mN/g for all, PC: 19.95 ± 3.34 vs. 22.02 ± 3.45; MC: 14.99 ± 1.97 vs. 17.48 ± 2.44; DC: 19.88 ± 2.82 vs. 23.50 mN/g ± 2.64; $p < 0.05$ for all), but was unable to modify Ang II-induced contraction in the ileum (92.58 ± 21.23 mN/g vs. 104.24 ± 23.50 mN/g), MC (146.13 ± 18.53 mN/g vs. 127.88 ± 21.89 mN/g) and DC of diabetic rats (236.37 ± 19.03 mN/g vs. 248.38 ± 25.64 mN/g; $p > 0.05$ for all), decreasing it in the PC (166.14 ± 20.49 vs. 108.45 ± 19.00; $p < 0.05$) (Figure 8b).

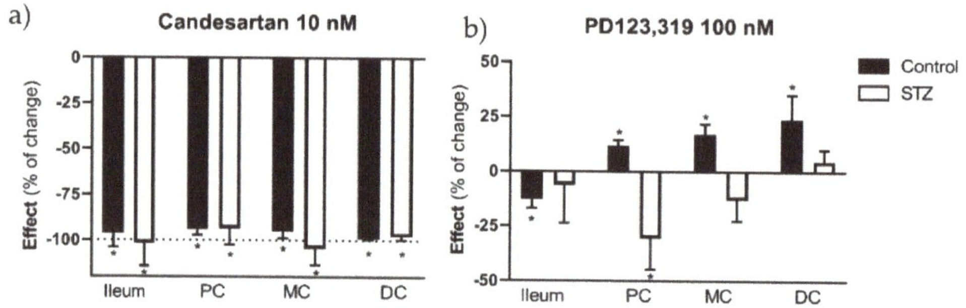


**Figure 8.** Angiotensin II contractile effect (expressed as percentage of change) in the ileum, proximal colon (PC), middle colon (MC) and distal colon (DC) of control (CTRL, $n$ = 5–8) and streptozotocin-induced diabetic rats (STZ, $n$ = 4–6) in the presence of the following antagonists: (**a**) candesartan ($AT_1R$ antagonist, 10 nM) and (**b**) PD123,319 ($AT_2R$ antagonist, 100 nM). Values are mean ± SEM. For statistical analysis we used a paired t test between the effect in the absence and presence of the antagonist.* $p < 0.05$ vs. the correspondent response to Angiotensin II in the absence of the antagonist.

## 3. Discussion

Our data show that the diabetic rat model chosen (DM chemically induced by an IP STZ injection, maintained for 14 days) presented all the typical signs of T1DM: body weight loss, polyphagia, polyuria and polydipsia [33–37]. In addition, diabetic rats gradually increased their fecal excretion whereas non-diabetic animals maintained a relatively stable fecal excretion during the entire experimental period. As pointed out before, this is the first study to quantify fecal excretion in STZ-induced diabetic animals. Besides the increase in mass, the fecal pellets from the diabetic group were well formed but were larger, wider and darker than those from the control group. These findings could eventually be attributed to polyphagia and intestinal distension, differing from Cuervas–Mon and collaborators' data, who described STZ-induced diabetic rats' feces as thick and amorphous, compared to those of control animals [38].

To our knowledge, this study is the first to show that the colon length and the perimeter of the ileum and colon are increased in this early DM model, and that the differences between control and STZ correlate to the different portions studied, in accordance with our visual observation of the marked dilatation of the intestine in STZ-induced animals.

Indeed, enlargement and increased length of the intestine and colon of STZ rats were already described by others, 10 and 8 weeks after DM induction, respectively [39,40]. A possible explanation for the increase in colon and intestine length described in these studies is the remodeling of the extracellular matrix (due to increased production of collagen type 1) and AGE accumulation [40]. In this study we decided to measure only the length of the colon, as it is macroscopically difficult to distinguish ileum boundaries. Our data also show that just 2 weeks after induction, STZ-induced rats present an increase in ileum and colon weight. Forrest and colleagues found that dry colon weight increased significantly in diabetic animals (8 weeks after induction) compared to controls and suggested that this could be related to increased colon length, since weight *per* length did not differ between the two experimental groups [35]. Others observed that weight, but not length, of insulin-treated diabetic rats was significantly higher compared to controls, thus contradicting Forrest and collaborators [41,42]. A possible explanation for the intestinal wall weight increase may be related to the tissue water content, which has been reported to be higher in diabetic animals [39]. However, we did not observe any difference between control and STZ-induced animals in the wet-to-dry ratio of the intestinal segments studied, results that are corroborated by other researchers [35]. For the time being, there is no clear answer as to which mechanisms are triggering the intestinal mass increase in diabetic animals, but Jervis and colleagues suggested that this enlargement could be an adaptation to polyphagia, a characteristic sign of the disease, since intestinal smooth muscle cells are plastic and adapt to functional demand, by remodeling [43]. Curiously, other causes of polyphagia such as lactation or hypothalamic lesions seem to induce similar intestinal consequences [44–46]. On the other hand, another study revealed that even when the food intake of diabetic rats was matched to that of controls, the intestinal weight of diabetic animals remained higher [9].

Our study innovatively uncovers several early histomorphometric alterations in the ileum and colon of T1DM rats and these alterations did not differ according to the different portions studied. Indeed, there are no previous histopathological data on the colon of STZ-induced rats just 2 weeks after induction, although a previous study showed similar results in the ileum 7 and 14 days after induction [9]. The same authors also studied histological characteristics of the middle colon, reporting increased intestinal wall thickness in longer STZ-induced models (4 and 8 weeks after induction) compared to controls [10].

Contrary to what happens when we look at the intestinal wall as a whole, the differences seen by layers are determined by the portion studied. This occurs since in the distal colon only the muscle layers are affected. So, the variation in thickness of the layers of the intestinal wall between diabetic and control animals becomes progressively less evident in the proximal–distal direction (from ileum to distal colon), in agreement with what was previously described by Fregonesi and collaborators [18]. This is a curious finding that reinforces the relevance of studying several intestinal segments to avoid generalizing phenomena that may occur in specific regions. Several studies indicate that increased intestinal thickness in diabetic animals may be due to: a) increased mucosa proliferation (due to higher food intake, increased expression of glucagon-like peptide 2, accumulation of AGE and/or suppression of apoptosis) and b) increased muscle layers (due to AGE mediated effects, collagen type I accumulation and/or smooth muscle cells hypertrophy) [33,40,47–51]. However, further studies are needed to understand if any of the possibilities mentioned above explain the histomorphometric alterations observed, or if there are other mechanisms involved.

The studies conducted on ileum and colon reactivity suggest that there are no changes in the intestinal function of STZ-induced rats just two weeks after induction, since the contractile response to KCl and ACh remained unchanged in all segments studied. Previous studies using rat ileum showed a decrease in the contractile response to ACh 30 days and 6 months after STZ-induction, but this change does not seem to be related to cholinergic innervation damage or acetylcholinesterase activity modification [21,38,52]. Concerning the colon, it was not possible to find differences between the contractile response to ACh in control and STZ-induced rats, injected 30 days previously [52]. However, in a genetic model of T2DM, after a long period of disease (60 weeks) the contractile response to carbachol (an ACh mimetic) in the PC was lower than that of controls, while the response in the DC appeared to be unaffected [22]. Thus, it seems that cholinergic activity in the colon and ileum of diabetic animals may depend on several factors, such as type of diabetes, intestinal segment affected and diabetes evolution time, suggesting that main alterations in diabetic intestinal motility are probably related to changes in smooth muscle layers and non-cholinergic innervation [21,22,38].

We therefore decided, in an innovative way, to evaluate the reactivity of the ileum and colon of diabetic animals to Ang II. The results presented in the functional studies suggest a loss of contractile force in response to Ang II in the ileum, PC and MC but not in the DC of STZ-induced rats, compared to controls, probably due to the fact that the distal segments of the GI tract are the last ones to be affected by diabetic complications [18]. To our knowledge this is the first time that an altered Ang II response is reported in diabetic animals, an effect that could be associated with the structural alterations observed, loss of specific neurons (mostly in the myenteric plexus) and changes in the local tissue levels of Ang II [17,27,53]. Ang II activates both receptors in the smooth muscle cells but also presynaptic receptors in other cells crucial for colonic function, an intricate network that has been reported to be altered in the diseased colon [27,29,54]. Regarding Ang II-mediated effects, it is known that contractile responses in intestinal smooth muscle occur mainly through the activation of $AT_1R$, while $AT_2R$'s role according to our group and others, seems to be more important under pathological conditions [27,55,56]. Not surprisingly, we observed that the $AT_1R$ antagonist (candesartan, 10 nM) completely abolished AngII-mediated contractile response in the ileum and all colon segments of both control and diabetic animals. However, the blockade of $AT_2R$ with PD123319 (100 nM) was more intriguing. In the colon of control rats, we observed that the $AT_2R$-associated counterbalance of Ang II $AT_1R$-mediated contractile effects was no longer present in the DC and MC of diabetic animals, and was even reversed in the PC, as we have reviewed previously [53]. Interestingly, the contractile effect of Ang II in the ileum of control rats was decreased in the presence of PD123319. This points to a putative contractile effect mediated by the $AT_2R$, which although uncommon was previously described in other studies [57,58]. Even so, in the ileum this is not observed, reinforcing the idea that under pathological conditions the effect mediated by the $AT_2R$ in the ileum and throughout the colon is loss/altered, as previously described by our group in an experimental model of colitis in rats [27,53,59].

## 4. Materials and Methods

### 4.1. Animals and Housing

Since female rats seem to be less sensitive to STZ [7], forty-seven male Wistar rats, 10 to 14 weeks in age (weighing 300–400 g), were used in this study, including control ($n$ = 24) and diabetic ($n$ = 23) animals, that were distributed between the different experimental protocols. All control animals were used in the experimental procedures (since we used the same intestinal portions in different functional studies), but only eight of these rats were daily monitored in the animal house facility. Control animals were used in collaboration with other groups that collected organs such as heart, muscle and brain, in a perspective of reducing animals used in experimental research. Sample size was decided using the free software Sample Size Calculator (©2022—ClinCalc LLC, https://clincalc.com/stats/samplesize.aspx). Animals were maintained at the ICBAS-UP rodent animal house facility and the project was approved by the animal welfare body (P311/2019). This work followed the ARRIVE guidelines for reporting experiments with animals [60] (see supplementary material). Animals were maintained in a 12 hours' light/dark cycle, with controlled ventilation, temperature (20–24 °C) and relative humidity (40–60%). All animals were housed in groups of two in Sealsafe Plus GR900 Tecniplast® cages with proper bedding (Corncob ultra 12, Ultragene), with free access to autoclaved tap water (two bottles per cage) and laboratory rodent food (4 RF21, Mucedola S.r.l., Italy). Environmental enrichment such as paper tunnels and nesting material was provided in all animal cages.

### 4.2. Diabetes Induction

On the day of DM induction (d0) animals were fasted for 4 h (food taken from the box where the animals were housed) with free access to water. The STZ solution (S0130, Sigma-Aldrich, St. Louis, MO, USA; 55 mg/mL in citrate buffer, pH 4.5) was prepared just prior to the injection, since a freshly prepared solution is considered to be more effective [4]. Diabetes was randomly induced by a single intraperitoneal injection of 55 mg/kg of STZ (a concentration that has proven successful in our group (data not published) and also by other authors [61]), under the analgesic effect of tramadol (Tramal® oral suspension, 100 mg tramadol/mL, Grünenthal, Portugal) (20 mg/kg, PO), administered moments before [4]. The total volume of STZ solution (55 mg/kg) administered to each animal depended on its weight on the day of induction, ranging from 0.3 to 0.4 mL. Rats had *ad libitum* access to water and food until the end of the protocol (day 14). Animals were considered diabetic if 48 h after STZ injection their blood glucose was ≥250 mg/dL, a situation that occurred in 23 of the 32 animal that were induced (diabetes induction success of 72%). These 23 hyperglycemic rats were included in the STZ group and used in the respective experimental protocols. Glycemia was evaluated using a FreeStyle Precision Neo (Abbott, Canada) glucometer. The blood glucose level of diabetic rats was measured by puncturing one of the tail veins at d0 (control value), d2 (to confirm or discard DM) and d7. On d14, animals were sacrificed by decapitation, using a guillotine suitable for rats (Small Guillotine, Harvard Apparatus) and blood glucose levels were obtained from blood samples collected from the abdominal aorta.

### 4.3. Animal Monitorization and Welfare Evaluation

The animals included in this project were daily monitored (11:00 h to 13:00 h) throughout the entire protocol (d0–d14), and all information was registered in an individual evaluation table (confounders were not controlled). The evaluation started in the maintenance room, assessing the coat's appearance, piloerection, animal's posture, abdominal discomfort and changes in the breathing pattern (welfare evaluation). Then, in the observation room and with the cage open, the same parameters were observed, and the animals' hydration status was evaluated. Monitoring proceeded by weighing the animal and water/food in order to calculate daily intake. The appearance of the feces was also evaluated, and fecal pellets were weighed 48 h after collection to assure uniform drying of all collected samples. The cages were changed every 2 days or whenever they became excessively wet due to diabetes-associated polyuria.

### 4.4. Intestinal Macroscopic Evaluation

On protocol d14 control and STZ-induced rats were euthanized. The abdomen was opened, and the overall appearance of the viscera was evaluated. The abdominal aorta was identified and punctured to collect blood to measure glycemia. The ileum and colon were collected and weighed intact and after cleaning gently their content using Krebs-Henseleit solution (in mM: 118 NaCl; 4.8 KCl; 2.5 $CaCl_2 \cdot 2H_2O$; 1.2 $NaH_2PO_4 \cdot H_2O$; 1.2 $MgSO_4 \cdot 7H_2O$; 25 $NaHCO_3$; 0.02 $Na_2EDTA$; 0.3 Ascorbic acid; 11 monohydrated glucose). The longitudinal length of the colon was measured and a 1 cm portion of the ileum and middle colon was opened through the non-mesenteric border and laid flat to measure the circumferential perimeter (mm).

### 4.5. Intestinal Microscopic Evaluation

Samples (0.5 cm long) of the ileum and colon of diabetic and control animals were collected for histological examination. More precisely, the portion of the ileum was collected 3 cm proximal to the ileocecal junction; the proximal colon (PC) was collected 3 cm distal from the cecum; the distal colon (DC) 3 cm proximal to the anus and the middle colon (MC) 3 cm proximal to where the DC was collected. Each sample was opened through the anti-mesenteric border and fixed in 4% formalin. Samples were routinely processed and paraffin-embedded, cut in 3 µm-thick sections and stained with hematoxylin-eosin (HE) for histological evaluation [10]. Each section was evaluated under an optical microscope (Nikon, model Eclipse E600, Nikon Instruments, Miami, FL, USA) and photographed in two or three different representative regions with objective lens of 4×, 10× and 20× (magnification of 40×, 100× and 200×). The images were used to measure the thickness of the mucosa, submucosa, circular muscle and longitudinal muscle, always by the same person, using the free ImageJ® software 1.53t. For each sample the layer thickness was measured in nine different locations and averaged. The measurements were only carried out in images where all the intestinal wall could be observed.

### 4.6. Intestinal Functional Evaluation

Four 1 cm long portions were collected from the ileum and colon of diabetic and control animals to evaluate smooth muscle contraction. The ileum was taken 2 cm proximal to the ileocecal junction; the PC 2 cm distal from cecum; DC 2 cm from anus and MC 2 cm proximal to the DC. Each sample was mounted in a vertical organ bath along its longitudinal axis, fixed to the bottom of the bath and to an isometric transducer (UGO BASILE S.R.I., Italy, Model 7004) using sewing threads. The bath was continuously aerated with carbogen (95% $O_2$ and 5% $CO_2$) and maintained at 37 ± 1 °C. Tissues were stretched to an initial resting tension of 1 g and mechanical responses were recorded using a PowerLab system (ADInstruments, Oxford, UK). All tissues were washed twice, every 15 min, and triggered with 10 µM of ACh. They were then washed and allowed to stabilize for 15 min more before starting one of the following protocols:

- a cumulative concentration–response curve to ACh (Sigma-Aldrich, USA; 1 nM to 10 mM)
- a non-cumulative concentration–response curve to Ang II (Sigma-Aldrich, USA), according to the range of concentrations that was previously determined in other studies of this research group: ileum, PC and MC: 300 pM to 100 nM; DC: 1 nM to 300 nM [27]. Between each Ang II concentration tissues were washed for 1 h (every 15 min), to avoid receptor desensitization.
- the response to a single concentration of Ang II (Ileum, PC and MC: 30 nM, DC: 100 nM) in the absence and presence of candesartan (a kind gift from Dr. Fredrik Palm, Uppsala University, Sweden; 10 nM, $AT_1R$ antagonist) or PD123319 (Sigma-Aldrich, USA; 100 nM, $AT_2R$ antagonist). Tissues were incubated for 20 min with the antagonists before the second stimulation with Ang II.

At the end of every protocol, the contractile response to potassium chloride (KCl, 125 mM) was recorded.

Finally, each portion used in the functional study was weighed immediately after the protocol (fresh weight) and after drying for 48 h, at room temperature (dry weight). The fresh weight was used to normalize the contractile response. Fresh and dry weight were used to calculate the wet-to-dry ratio, as an index of edema, according to the following equation: WtDr = (WetWeight − DryWeight)/DryWeight.

*4.7. Statistical Analysis*

The GraphPad Prism©8.1.2 software was used for statistical analysis of data. The unpaired Student's t-test was used to analyze animal monitorization and macroscopic evaluation. For comparison between two experimental groups (CTRL and STZ) the Student's t test was used for variables with a Gaussian distribution and the Mann–Whitney test for those with a non-Gaussian distribution. The two-way ANOVA was used to look for interaction in the data from histological evaluation and functional data. Accordingly, data were expressed as mean ± SEM for the Student's *t*-test and median [95% CI] for the Mann–Whitney test where "*n*" indicates the number of animals per group. In all cases, a $p$ value of less than 0.05 was considered to denote a statistically significant difference.

## 5. Conclusions

The results presented in this study demonstrate that it is possible to refine a classic animal model of T1DM, improving animal welfare. In this early (two-week evolution) STZ-induced T1DM model we observed (Figure 9): (1) all the characteristic signs of T1DM (polydipsia, polyuria, polyphagia and body weight loss) and increased fecal excretion; (2) increased length, perimeter and weight in the ileum and colon; (3) increased thickness of several histological intestinal layers (less evident in CD) of the ileum and colon, and (4) decreased Ang II-induced smooth muscle contraction (less evident in the DC) associated with altered balance between the function of Ang II receptors. These reported histomorphometric differences and altered reactivity may help to explain diabetic enteric dysmotility and will be deepened in future studies.

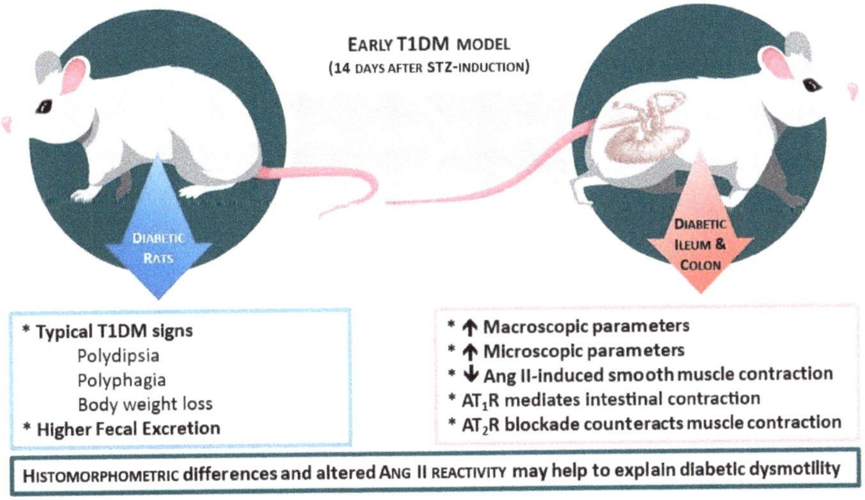


**Figure 9.** Schematic representation of the major findings observed in this early T1DM model. Diabetic rats showed typical DM signs (**left** part of the scheme). The ileum and colon revealed an increase in both macro/microscopic parameters and a decrease in Ang II-induced smooth muscles contraction, mediated by both $AT_1R$ and $AT_2R$ activation (**right** part of the scheme).

**Supplementary Materials:** The supporting information can be downloaded at: https://www.mdpi.com/article/10.3390/ijms232113233/s1

**Author Contributions:** Conceptualization: M.M. and M.D.-A.; methodology: M.E.-M., D.M.-P., M.F.-D., P.D.-P., M.M. and M.D.-A.; data analysis: M.E.-M., M.F.-D., P.D.-P., M.M. and M.D.-A.; investigation: M.E.-M., M.F.-D., M.M. and M.D.-A.; writing—original draft preparation, M.E.-M.; writing—review and editing: M.F.-D., M.M. and M.D.-A.; supervision: M.M. and M.D.-A.; project administration: M.M. and M.D.-A.; funding acquisition: M.M. and M.D.-A. All authors have read and agreed to the published version of the manuscript.

**Funding:** This work was supported by Fundação para a Ciência e Tecnologia (FCT) by a PhD scholarship (2020.06502.BD) to M.E.M. This research was funded through the project UIDB/50006/2020, funded by FCT/MCTES through national funding.

**Institutional Review Board Statement:** All animal procedures were performed according to the Portuguese DL n° 113/2013 and European Guidelines for humane and responsible animal care (European Directive 2010/63). All protocols were approved by ICBAS-UP animal welfare body (P311/2019) in accordance with the ARRIVE guidelines for reporting experiments. Animals were housed at the ICBAS-UP rodent animal facility, which is approved by the national competent authority (024159/2017 DGAV).

**Informed Consent Statement:** Not applicable.

**Data Availability Statement:** Data is available upon request.

**Acknowledgments:** The authors thank Céu Pereira for excellent technical support.

**Conflicts of Interest:** The authors declare no conflict of interest.

## References

1. Krishnan, B.; Babu, S.; Walker, J.; Walker, A.B.; Pappachan, J.M. Gastrointestinal complications of diabetes mellitus. *World J. Diabetes* **2013**, *4*, 51–63. [CrossRef] [PubMed]
2. Radenković, M.; Stojanović, M.; Prostran, M. Experimental diabetes induced by alloxan and streptozotocin: The current state of the art. *J. Pharmacol. Toxicol. Methods* **2016**, *78*, 13–31. [CrossRef] [PubMed]
3. Sharma, R.; Dave, V.; Sharma, S.; Jain, P.; Yadav, S. Experimental Models on Diabetes: A Comprehensive Review. *Int. J. Adv. Pharm. Sci.* **2013**, *4*, 1–8.
4. Eleazu, C.O.; Eleazu, K.C.; Chukwuma, S.; Essien, U.N. Review of the mechanism of cell death resulting from streptozotocin challenge in experimental animals, its practical use and potential risk to humans. *J. Diabetes Metab. Disord.* **2013**, *12*, 60. [CrossRef] [PubMed]
5. Goyal, S.N.; Reddy, N.M.; Patil, K.R.; Nakhate, K.T.; Ojha, S.; Patil, C.R.; Agrawal, Y.O. Challenges and issues with streptozotocin-induced diabetes—A clinically relevant animal model to understand the diabetes pathogenesis and evaluate therapeutics. *Chem. Interactions* **2016**, *244*, 49–63. [CrossRef] [PubMed]
6. Gajdosík, A.; Gajdosíková, A.; Stefek, M.; Navarová, J.; Hozová, R. Streptozotocin-induced experimental diabetes in male Wistar rats. *Gen Physiol. Biophys.* **1999**, *18*, 54–62.
7. Furman, B.L. Streptozotocin-Induced Diabetic Models in Mice and Rats. *Curr. Protoc. Pharmacol.* **2015**, *70*, 5.47.1–5.47.20. [CrossRef]
8. Zhao, M.; Liao, D.; Zhao, J. Diabetes-induced mechanophysiological changes in the small intestine and colon. *World J. Diabetes* **2017**, *8*, 249–269. [CrossRef]
9. Zhao, J.; Yang, J.; Gregersen, H. Biomechanical and morphometric intestinal remodelling during experimental diabetes in rats. *Diabetologia* **2003**, *46*, 1688–1697. [CrossRef]
10. Zhao, J.; Nakaguchi, T.; Gregersen, H. Biomechanical and Histomorphometric Colon Remodelling in STZ-Induced Diabetic Rats. *Am. J. Dig. Dis.* **2008**, *54*, 1636–1642. [CrossRef]
11. Chandrasekharan, B.; Srinivasan, S. Diabetes and the enteric nervous system. *Neurogastroenterol. Motil.* **2007**, *19*, 951–960. [CrossRef] [PubMed]
12. Horváth, V.J.; Putz, Z.; Izbéki, F.; Körei, A.E.; Gerő, L.; Lengyel, C.; Kempler, P.; Várkonyi, T. Diabetes-Related Dysfunction of the Small Intestine and the Colon: Focus on Motility. *Curr. Diabetes Rep.* **2015**, *15*, 94. [CrossRef] [PubMed]
13. Dedeli, O.; Pakyuz, S.C.; Daban, U.K.; Kipcak, S.; Sari, D. Prevalence of Gastrointestinal Symptoms and Its Effect on Quality of Life among Patients with Diabetes Mellitus. *Am. J. Nurs. Res.* **2015**, *3*, 48–53. [CrossRef]
14. Talley, N.J.; Young, L.; Bytzer, P.; Hammer, J.; Diet, M.L.M.N.; Jones, M.; Horowitz, M. Impact of chronic gastrointestinal symptoms in diabetes mellitus on health-related quality of life. *Am. J. Gastroenterol.* **2001**, *96*, 71–76. [CrossRef]
15. He, C.-L.; Szurszewski, J.H.; Farrugia, G.; Soffer, E.E.; Walsh, R.; Ferris, C.D. Loss of interstitial cells of cajal and inhibitory innervation in insulin-dependent diabetes. *Gastroenterology* **2001**, *121*, 427–434. [CrossRef]

16. Furlan, M.M.D.; Molinari, S.L.; Neto, M.H.D.M. Morphoquantitative effects of acute diabetes on the myenteric neurons of the proximal colon of adult rats. *Arq. de Neuro-Psiquiatria* **2002**, *60*, 576–581. [CrossRef]
17. Honoré, S.M.; Zelarayan, L.C.; Genta, S.B.; Sánchez, S.S. Neuronal loss and abnormal BMP/Smad signaling in the myenteric plexus of diabetic rats. *Auton. Neurosci.* **2011**, *164*, 51–61. [CrossRef]
18. Fregonesi, C.E.P.T.; De Miranda-Neto, M.H.; Molinari, S.L.; Zanoni, J.N. Quantitative study of the myenteric plexus of the stomach of rats with streptozotocin-induced diabetes. *Arq. Neuro-Psiquiatria* **2001**, *59*, 50–53. [CrossRef]
19. Brasileiro, A.D.; Garcia, L.P.; Silva, S.D.C.D.; Rocha, L.B.; Pedrosa, A.L.; Vieira, A.S.; da Silva, V.J.D.; Rodrigues, A.R.A. Effects of diabetes mellitus on myenteric neuronal density and sodium channel expression in the rat ileum. *Brain Res.* **2018**, *1708*, 1–9. [CrossRef]
20. Zanoni, J.; Miranda, N.; Bazzote, R. Morphological and quantitative analysis of the neurons of the myenteric plexus of the cecum of streptozotocin-induced diabetic rats. *Arq. Neuropsiquiatr.* **1998**, *55*, 696–702. [CrossRef]
21. Lucas, P.D.; Sardar, M. Effects of Diabetes on Cholinergic Transmission in Two Rat Gut Preparations. *Gastroenterology* **1991**, *100*, 123–128. [CrossRef]
22. Kim, S.J.; Park, J.H.; Song, D.K.; Park, K.S.; Lee, J.E.; Kim, E.S.; Cho, K.B.; Jang, B.K.; Chung, W.J.; Hwang, J.S.; et al. Alterations of Colonic Contractility in Long-term Diabetic Rat Model. *J. Neurogastroenterol. Motil.* **2011**, *17*, 372–380. [CrossRef] [PubMed]
23. Frøkjær, J.B.; Andersen, S.D.; Ejskjær, N.; Funch-jensen, P.; Drewes, A.M. Impaired contractility and remodeling of the upper gastrointestinal tract in diabetes mellitus type-1. *World J. Gastroenterol.* **2007**, *13*, 4881–4890. [CrossRef] [PubMed]
24. Nehme, A.; Zouein, F.A.; Zayeri, Z.D.; Zibara, K. An Update on the Tissue Renin Angiotensin System and Its Role in Physiology and Pathology. *J. Cardiovasc. Dev. Dis.* **2019**, *6*, 14. [CrossRef]
25. Garg, M.; Angus, P.W.; Burrell, L.M.; Herath, C.; Gibson, P.R.; Lubel, J.S. Review article: The pathophysiological roles of the renin–angiotensin system in the gastrointestinal tract. *Aliment. Pharmacol. Ther.* **2012**, *35*, 414–428. [CrossRef]
26. Crowley, S.D.; Coffman, T.M. Recent advances involving the renin–angiotensin system. *Exp. Cell Res.* **2012**, *318*, 1049–1056. [CrossRef]
27. Ferreira-Duarte, M.; Rodrigues-Pinto, T.; Sousa, T.; Faria, M.A.; Rocha, M.S.; Menezes-Pinto, D.; Esteves-Monteiro, M.; Duarte-Araújo, M.; Morato, M. Interaction between the Renin–Angiotensin System and Enteric Neurotransmission Contributes to Colonic Dysmotility in the TNBS-Induced Model of Colitis. *Int. J. Mol. Sci.* **2021**, *22*, 4836. [CrossRef]
28. Gunn, A. The action of angiotensin on the human colon in vitro. *Br. J. Pharmacol.* **1970**, *39*, 34–39.
29. Serio, R.; Mastropaolo, M.; Zizzo, M.G.; Mul, F. Angiotensin II contractile effects in mouse colon: Role for pre- and post-junctional AT 1A receptors. *Acta Physiol.* **2012**, *207*, 337–345. [CrossRef]
30. Ewert, S.; Spak, E.; Olbers, T.; Johnsson, E.; Edebo, A.; Fa, L. Angiotensin II induced contraction of rat and human small intestinal wall musculature in vitro. *Acta Physiol.* **2006**, *188*, 33–40. [CrossRef]
31. Kubíčková, R. Differential Reactivity of the Longitudinal and Circular Muscle of the Rat Distal Colon. Diploma Thesis, Univerzita Karlova, Prague, Czechia, 2016.
32. Malik, A.; Saha, S.; Morya, R.K.; Bhadada, S.K.; Singh, P.K.; Rana, S.V. Angiotensin—Converting Enzyme Gene Polymorphism Influences Gastrointestinal Motility in Type 2 Diabetes Mellitus. *Biochem. Genet.* **2020**, *59*, 335–345. [CrossRef] [PubMed]
33. Chen, P.; Zhao, J.; Gregersen, H. Up-Regulated Expression of Advanced Glycation End-Products and Their Receptor in the Small Intestine and Colon of Diabetic Rats. *Dig. Dis. Sci.* **2011**, *57*, 48–57. [CrossRef] [PubMed]
34. Akbarzadeh, A.; Norouzian, D.; Mehrabi, M.R.; Jamshidi, S.; Farhangi, A.; Verdi, A.A.; Rad, B.L. Induction of diabetes by streptozotocin in rats. *Indian J. Clin. Biochem.* **2007**, *22*, 60–64. [CrossRef]
35. Forrest, A.; Huizinga, J.D.; Wang, X.-Y.; Liu, L.W.C.; Parsons, M. Increase in stretch induced rhythmic motor activity in the diabetic rat colon is associated with loss of ICC of the submuscular plexus. *Am. J. Physiol. Liver Physiol.* **2008**, *294*, G315–G326. [CrossRef] [PubMed]
36. Rees, D.A.; Alcolado, J.C. Animal models of diabetes mellitus. *Diabet. Med.* **2005**, *22*, 359–370. [CrossRef]
37. Davis, F.B.; Davis, P.J. Water metabolism in diabetes mellitus. *Am. J. Med.* **1981**, *70*, 210–214. [CrossRef]
38. Cuervas-Mons, M.; Morte, L.; Junquera, C.; Cajal, S.R.Y. Effects of experimental diabetes in the noradrenergic and cholinergic nerves of the rat small intestine. *Histol. Histopathol.* **1990**, *5*, 193–198.
39. Carson, F.L. Mechanisms of adaptation in rat small intestine: Regional differences in quantitative morphology during normal growth and experimental hypertrophy. *J. Anat.* **1989**, *164*, 189–200.
40. Siegman, M.J.; Eto, M.; Butler, T.M. Remodeling of the rat distal colon in diabetes: Function and ultrastructure. *Am. J. Physiol. Physiol.* **2016**, *310*, C151–C160. [CrossRef]
41. Izbéki, F.; Wittman, T.; Rosztóczy, A.; Linke, N.; Bódi, N.; Fekete, E.; Bagyánszki, M. Immediate insulin treatment prevents gut motility alterations and loss of nitrergic neurons in the ileum and colon of rats with streptozotocin-induced diabetes. *Diabetes Res. Clin. Pr.* **2008**, *80*, 192–198. [CrossRef]
42. Sukhotnik, I.; Shamir, R.; Bashenko, Y. Effect of Oral Insulin on Diabetes-Induced Intestinal Mucosal Growth in Rats. *Dig. Dis. Sci.* **2011**, *56*, 2566–2574. [CrossRef] [PubMed]
43. Jervis, L.; Levim, R. Anatomic adaptation of the alimentary tract of the rat to the hyperphagia of chronic alloxan-diabetes. *Nature* **1996**, *210*, 391–393. [CrossRef] [PubMed]
44. Cripps, A.W.; Williams, V.J. The effect of pregnancy and lactation on food intake, gastrointestinal anatomy and the absorptive capacity of the small intestine in the albino rat. *Br. J. Nutr.* **1975**, *33*, 17–32. [CrossRef] [PubMed]

45. Brobeck, J.R.; Tepperman, J.; Long, C.N.H. Experimental Hypothalamic Hyperphagia in the Albino Rat. *Yale J. Biol. Med.* **1943**, *15*, 831. [PubMed]
46. Kageyama, H.; Kageyama, A.; Endo, T.; Osaka, K.; Nemoto, T.; Hirano, Y.; Inoue, S. Ventromedial hypothalamus lesions induce jejunal epithelial cell hyperplasia through an increase in gene expression of cyclooxygenase. *Int. J. Obes.* **2003**, *27*, 1006–1013. [CrossRef]
47. Fischer, K.D.; Dhanvantari, S.; Drucker, D.J.; Brubaker, P.L.; Kirk, D.; Dhanvantari, S.; Brubaker, P.L. Intestinal growth is associated with elevated levels of glucagon-like peptide 2 in diabetic rats. *Am. J. Physiol.* **1997**, *273*, E815–E820. [CrossRef]
48. Hartmann, B.; Thulesen, J.; Juul, K.; Kissow, H. Immunoneutralization of endogenous glucagon-like peptide-2 reduces adaptive intestinal growth in diabetic rats. *Regul. Pept.* **2002**, *105*, 173–179. [CrossRef]
49. Ahmed, N.; Thornalley, P.J. Advanced glycation endproducts: What is their relevance to diabetic complications? *Diabetes Obes. Metab.* **2007**, *9*, 233–245. [CrossRef]
50. Stern, D.M.; Du, S.; Fang, S.; Marie, A. Receptor for advanced glycation endproducts (RAGE) and the complications of diabetes. *Ageing Res. Rev.* **2002**, *54*, 1615–1625. [CrossRef]
51. Noda, T.; Iwakiri, R.; Fujimoto, K.; Yoshida, T.; Utsumi, H.; Sakata, H.; Hisatomi, A.; Aw, T.Y. Suppression of apoptosis is responsible for increased thickness of intestinal mucosa in streptozotocin-induced diabetic rats. *Metabolism* **2001**, *50*, 259–264. [CrossRef]
52. Liu, H.S.; Karakida, T.; Homm, S. Acetylcholine Smooth and Substance P Responsiveness muscles in Streptozotocin Diabetic of Intestinal Rats. *Jpn. J. Physiol.* **1988**, *38*, 787–797. [CrossRef] [PubMed]
53. Ferreira-Duarte, M.; Sousa, J.B.; Diniz, C.; Sousa, T.; Duarte-Araújo, M.; Morato, M. Experimental and Clinical Evidence of Endothelial Dysfunction in Inflammatory Bowel Disease. *Curr Pharm. Des.* **2020**, *26*, 3733–3747. [CrossRef] [PubMed]
54. Zizzo, M.G.; Auteri, M.; Amato, A.; Caldara, G.; Nuzzo, D.; Di Carlo, M.; Serio, R. Angiotensin II type II receptors and colonic dysmotility in 2,4-dinitrofluorobenzenesulfonic acid-induced colitis in rats. *Neurogastroenterol. Motil.* **2017**, *29*, e13019. [CrossRef] [PubMed]
55. Zizzo, M.G.; Caldara, G.; Bellanca, A.; Nuzzo, D.; Di Carlo, M.; Serio, R. PD123319, angiotensin II type II receptor antagonist, inhibits oxidative stress and inflammation in 2, 4-dinitrobenzene sulfonic acid-induced colitis in rat and ameliorates colonic contractility. *Inflammopharmacology* **2020**, *28*, 187–199. [CrossRef] [PubMed]
56. da Silva, A.C.A.; Severo, J.S.; dos Santos, B.L.B.; Mendes, P.H.M.; Nobre, L.M.S.; de Oliveira, A.P.; Ferreira, F.C.S.; Medeiros, J.V.R.; Lima-Junior, R.C.; Havt, A.; et al. Moderate Physical Exercise Activates ATR2 Receptors, Improving Inflammation and Oxidative Stress in the Duodenum of 2K1C Hypertensive Rats. *Front. Physiol.* **2021**, *12*, 734038. [CrossRef]
57. Touyz, R.M.; Schiffrin, E.L. Ang II-stimulated Superoxide Production Is Mediated via Phospholipase D in Human Vascular Smooth Muscle Cells. *Hypertension* **1999**, *34*, 976–982. [CrossRef]
58. Pelet, C.; Mironneau, C.; Rakotoarisoa, L.; Neuilly, G. Angiotensin II receptor subtypes and contractile responses in portal vein smooth muscle. *Eur. J. Pharmacol.* **1995**, *279*, 15–24. [CrossRef]
59. Ferreira-Duarte, M.; Rodrigues-Pinto, T.; Menezes-Pinto, D.; Esteves-Monteiro, M.; Gonçalves-Monteiro, S.; Capas-Peneda, S.; Morato, M.; Duarte-Araújo, M. 2,4,6-trinitrobenzenesulfonic acid-induced colitis in Rattus norgevicus: A categorization proposal. *Exp. Anim.* **2021**, *70*, 245–256. [CrossRef]
60. du Sert, N.P.; Hurst, V.; Ahluwalia, A.; Alam, S.; Avey, M.T.; Baker, M.; Browne, W.J.; Clark, A.; Cuthill, I.C.; Dirnagl, U.; et al. The ARRIVE guidelines 2.0: Updated guidelines for reporting animal research. *PLoS Biol.* **2020**, *18*, e3000410. [CrossRef]
61. Saleh, D.O.; Bayoumi, A.R.; El-Eraky, W.I.; El-Khatib, A.S. Streptozotocin-induced vascular and biochemical changes in rats: Effects of rosiglitazone vs. metformin. *Bull. Fac. Pharm.*. **2013**, *51*, 131–138. [CrossRef]

MDPI
St. Alban-Anlage 66
4052 Basel
Switzerland
www.mdpi.com

*International Journal of Molecular Sciences* Editorial Office
E-mail: ijms@mdpi.com
www.mdpi.com/journal/ijms

www.ingramcontent.com/pod-product-compliance
Lightning Source LLC
Chambersburg PA
CBHW061445210926
51850CB00008B/1616
* 9 7 8 3 0 3 6 5 9 2 1 8 3 *